Water science reviews 5
The molecules of life

# Water Science Reviews 5

## The molecules of life

EDITED BY

FELIX FRANKS

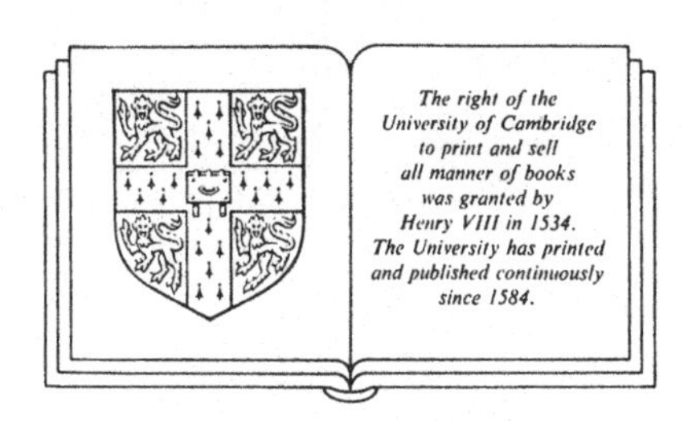


CAMBRIDGE UNIVERSITY PRESS

CAMBRIDGE

NEW YORK PORT CHESTER

MELBOURNE SYDNEY

CAMBRIDGE UNIVERSITY PRESS
Cambridge, New York, Melbourne, Madrid, Cape Town, Singapore, São Paulo, Delhi

Cambridge University Press
The Edinburgh Building, Cambridge CB2 8RU, UK

Published in the United States of America by Cambridge University Press, New York

www.cambridge.org
Information on this title: www.cambridge.org/9780521365772



First published 1990
This digitally printed version 2008

*A catalogue record for this publication is available from the British Library*

ISBN 978-0-521-36577-2 hardback
ISBN 978-0-521-10053-3 paperback

# Contents

# Foreword

Water existed on this planet long before life evolved. The developing chemistry of life processes therefore had to take account of the eccentric physical properties of the universal solvent medium. Water participates in most biochemical reactions, *viz.* condensation, hydrolysis, oxidation and reduction, either as a reactant or a product. The conformational properties of complex organic molecules and their interactions are also affected by the stereochemistry of hydration effects which are dominated by the tetrahedral structure of the $H_2O$ molecule and its hydrogen-bonding capability.

The influence of hydration on macromolecular structures is a subject of intense study by experimental, theoretical and, of late, computer simulation methods. It is very much a case of running before we can walk. The hydration interactions of monomers and oligomers have been neglected, although they can probably shed light on some important aspects of macromolecular behaviour.

This volume is devoted to the solution behaviour of the molecules of life: lipids, nucleotide bases, amino acids and sugars. Theory is confronted by experiment, and comparisons are made between the properties of molecules in crystals and in solution. The diversity in the various types of molecular behaviour is impressive, with hydrophobic effects dominant in controlling lipid interactions, while complex three-dimensional hydrogen bonding patterns regulate the interactions between sugars and also their conformational equilibria in solution.

The general conclusion is that molecular shapes and interactions in solution, and presumably also *in vivo*, are governed by weak, non-covalent effects, some spherically symmetrical and others orientation-specific. While there is a multitude of sensitive and refined experimental techniques by which such effects can be studied, their calculation and prediction still pose severe problems: vacuum energy minimization calculations are of limited use, as are also analogies with crystal structures, at least in the case of carbohydrates. The incorporation of solvent effects into theoretical or computer treatments in any but a naive manner is in its early stages but remains the goal. Until it has been achieved, computer modelling of molecules and their interactions must remain a sterile exercise.

Advances are being made and it is the hope of the authors and editors that the reviews collected in this volume will be a timely stimulus to further efforts.

# Lyotropic effects of water on phospholipids

JOHN H. CROWE AND LOIS M. CROWE

*Department of Zoology, University of California, David, CA 95616*

## 1. Introduction

Water has profound lyotropic effects on membrane phospholipids (reviewed in refs 1–3), including depression of the temperature at which they pass from the gel to the liquid crystalline phase. [**4–9**] This effect has enormous consequences for the structure and function of phospholipid bilayers both *in vitro* and *in vivo*. In fact, studies on natural membranes and pure phospholipids have demonstrated over the past two decades that dramatic alterations in the organization of membrane lipids are induced as a result of dehydration-induced phase transitions (reviewed in refs 10–12). Among these alterations are: phase separations of membrane constituents [**13, 14**] leading to such phenomena as aggregation of membrane proteins [**15**] and formation of non-bilayer phases, [**16**] fusion between adjacent bilayers; [**14, 17**] and transient permeability changes, resulting in leakage of the contents of vesicles or cells to the surrounding medium (reviewed in refs 18, 19).

The impetus for our own work in this field over the past two decades has been a search for the mechanism by which certain organisms such as seeds of plants, yeast cells, and even some lower animals are capable of surviving more or less complete dehydration (see refs 9, 10, 18, 20 for background). We will not deal extensively with this phenomenon in this review since we have recently done so elsewhere, [**9, 10, 23**] but it is of such interest that we wish to comment briefly on what is known about the mechanism. According to the available evidence, all such organisms contain large quantities of disaccharides (usually trehalose or sucrose). These sugars have the ability to form hydrogen bonds with the polar headgroups of membrane phospholipids, and in effect replace the water that is hydrogen bonded to the same polar groups in fully hydrated bilayers. The result is that many of the physical properties of the dry phospholipid bilayers closely resemble those of fully hydrated bilayers. Most of the effects of water to be discussed here are mimicked to a great extent by the presence of these disaccharides.

We have organized this review around increasing levels of molecular complexity. In the first sections we review the evidence that water alters many

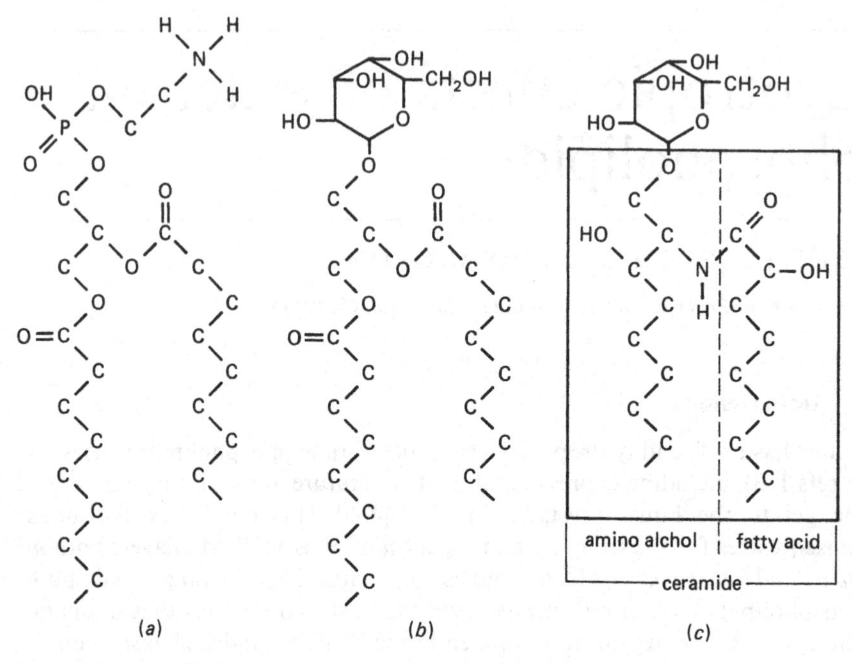


Figure 1. Structures of some representative polar membrane lipids. (*a*) Phosphatidylethanolamine: the polar portion includes an amino group. In phosphatidylcholine, the amino nitrogen is trimethylated. (*b*) Monogalactosylidiacylglycerol: the polar headgroup attached to the glycerol backbone is the sugar galactose. Other sugar headgroups are digalactose and mono- and diglucose. (*c*) A cerebroside: this sphingolipid has a galactose headgroup attached to the ceramide portion. Other headgroups are phosphorylcholine, phosphorylethanolamine, and mono-, di- and oligosaccharides. Sphingolipids do not have a glycerol backbone, but a relatively planar amide bond region linking the amino alcohol and fatty acid portions of the ceramide. They are further modified by the number and position of the –OH groups on the amino alcohol and fatty acid.

physical properties of pure phospholipids. We then show that it has similar effects on natural membranes isolated from cells and on intact cells.

## 2. Effects of water on polar membrane lipids

Phospholipids and other polar lipids exist in several phases, and transition from one to another is affected by both temperature and hydration. **[4, 21, 22]** Since the various polar lipids commonly found in cells display different hydration-dependent phase behavior, we have divided the first portion of this review into sections that deal with the major classes for which data are available. The chemical structures of these classes are represented in Figure 1. For the benefit of the reader unfamiliar with the polymorphism exhibited by these lipids as they undergo thermotropic phase transitions, we illustrate these properties in Figure 2.

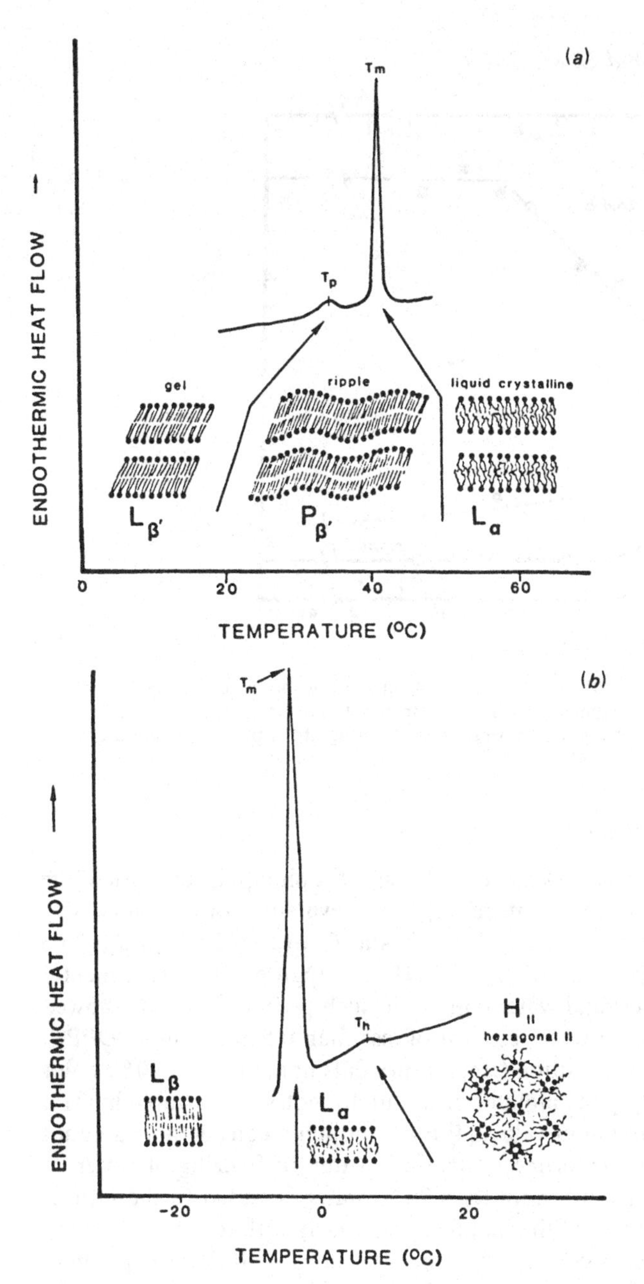


Figure 2. (*a*) Calorimetric scan of hydrated dipalmitoylphosphatidylcholine (DPPC) multilamellar vesicles with a diagram of the bilayer structures associated with each phase. As the temperature of the DPPC is raised a transition from the gel ($L_{\beta}'$) phase to the ripple (pretransition, $P_{\beta'}$) phase occurs at a characteristic transition temperature, $T_p$. In both the $L_{\beta'}$ and $P_{\beta'}$ phases, the acyl chains of the lipid are tilted with respect to the bilayer normal. A further increase in temperature causes a transition to the liquid crystalline phase at a characteriestic temperature, $T_m$. In the liquid crystalline phase, the acyl chains are disordered. (*b*). Calorimetric scan of dioleoylphosphatidylethanolamine (DOPE). The acyl chains of phosphatidylethanolamines are parallel to the bilayer normal. The phosphatidylethanolamine undergoes a gel to liquid crystalline transition at $T_m$, and a second, liquid crystalline to inverted hexagonal ($H_{II}$) transition at $T_h$. In the inverted hexagonal phase the lipid forms long, water-filled cylinders with the headgroups facing the aqueous core. The cylinders are in a three-dimensional hexagonal array (from ref. 3).

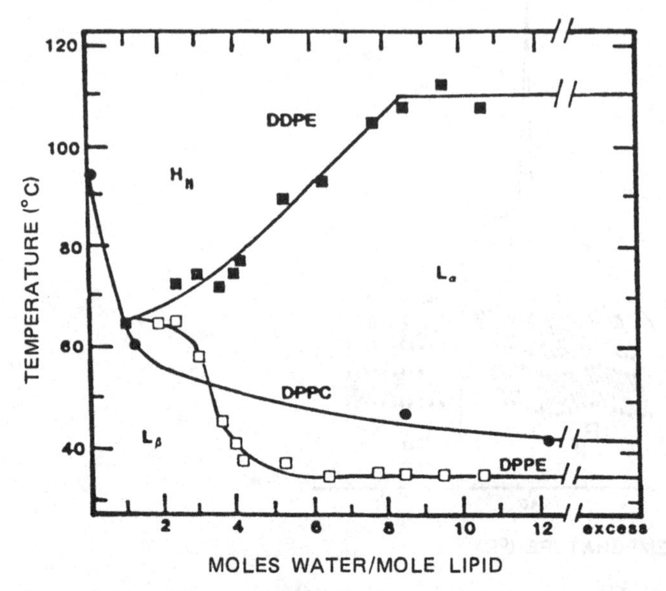


Figure 3. Phase diagram of DPPC (data from ref. 4) shown together with that for didodecylphosphatidylethanolamine (DDPE) (data from ref. 36) at various water contents. Changing water content or changing temperature could bring about phase separation of a mixture of these two phospholipids.

## 2.1. *Phosphatidylcholines*

*Effects of water on phase behavior* When phospholipids are dried the calorimetric transition temperature ($T_m$) is elevated. For example, dry dipalmitoylphosphatidylcholine (DPPC) has a $T_m$ over 105 °C, [**5–9**] while fully hydrated DPPC has a $T_m$ of 41 °C (Figure 3). Experimental measurement of the residual water content in such preparations [**24**] showed that the dry DPPC has a water content of less than 0.5 moles/mole DPPC. As water is added, this transition temperature falls until at about 20 % water content (0.25 g $H_2O$/g phospholipid or about 10 moles water/mole lipid) it reaches its limiting minimum value. [**4**] Even though a considerable amount of water is present, no calorimetric transition due to freezing of water or melting of ice appears until more than 20 % water is added to the DPPC, after which the ice content of the sample rises linearly with the water content. This unfreezable water is considered to be associated with the phospholipid headgroup. [**4**]

Studies with other physical techniques have yielded similar results. For example, $^2H_2O$ NMR shows that in the gel phase about 21 moles deuteron/mole DPPC showed restricted motion, were bound and partially oriented. [**24**] Above 21 moles/mole DPPC the water molecules showed rapid tumbling, and between 9 and 21 moles/mole, there appeared to be exchange between the bound and free environments. X-ray diffraction studies [**25**]

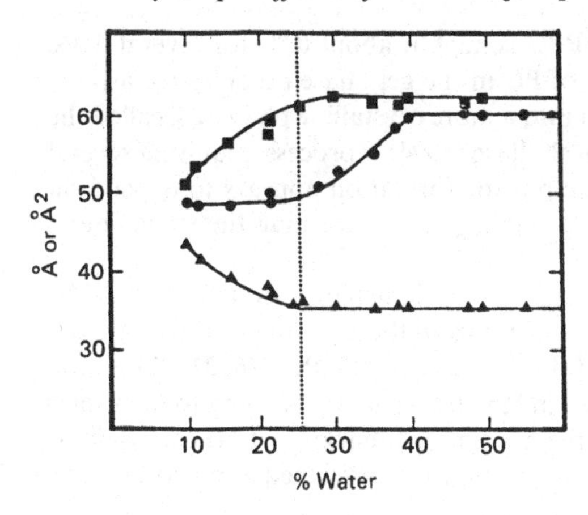


Figure 4. Lamellar repeat distance ($d$) (circles), bilayer thickness ($d_l$) (triangles), and area/lipid molecule ($S$) (squares) of DMPC as a function of water content as determined by X-ray diffraction. The line drawn at 25% water content shows that at that hydration level, both the area/molecule and the thickness of the lipid bilayer have reached limiting values while the bilayer spacing continues to increase to 40% water content. The water incorporated from 25–40% is intercalated between the bilayers, increasing the water thickness ($d_w$), while water above 40% exists as a separate phase. (Data from ref. 26).

showed that the 2 moles $H_2O$/mole DMPC are hydrogen bonded to the phosphate head-groups in the same half bilayer to form a phosphate–water ribbon linking adjacent phosphatidylcholine (PC) molecules to each other, and are also hydrogen bonded across the bilayer interface to other symmetrically related water molecules associated with the adjacent bilayer. The location in the PC bilayer of the other 8–10 water molecules is unknown.

X-ray diffraction of dimyristoylphosphatidylcholine (DMPC) in the liquid crystalline phase shows that the lamellar repeat distance is constant until 20% water content, while at the same time the bilayer thickness decreases to a minimal value of 35 Å. **[26]** At greater water contents, the bilayer thickness is constant, but the lamellar repeat distance increases with increasing hydration up to 40% water (27 moles $H_2O$/mole lipid). Additional water exists as a separate phase. **[4, 22, 26]**

Thus, as water is added to anhydrous PCs it initially binds to the headgroup region. **[27]** This has the effect of increasing the area/headgroup and the distance between the acyl chains, thus decreasing the attractive van der Waals' forces between adjacent lipid molecules (Figure 4). As a result, the lipid undergoes chain melting more readily, i.e., at lower temperatures. PC does not show unlimited swelling in water, indicating that there are attractive forces between these neutral bilayers.

*Metastability in phosphatidylcholines* Until recently, it was thought that PCs were stable in the gel phase ($L_\beta$) at temperatures below the pretransition.

It has now been shown that if DPPC is kept at about 0 °C for several days, the usual quasi-hexagonal lattice of PC in the gel phase becomes metastable and will spontaneously transform into a more crystalline phase [**28**] called the subgel. [**29**] Formation of the subgel phase is a slow process, requiring several days, [**6, 28, 30**] and the time required for formation appears to depend on acyl chain length with both $C_{12-16}$ and $C_{18-20}$ PCs forming the subgel more slowly than $C_{16}$ (DPPC). [**28**]

X-ray diffraction [**31, 32**] shows that the lamellar periodicity of DPPC decreases by about 4.2–4.5 Å upon going from the $L_{\beta}$ to the subgel phase. By analogy with hydration studies of the $L_{\beta}$ phase of DMPC, [**26, 31 ,32**] several groups suggested that the decrease in lamellar spacing on going to the subgel phase was due to dehydration of the headgroups. Packing of the chains in the subgel phase is uncertain, but has been tentatively assigned as orthorhombic. [**32**]

IR spectroscopy shows a general reduction in bandwidth and increase in peak height, indicating a reduction in the mobility of various functional groups, especially ones associated with the acyl chains or the polar–hydrophobic interface. [**30**] In particular, the C—O stretching band of DPPC (between 1741 and 1727 $cm^{-1}$) in the subgel phase resembles that of DPPC monohydrate or anhydrous DPPC. [**33**] Although the packing mode cannot be assigned unambiguously from IR data, changes in the $CH_2$ scissoring band also indicate that there are changes in the acyl chain packing of DPPC in the subgel phase. [**30**] Other Fourier transform IR (FT-IR) results, using DPPC with one or both chains totally deuterated, show a reduction in the transition temperature for both the pre- and main transitions with increasing deuteration, consistent with previous results, [**34**] but the temperature of the subtransition is independent of the degree of deuteration. [**35**] Since the subtransition results in changes in acyl chain packing as shown by X-ray diffraction and FT-IR, [**30–32**] and deuteration changes the packing and therefore the temperature of the pre- and main transitions but not the subtransition, it appears that the subtransition is dominated by changes in hydration of the headgroup rather than changes in chain packing as the PC goes from the subgel to the $L_{\beta}$ phase.

The relationship between the subgel phase and the water content has been studied by differential scanning calorimetry (DSC). [**6**] Measured amounts of water were added to very dry DPPC, and calorimetric scans were run with and without incubation to form the subgel. A close examination of the ice melting endotherm at several water contents shows that in the annealed samples, where DPPC is in the subgel phase, a slight shoulder appears on the broad melting endotherm, which the authors suggest indicates that part of the bound water has become more mobile. Concomitant with the appearance of the shoulder there is a diminution of the broad endotherm at the low temperature end. Kodama, Hashigami and Seki [**6**] thus detected in DPPC at low levels of hydration (about 19 %) three types of water: water that does not freeze, a second type that has a depressed freezing and melting point and bulk water. When these samples are incubated to form a subgel phase, part of the second type of water appears to be converted to bulk water, which melts at a slightly higher temperature and appears as a shoulder on the broad low endotherm.

## 2.2. *Phosphatidylethanolamines*

*Effects of water on phosphatidylethanolamines* Phosphatidylethanolamines (PEs), like PCs, undergo a main chain-melting transition at temperatures that depend on acyl chain length and degree of unsaturation. Unlike PC, the acyl chains of PE are perpendicular to the plane of the bilayer in the gel phase ($L_\beta$) and do not undergo a pretransition to give the periodic ripple phase ($P_\beta'$) of saturated PC (Figure 2). The headgroup of PE is also smaller than that of PC since ther terminal amino group is not methylated (Figure 1). This free amino group of PE has important consequences for the structure of this group of phospholipids. In the gel phase the space requirement of the headgroup equals the cross-sectional area of close-packed hydrocarbon chains. When PE (especially unsaturated PE) undergoes the chain-melting transition, the cross-sectional area of the chains increases, and if the small headgroups maintain their tight packing the layer will curve. This tendency to curve can be counteracted by inserting polar molecules such as water to fill spaces between the headgroups. Many PEs, especially unsaturated ones, will form a phase where this tendency to curve results in the lipid molecules being arranged in long cylinders with their headgroups at the core of the cylinder (Figure 2). This phase, called the inverted hexagonal or Hex II phase ($H_{II}$), forms at relatively high temperatures and low hydrations in PEs with saturated hydrocarbon chains, **[36, 37]** but can form at lower temperatures in fully hydrated PEs with unsaturated chains. For example, the $H_{II}$ transition temperature ($T_h$) of dioleoylphosphatidylethanolamine (DOPE) is about 8 °C, **[38, 39]** while for egg yolk PE the $T_h$ is about 30 °C. **[39, 40]**

The amine in the headgroup of PE forms hydrogen bonds directly with the phosphate of the headgroup of adjacent molecules, either laterally or across the bilayer. **[41–44]** Thus, PE has stronger lipid–lipid interactions than lipid–water interactions. As a result, it binds far less water than does PC. There is not complete agreement in the literature concerning exactly how much water is present in fully hydrated PE, however. Lis *et al.*, **[45]** using X-ray difffraction to determine the structural parameters of egg yolk PE at varying degrees of hydration, found an equilibrium water distance between bilayers ($d_w$) of about 20 Å occurring at a water:PE mole ratio of 25:1. This large water space between bilayers implies that the attractive forces between PE bilayers (van der Waals', electrostatic, and hydrogen bonding) are rather weak, a conclusion that has been challenged by others. **[44]** On the other hand, Ladbrooke and Chapman **[46]** found that egg yolk PE binds only about

7–8 moles of water. A $^2$H NMR study of the mobility of $^2H_2O$ in egg PE bilayers at various degrees of hydration found that the egg PE bound no more than 12 moles water/mole lipid. [47] For a series of saturated acyl chain PEs ($C_{12}$–$C_{20}$) the gel–liquid crystalline transition temperature fell to a limiting minimum value at around 7 moles of water, indicating there was sufficient water present to cause the maximum of structural changes in the PE bilayers. [48] X-ray diffraction of dilauroylphosphatidylethanolamine (DLPE) when fully hydrated and almost anhydrous gave a $d_w$ of 12 Å for bilayers in the liquid crystalline phase and 5 Å in the gel phase and corresponding limiting hydrations of about 11 moles $H_2O$/mole lipid in the liquid crystalline phase and 3–6 moles/mole in the gel phase. [48] In the liquid crystalline phase of DLPE, the water-order correlation length (the 'hydration repulsion' decay distance of Lis *et al.* [45]) was calculated to be 2.5 Å. [48] Since Cevc and Marsh [48] found a water layer of 12 Å for liquid crystalline bilayers of DLPE, this indicates considerably more attractive force between the PE bilayers than suggested by Lis *et al.* [45]

More recently, McIntosh and Simon [44] have calculated the structural parameters for gel and liquid crystalline phase DLPE using X-ray diffraction, electron density profiles, and previously published dilatometry data of Wilkinson and Nagle [49] for the specific volume of DLPE. These calculations indicate that there are about 7 and 9 waters per DLPE in the gel and liquid crystalline states respectively, similar to previously reported values. [46–8] The water layer between fully hydrated bilayers is the same in both phases (5 Å, about the diameter of two water molecules) and contains about half the water of hydration, with the rest being associated with the headgroup. Further calculations using the area/molecule and the bilayer (lipid) thickness indicate that the two 'extra' water molecules present in the liquid crystalline phase are incorporated into the headgroup region. [44] The above measurements can be compared to the $d_w$ of 27 Å for DMPC in the liquid crystalline phase. [26] All of these data support the conclusion that PE is significantly less hydrated than PC, and the work of Cevc and Marsh [48] and McIntosh and Simon [44] suggest that intermolecular attractive forces acting between adjacent bilayers are rather strong and counteract the hydration repulsive force present in PC bilayers. [45, 50]

*Effects of water on the $H_{II}$ phase* The effect of hydration on $H_{II}$ formation has been studied by DSC and $^{31}$P NMR for didodecylphosphatidylethanolamine (DDPE). Decreasing the water content below 6 moles/mole DDPE causes a rise in the main transition temperature similar to the effect of hydration on PC (Figure 3). The hexagonal transition temperature begins to decrease below 9 waters/lipid, indicating either greater dehydration in the hexagonal phase or greater hydration of the $L$ phase with increasing temperature (Figure 3). At the lowest water content, the transitions from gel to liquid crystalline to hexagonal phase are not separated in temperature. [36]

Increasing the sodium chloride concentration causes an increase in $T_m$ and a decrease in $T_h$ of both diacyl and dialkyl PE, [**36, 51**] while a direct transition of $C_{16}$ acyl and alkyl PE from gel to $H_{II}$ occurs in saturated sodium chloride. [**52**] How the salt causes this effect is not known. More recently, similar effects of other solutes on the phase behavior of PEs in the presence of excess water have been reported. [**39**]

*Effects of lipid–lipid interactions on phase behavior* The relative contributions of steric bulk, hydration and hydrogen bonding capacity, have been assessed through studies on PE analogs that have had headgroup bulk increased either by N-methylation, N-ethylation, addition of extra methylene groups between the phosphoryl and amino groups, or $C_2$-alkylation. [**36, 53, 54**] The last modification is the only one to increase steric bulk without interfering with hydrogen bonding between PE molecules. It was found that $C_2$-alkylated PE analogs had *lower* $T_h$ values, showing that they formed $H_{II}$ more easily. This might be accounted for by a relative dehydration due to the addition of hydrophobic substituents, while maintaining normal hydrogen bonding capacity between the amino and phosphoryl groups. [**54**] Lengthening the distance between the amino and phosphoryl groups made formation of $H_{II}$ more difficult, [**36, 54**] while adding methyl or ethyl groups to the amino group abolished the $H_{II}$ transition in the temperature range studied. [**54**]

Modifications that interfere with lipid–lipid hydrogen bonding also will increase the degree of lipid–water interactions and the effective hydration of the headgroup. The relative contributions of each of these effects is not clear. The effects on the gel–liquid crystalline transition are more subtle, but in general the largest part of the reduction in $T_m$ results from the addition of carbons to the PE headgroup. For an equivalent amount of steric bulk, modifications that interfered with hydrogen bonding between the adjacent PE molecules lowered the $T_m$ slightly more. [**53, 54**] Thus, it appears that the amount of hydration and the opportunity for lipid–lipid interactions through hydrogen bonding as well as the steric bulk of the headgroup are important factors in determining the stability of the lamellar phase.

*Metastability of PE bilayers* Besides the gel, liquid crystalline, and $H_{II}$ phases, hydrated PE can exist in crystalline phases. [**53, 56–8**] The presence of crystalline phases depends on the thermal history and the method of preparation of the sample: if the PE is hydrated and stored below its $T_m$, it will undergo a cooperative thermotropic transition with a large calorimetric enthalpy, $H_{cal}$, at a temperature above the normal chain-melting. If a second calorimetric scan is run soon after the first, this 'high-melt' transition [**53**] is absent. When the dry PE samples are carefully hydrated by repeated cycling at temperatures above the 'high melt' transition they undergo a chain-melting transition at a lower temperature, but upon storage at low temperatures a crystalline phase with a high melt transition appears. [**53–5, 57**]

X-ray diffraction [**55, 56, 58**] demonstrates the crystalline chain packing and decreased lamellar spacing of this phase. FT-IR spectra of the high melt samples of PE are almost identical to those of the anhydrous solid. [**59**] The spectroscopic and diffraction results suggest that the high enthalpy of the high melt transition is due to simultaneous chain-melting and hydration of the headgroups. Other crystalline phases of PE have been demonstrated with different hydration or storage methods [**56, 58**] or at high pH and in the presence of $Ca^{2+}$. [**51**] All of these studies show that fully hydrated PE is metastable in both the gel and liquid crystalline phases and can form dehydrated crystalline phases upon storage at physiological temperatures. At present, it is not known whether phase separation and formation of dehydrated crystalline phases of PE can occur in mixtures with other lipids, so any biological significance is an open question. However, these results illustrate how important it is to specify carefully the conditions of sample preparation and experimental design.

### 2.3. *Glycoglycerolipids*

*Effects of water on phase behavior* PCs and PEs are the most common glycerophospholipids and have been widely studied. Less is known about other glycerolipids, although many occur in high concentration in some membranes. In this section, we will consider the interactions of amphiphilic gluco- and galactoglycerolipids in which the various phosphate headgroups are replaced by sugar moieties.

Mono- and digalactosyldiacylglycerols (MGalDG and DGalDG) are the major polar lipids of chloroplast membranes of higher plants, comprising about 70% of the polar lipids. [**60, 61**] Bacterial membranes contain about 70% mono- and diglucodiacylglycerol (MGluDG and DGluDG). [**62**]

The hydration and phase behavior of galactosylglycerols were first studied by Rivas and Luzzati [**60**] in extracts from corn leaf chloroplasts. The galactolipids incorporated only a small amount of water between the bilayers, a behavior similar to that seen in PE. These mixed galactolipids also formed the $H_{II}$ phase at all temperatures at lower water contents and a cubic phase at high temperatures and intermediate water content.

When purified MGalDG and DGalDG isolated from *Pelargonium* leaves were studied by X-ray diffraction and freeze-fracture, it was found that the formation of the $H_{II}$ phase was due to the monoglucolipids, since they were in that phase at all water contents down to at least −15 °C, while the diglucolipids were always present in the lamellar phase. Both glucolipids had high degrees of acyl chain unsaturation, with about 99% of the MGalDG chains being polyunsaturated, and both reached limiting hydration at 22% water (13 moles/mole MGalDG and about 16 moles/mole DGalDG); [**63**] these limiting hydrations are similar to those of PE and only about half those of PC (see above).

Corresponding studies of hydration and phase behavior of more saturated pure MGluDG and DGluDG using X-ray diffraction and $^2H_2O$ NMR showed a maximum hydration of about 7 moles water/mole lipid in the gel phase and about 8–9 moles/mole in the liquid crystalline phase for both lipids. [**62**] Considering the larger number of –OH groups present in DGluDG, it is surprising that both glucolipids have the same maximum hydration (see also ref. 63). Wieslander, Ulmius, Lindblom and Fontell [**62**] suggested that this effect could be explained if the polar headgroup were more or less parallel to the plane of the bilayer; this conformation was later found to be present in the analogous galactolipid, DGalDG, [**64**] while in a synthetic $C_{14}$ monoglucosyldialkylglycerol the headgroup is extended away from the bilayer surface. [**65**]

MGluDG was found to be in the $H_{II}$ phase both in the dry state and in excess water, while DGluDG was lamellar. However, the chain-melting transition of both glucodiacylglycerols was about 30 °C higher than those of PCs with the same acyl chain composition, [**62**] implying that both MGluDG and DGluDG have strong intermolecular interactions due to the large numbers of hydrogen bond donating and accepting OH groups. Since the hydration of both glucolipids is the same, and both have strong intermolecular interactions, the absence of $H_{II}$ in DGluDG (and DGalDG) appears to be due to the greater steric bulk of the saccharide headgroup.

## 2.4. *Sphingolipids*

*Hydration and phase behavior of sphingolipids* Sphingolipids are found in relatively small amounts in the membranes of both plants and animals, but are present in especially large amounts in brain and nervous tissue. Unlike glycerolipids, sphingolipids do not have a glycerol backbone, but are composed of an amino alcohol (sphingosine or dihydrosphingosine) linked to a long-chain fatty acid through an amide bond (Figure 1). The hydrophilic headgroup, which is also linked to the amino alcohol, may be composed of a phosphorus-containing moiety such as PC or PE or may consist of the sugar groups glucose or galactose. These latter glycosphingolipids are also called cerebrosides. The amino alcohol–fatty acid hydrocarbon chain portion of all of the sphingolipids (ceramide) is hydrophobic, and may be modified in different sphingolipids by the length and unsaturation of the hydrocarbon chains. Sphingolipids are also modified by various numbers of OH groups on the carbons of either chain near the amide linkage (Figure 1). Since natural sphingolipids often have unusually long fatty acids, their two hydrocarbon chains frequently differ in length by more than seven $CH_2$ residues and there may be considerable chain interdigitation in the center of the bilayer. [**66**]

The amide group and the hydroxyls of the amino alcohol and fatty acid portions of the ceramide contribute to intermolecular hydrogen bonding in the region below the polar headgroup. [**67**] Additionally, one might expect intermolecular hydrogen bonding contributions by ethanolamine, glucose and galactose headgroups. Cerebrosides have limiting hydrations of about 20%, or 11 moles water/mole lipid. [**22, 68**] The glucose moiety of a fully hydrated synthetic monoglucosylcerebroside has been shown to extend away from the bilayer, [**69**], in a similar conformation to that of a synthetic monoglucosyldialkylglycerol [**65**] and may possibly form hydrogen bonds with adjacent hexose units, as suggested by the low limiting hydration. On the other hand, sphingomyelin with a phosphocholine headgroup has the same limiting hydration as PC, [**22, 70**] suggesting that the intermolecular hydrogen bonding network of the *ceramide* portion does not affect the hydration behavior of the sphingolipids, but that hydration behavior is dependent on the headgroup.

The structural evidence available indicates that sphingolipids exist only in lamellar phases [**22, 67, 70–2**] except for anhydrous sphingomyelin at very high temperatures. [**70**] The structural and hydration similarities between monoglycosylceramides and monoglycosylglycerolipids would suggest that monoglycosylceramides have the potential for forming the $H_{II}$ phase. That they do not may be due to stabilization of the bilayer by the intermolecular bonding in the ceramide portion of the lipid at the interface between the acyl chains and the polar headgroup. Sphingolipids from tissues that are subjected to high mechanical stress have been found to contain a higher proportion of more hydroxylated ceramides and therefore more lateral hydrogen bonding. [**67, 73**] In addition, the hydrogen donors of the ceramides could interact with free electron pairs of ester and ether groups in glycerolipids and may link most of the lipid layers of the membrane in a network of hydrogen bonds, thus increasing the stability of the membrane. [**67**] The phase diagram of natural brain sphingomyelin shows a drop in $T_m$ with increasing hydration with maximum changes occurring at about 15% water content [**70**] in a manner analogous to PC.

*Metastability of sphingolipids* Various natural and synthetic sphingomyelins and cerebrosides have exhibited a non-equilibrium gel phase. [**68, 71, 72, 74**] The first, metastable, gel form appears on rapid cooling from above the chain-melting transition. On subsequent heating the cerebroside or sphingomyelin undergoes an exothermic reaction which leads to the second, stable gel state, and finally an endothermic reaction with a high $H_{cal}$ (gel–liquid crystalline transition). The stable gel phase also forms upon slow cooling of the sample or incubation at room temperature. [**68, 71, 72**] X-ray diffraction measurements of the lamellar spacings show that the metastable gel form has a larger repeat distance than the stable gel. [**71, 72**] Ruocco *et al.* [**72**] suggest that the reduction in spacing during the metastable–stable gel transition is due to an increase in chain tilt. However, the decreased spacing could also be explained by a transition from a hydrated to dehydrated gel, as has been demonstrated for metastable phases in both PE and PC. **[105]** A reduction in lamellar spacing due to interdigitation of the hydrocarbon chains would not apply in this case since the two chain lengths are approximately equal in these synthetic cerebrosides and sphingomyelin. The high enthalpy transition at higher temperature was suggested to be due to a concomitant hydration and chain-melting transition. **[68]** X-ray diffraction of synthetic sphingomyelin shows that the metastable gel pattern is typical of a hydrated gel phase phospholipid, while the stable gel phase has a number of sharp diffraction lines consistent with greater ordering of the hydrocarbon chains, **[71]** again similar to the stable, dehydrated gel phases of PC and PE.

## 3. Hydration forces between bilayers

### 3.1. *Nature of the hydration force*

The physical forces determining interactions between surfaces in water have traditionally been accounted for by electrostatic double layers and van der Waals' forces. However, there is a large body of evidence now that an additional force is involved which is related to the hydration of the surface (reviewed in refs 75 and 76). This extraordinary force, which has been thought to be due to dynamic structuring of the solvent by the surface and has been called a 'hydration force', was first discovered in phospholipid bilayers in multilamellar arrays. **[50, 77]** Using X-ray diffraction to measure the bilayer separation Rand, Parsegian and their colleagues have applied an externally applied osmotic dehydration to multilamellar vesicles of PC and deduced the repulsive pressure between the opposing bilayers. They found hydration repulsion dominates all bilayer interactions in the range of 15–25 Å separation, depending on the lipid. **[45]** At larger separations the repulsive force is matched by van der Waals attractions (at least in the case of PCs), which as a result attain finite separations in water. A second repulsive force that operates in this same range of bilayer separations, known as the undulation or fluctuation pressure, arises from thermally induced fluctuations in the bilayer surface. **[78, 79]** At closer separations steric effects of the headgroups on each other dominate. **[80]** Similar measurements have now been done with large polymers in solutions such as DNA **[81]** and with mica sheets, **[82]** sometimes coated with phospholipids. **[83]** By contrast with PCs, charged phospholipids separate in water indefinitely, with a decay rate given by Debye–Huckel theory, **[84]** but modified somewhat by steric interactions.

Hydration repulsion appears to occur at all surfaces interacting at distances where perturbations of the solvent structure by each surface overlap. **[76]** On a molecular scale the repulsive forces are small, but when they are summed over the large number of water molecules associated with each surface they result in large forces that are easily measured. This force is the major barrier to contact and fusion between bilayers. [**76**]

### 3.2. *Modification of hydration repulsion*

*Divalent cations* Unilamellar vesicles of acidic phospholipids such as phosphatidylserine can in the presence of $Ca^{2+}$ be caused to fuse into multilamellar vesicles, accompanied by almost complete dehydration. [**85**] The surface water that is normally responsible for the repulsion between bilayers is completely displaced when $Ca^{2+}$ binds to the negatively charged headgroup, leading to molecular contact between bilayers and ultimately to fusion. The affinity of the interaction bilayers for $Ca^{2+}$ is apparently much higher than that of isolated bilayers, [**86**] and this maintains the closely opposed bilayers in a dehydrated state. [**75**] This phenomenon is often used as a model for fusion between bilayers. [**85**]

*Cholesterol* Cholesterol is known to affect the hydration of bilayers, [**87**] the depth of penetration of water into bilayers [**88**] and the association of adjacent PC bilayers. [**89**] One might expect this molecule to alter the hydration pressure significantly, a prediction that has recently been demonstrated to be correct. [**90**] Addition of cholesterol significantly alters the van der Waals' attraction between bilayers, both by increasing the polarizability of the hydrocarbon chain region [**91**] and changing the contribution of the polar headgroup region to repulsion, [**92**] effects due apparently to alteration of the hydration state of the headgroups. [**90**]

## 4. Native membranes and intact cells

### 4.1. *Measurement of phase transitions*

DSC has only limited uses for measurements on phase transitions in native membranes and intact cells. In such mixed lipid systems, where the cooperativity of melting is low, DSC yields either broad endotherms with very low enthalpy or small isolated endotherms that we have suggested represent melting of phase separated lipid domains. [**93**] A superior way of measuring these phase transitions is with FT-IR spectroscopy, which provides a fast, accurate measurement (reviewed in ref. 94). We have simply adapted the methods used by others for measurements of $T_m$ in phospholipids (reviewed in ref. 95) and intact bacterial cells [**96**] for use in this context. We believe these are the first such measurements with eukaryotic cells. We have recently used this method to record effects of water on lipid phase transitions in native membranes and in intact cells.

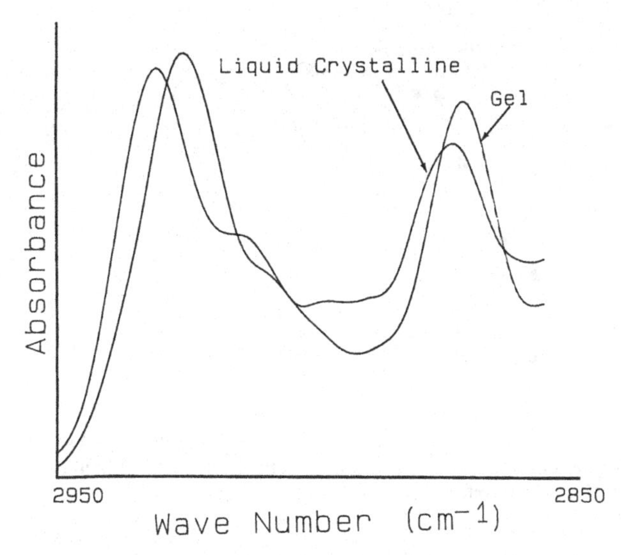


Figure 5. IR spectra in the hydrocarbon chain stretching region taken from native membrane vesicles prepared from pollen grains.

## 4.2. *Effects of water on native membranes*

Typical spectra taken from membrane vesicles isolated from pollen grains (chosen for study because the pollen survives extreme dehydration) is shown in Figure 5. All the features characteristic of hydrocarbons are evident in these spectra. When the membranes were cooled and then rewarmed, vibrational frequency increased over about a 10 °C temperature range (Figure 5). We interpret this frequency shift to represent increasing disorder in membrane phospholipids. As we explained above, it may be due to melting of several populations of phospholipids. When the membranes were dried, the temperature at which the melting occurred shifted upwards from about −6 °C to over 60 °C, an effect similar to that seen with pure phospholipids (Figure 5).

## 4.3. *Effects of water on intact cells*

The first measurements of $T_m$ in intact cells have been made on the same pollen grains as those used above for membrane preparations. IR spectra can be obtained easily from the intact organisms and portions of the spectra can be assigned to phospholipids. **[93, 97]** As with phospholipids and membranes, shifts in frequency can be seen with temperature as the hydrocarbons undergo their phase transition from the gel to the liquid crystalline phase. **[93, 97]** A detailed description of the procedure used to obtain these spectra and to assign them to phospholipids is presented elsewhere. **[93]** Care must

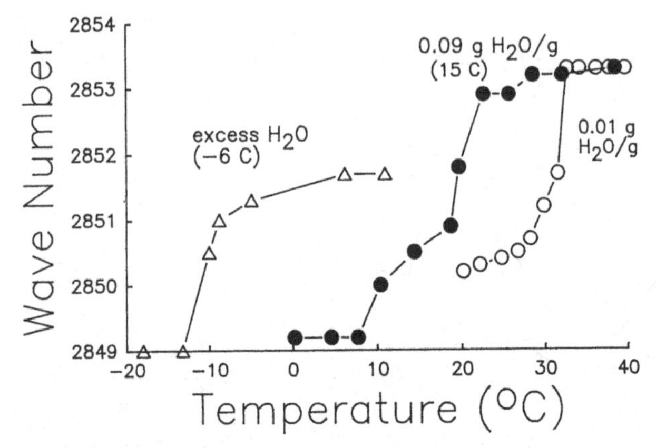


Figure 6. Frequency of the $CH_2$ symmetric stretch in pollen grains containing the indicated amounts of water and as a function of temperature. From the midpoint of the transition a value for $T_m$ can be extracted. (Data from ref. 97).

be taken to minimize light scattering by distributing the experimental material as uniformly as possible on the windows of the sample holder. In assigning the phase transitions evident in the data, we conducted biochemical analysis of the contents of the pollen. The results showed that these organisms contain three major classes of hydrocarbons: (1) An extracellular hydrocarbon, which can be removed by washing the pollen in organic solvents, thus eliminating this fraction from consideration. This treatment does not alter the viability of the pollen. (2) Neutral lipids, which we found to have a major transition centered on about 15 °C. This transition does not change with water content of the pollen. (3) Phospholipids, which have $T_m$s varying from −6 °C in the fully hydrated pollen to 32 °C in the dry pollen. Accordingly, the hydration sensitivity of $T_m$ in the phospholipids can be used to distinguish their transition from that of the interfering neutral lipids. A representative plot of the change in frequency with temperature for the $CH_2$ symmetric stretch is shown in Figure 6. From the midpoint of the frequency change an average value for $T_m$ can be extracted, and the data show clearly that this value shifts with water content, much as would be expected for pure phospholipids. Similar measurements have been made on a variety of cells, including newly emerged hypocotyls from seeds of several plant species and the sperm of several animal species, **[93]** indicating that such measurements are possible with a wide variety of cell types. The important point we wish to make here is that these transitions have hydration-dependent behavior similar to that seen in phospholipids and isolated membranes.

From measurements of $T_m$ extracted from frequency–temperature plots taken for pollen at different water contents, it was possible to construct the phase diagram shown in Figure 7. The data show that $T_m$ falls sharply when the pollen grains are hydrated, reaching a minimal value at a water content

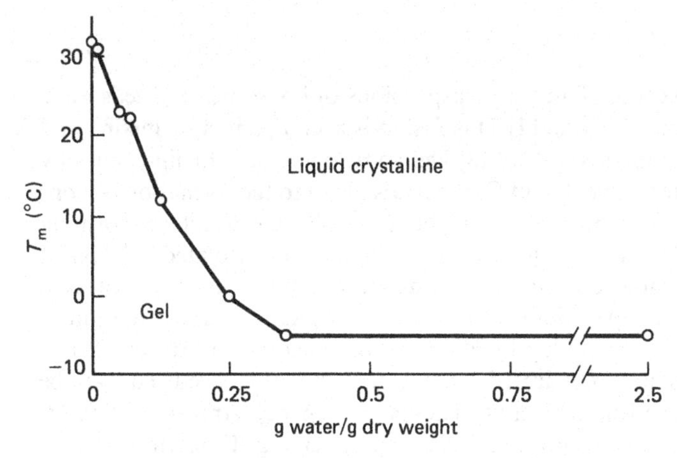


Figure 7. A hydration-dependent phase diagram for pollen grains, extracted from measurements of $T_m$ on pollen at the indicated water contents. (Data from ref. 97).

of about of 0.25 g water/g dry weight. Although this phase diagram differs in detail from that for a pure phospholipid (cf. Figure 3), the probable reasons for which are discussed elsewhere, **[19]** we suggest that this comparison supports the proposition that we are measuring the hydration-dependent phase transition in membrane phospholipids in the pollen. The phase diagram for the pollen shown here is the first such phase diagram for an intact cell.

## 5. Physiological consequences of dehydration

### 5.1. *Phase separations*

Although few of the studies reviewed above were done with mixtures of lipids, many workers in this area have suggested that thermotropic phase transitions and dehydration have damaging consequences for cell membranes. Thermotropic or dehydration-induced transitions can induce phase separations in membranes: as a membrane is cooled (or dehydrated), the lipid(s) with a higher $T_m$ will go into the gel phase first and separate from lipids with lower $T_m$ (Figure 3), **[12, 98, 99]** and the resulting gel phase lipids will undergo lateral separation from membrane proteins. **[100]** Many studies have also shown that during thermotropic phase transitions membranes exhibit increased leakage of trapped solutes, the rate of which may be increased in the presence of lateral phase separation of membrane components. **[101–4]** Theoretical treatments of some of these results **[102, 106]** correlate permeability characteristics of liposomes with the proportion of interfacial lipid at the boundary between gel and liquid crystalline regions.

### 5.2. *Fusion*

Dehydration or freezing of aqueous suspensions of liposomes and cells leads to membrane fusion. [**17, 106–11**] This fusion is accompanied by mixing and leakage of vesicle contents. [**112, 113**] Dehydration of cells or liposomes by glycerol, polyethylene glycol, and $Ca^{2+}$ can also lead to the formation of non-bilayer structures in membranes with the appropriate non-bilayer forming lipids (e.g. refs 114–16). Dehydration of liposomes, isolated biological membranes and whole cells by air-drying or lyophilization also leads to fusion, lateral phase separations of proteins and lipids and the formation of lipid crystals and $H_{II}$ phase lipid (Figure 5) (reviewed in refs 10, 11). When such preparations are rehydrated, the fusion that has occurred can be assessed by freeze-fracture and resonance energy transfer. [**10, 11**] Rehydrated biological membranes also show changed distributions of intramembrane particles and are leaky, presumably due to the inability of the membrane to reestablish the proper protein and lipid distribution. [**10, 11**]

### 5.3. *Leakage*

*Leakage during thermotropic phase transitions* As phospholipid bilayers pass through the thermotropic transition from gel ($L_\beta$) to liquid crystalline ($L_\alpha$) phase they undergo transient changes in permeability (e.g. refs 117, 118). We illustrate this point through a specific example. Small (100 nm in diameter) unilamellar vesicles composed of DMPC were prepared with carboxyfluorescein trapped in the aqueous interior, and leakage of the fluorescent probe was then monitored as the vesicles passed through the main calorimetric phase transition, which, in the case of DMPC, is centered on 23 °C. The results (Figure 8) show that leakage of the probe commences at about 20 °C and is complete by 27 °C. Thus, the midpoint of the leakage curve is at 23.5 °C, occurring simultaneously with the calorimetric transition.

It is still not entirely clear why the permeability of bilayers increases as they pass through the phase transition, but the most predominant idea found in the current literature is that packing defects exist at boundaries between liquid crystalline and gel phase domains. Leakage would occur, according to this notion, through the defects. [**95**] Such defects would exist even in phospholipid vesicles composed of single species of phospholipid (e.g., the DMPC vesicles discussed here); since the phase transition is not completely cooperative, occurring over a temperature range, liquid crystalline and gel phase domains must coexist during the transition. With complex mixtures of phospholipids this effect may be exaggerated by the following sequence of events. There is good evidence that lateral phase separations of at least some phospholipid classes may occur in the plane of the bilayer during cooling [**12**] or drying, [**10, 11**] mainly due to the immiscibility of some gel phase phospholipid classes. When such a phase separated mixture composed of

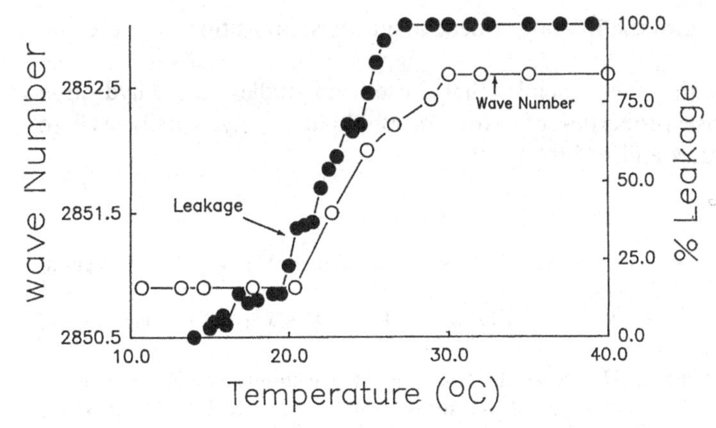


Figure 8. Leakage of carboxyfluorescein from DMPC vesicles heated through the gel–liquid crystalline transition temperature. For comparison, IR data showing melting of the hydrocarbon chains are also shown. (Data from ref. 19).

phospholipids with different transition temperatures is rewarmed, liquid crystalline and gel phase domains will unquestionably be found in the bilayer simultaneously until the domains are melted.

*Leakage during rehydration* Vesicles of many phospholipids would not be expected to lose their contents during dehydration since the liquid crystalline – gel transition would occur after bulk water into which the contents could be leaked was removed. However, during rehydration, when the vesicles undergo the gel–liquid crystalline transition they would be expected to leak into the surrounding medium. It follows that the gel–liquid crystalline phase transition is likely to be a primary vector for leakage across dry bilayers as they are rehydrated, depending on the respective hydrated and dry $T_m$s.

*The relationship of $T_m$ to physiological measurements* When dry cells are placed in water, particularly at low temperatures, they often show a transient leakage of intracellular solutes. In the case of dry pollen grains, Hoekstra [**118**] showed that this leakage is alleviated by heating above a species-specific temperature before the pollen grains are placed in water. He also showed that germination of the pollen grains was enhanced if they were heated before they were placed in water. Leakage declined in concert with the increase in germination. We subsequently found that the increase in germination coincides also with the increase in vibrational frequency for the $CH_2$ bands in the pollen. [**97**] Based on these findings, we suggest that at low temperature gel phase domains exist in the plasma membranes of cells in these pollen grains. As a result, when they are rehydrated, these gel phase domains undergo a transition to the liquid crystalline phase, and the pollen grains leak. Heating them above the transition temperature before they are rehydrated obviates the hydration-dependent phase transition, and they do not leak.

We conclude that the principles that arose from studies on the hydration dependent phase properties of pure phospholipids apply equally well to native membranes and to intact cells.

## References

1. H. Hauser. In *Water: A Comprehensive Treatise* (ed. F. Franks), Vol. 4. Plenum Press: New York, 1975.
2. H. Hauser. In *Water Relations of Foods* (ed. R. B. Duckworth). Academic Press: London, 1975.
3. L. M. Crowe and J. H. Crowe. In *Advances in Membrane Fluidity* (eds R. C. Aloia, C. C. Curtain and L. M. Gordon), Vol. 3. Alan R. Liss: New York, 1988.
4. D. Chapman, R. M. Williams and B. D. Ladbrooke. *Chem. Phys. Lipids* **1**, 445(1967).
5. M. Kodama, M. Kuwabara and S. Seki. *Biochim. Biophys. Acta* **689**, 567(1982).
6. M. Kodama, H. Hashigami and S. Seki. *Thermochimica Acta* **88**, 217(1985).
7. M. Kodama, H. Hashigami and S. Seki. *J. Colloid Interface Sci.*, in press.
8. J. H. Crowe, L. M. Crowe and D. Chapman. *Science* **223**, 701(1984).
9. L. M. Crowe and J. H. Crowe. *Biochim. Biophys. Acta* **946**, 193(1989).
10. J. H. Crowe, L. M. Crowe, J. F. Carpenter and C. Aurell Wistrom. *Biochem. J.* **242**, 1(1987).
11. J. H. Crowe, L. M. Crowe, J. F. Carpenter, A S. Rudolph, C. Aurell Wistrom, B. J. Spargo and T. J. Anchordoguy. *Biochim. Biophys. Acta Biomem. Rev.* **947**, 367(1988).
12. P. J. Quinn. *Cryobiology* **22**, 128(1985).
13. L. M. Crowe, J. H. Crowe and D. Chapman. *Arch. Biochem. Biophys.* **236**, 289(1985).
14. L. M. Crowe, C. Wormersley, J. H. Crowe, D. Reid, L. Appel and A. S. Rudolph. *Biochim. Biophys. Acta* **861**, 131(1986).
15. J. H. Crowe and L. M. Crowe. In *Biological Membranes* (ed. O. Chapman), Vol. 5. Academic Press: London, 1984, pp. 58–103.
16. L. M. Crowe and J. H. Crowe. *Arch. Biochem. Biophys.* **217**, 582(1982).
17. C. Womersley, P. S. Uster, A. S. Rudolph and J. H. Crowe. *Cryobiology* **23**, 245(1986).
18. L. M. Crowe and J. H. Crowe. In *Membranes, Metabolism, and Dry Organisms* (ed. A. C. Leopold). Cornell University Press: Ithaca, NY, 1986, pp. 210–30.
19. J. H. Crowe, L. M. Crowe and F. A. Hoekstra. *J. Bioenergetics Biomem.* **21**, 77(1989).
20. A. C. Leopold (ed.). *Membranes, Metabolism, and Dry Organisms.* Cornell University Press: Ithaca, NY, 1986.
21. V. Luzzati and F. J. Husson. *J. Cell Biol.* **12**, 207(1962).
22. F. Reiss-Husson. *J. Mol. Biol.* **25**, 363(1967).
23. J. H. Crowe, B. J. Spargo and L. M. Crowe. *Proc. Nat. Acad. Sci. USA* **84**, 1537(1987).
24. N. J. Salsbury, A. Darke and D. Chapman. *Chem. Phys. Lipids* **8**, 142(1972).

**25.** H. Hauser, I. Pascher, R. H. Pearson and S. Sundell. *Biochim. Biophys. Acta* **650**, 21(1981).
**26.** M. J. Janiak, D. M. Small and G. G. Shipley. *J. Biol. Chem.* **254**, 6068(1979).
**27.** A. Mellier and A. Diaf. *Chem. Phys. Lipids.* **46**, 51(1987).
**28.** S. C. Chen, J. M. Sturtevant and B. J. Gaffney. *Proc. Nat. Acad. Sci. USA* **77**, 5060(1980).
**29.** L. Finegold and M. A. Singer. *Biochim. Biophys. Acta* **855**, 417(1986).
**30.** D. G. Cameron and H. H. Mantsch. *Biophys. J.* **38**, 175(1982).
**31.** H. H. Fuldner. *Biochemistry* **20**, 5707(1981).
**32.** M. J. Ruocco and G. G. Shipley. *Biochim. Biophys. Acta* **684**, 59(1982).
**33.** S. F. Bush, H. Levin and I. W. Levin. *Chem. Phys. Lipids* **27**, 101(1980).
**34.** N. O. Petersen, P. A. Kroon, M. Kainosho and S. I. Chan. *Chem. Phys. Lipids* **14**, 343(1975).
**35.** H. L. Casal, H. H. Mantsch, D. G. Cameron and B. P. Gaber. *Chem. Phys. Lipids* **33**, 109(1983).
**36.** J. M. Seddon, G. Cevc and D. Marsh. *Biochemistry* **22**, 1280(1983).
**37.** M. Caffrey. *Biochemistry* **24**, 4826(1985).
**38.** R. M. Epand. *Chem. Phys. Lipids* **36**, 387(1985).
**39.** C. Aurell Wistrom, L. M. Crowe, B. Spargo and J. H. Crowe. *Biochim. Biophys. Acta*, in press.
**40.** P. R. Cullis and B. de Kruijff. *Biochim. Biophys. Acta* **513**, 31(1978).
**41.** J. M. Boggs. *Can. J. Biochem.* **58**, 755(1980).
**42.** J. M. Boggs. In *Membrane Fluidity* (eds M. Kates and L. A. Manson), Vol. 12. Plenum Press: New York, 1984, pp. 3–53.
**43.** J. Gagne, L. Stamatatos, T. Diacovo, S. W. Hui, P. L. Yeagle and J. R. Silvius. *Biochemistry* **24**, 4400(1985).
**44.** T. J. McIntosh and S. A. Simon. *Biochemistry* **25**, 4948(1986).
**45.** L. J. Lis, M. McAlister, N. Fuller, R. P. Rand and V. A. Parsegian. *Biophys. J.* **37**, 657 (1982).
**46.** B. D. Ladbrooke and D. Chapman. *Chem. Phys. Lipids* **3**, 304(1969).
**47.** E. G. Finer and A. Darke. *Chem. Phys. Lipids* **12**, 1(1974).
**48.** G. Cevc and D. Marsh. *Biophys. J.* **47**, 21(1985).
**49.** D. A. Wilkinson and J. F. Nagle. In *Liposomes: From Physical Structure to Therapeutic Application* (ed. A. Knight). Elsevier/North-Holland Biomedical Press: Amsterdam, 1981, pp. 273–97.
**50.** D. M. LeNeveu, R. P. Rand, V. A. Parsegian and D. Gingell. *Biophys. J.* **18**, 209(1977).
**51.** K. Harlos and H. Eibl. *Biochemistry* **20**, 2888(1981).
**52.** D. Marsh and J. M. Seddon. *Biochim. Biophys. Acta* **690**, 117(1982).
**53.** J. R. Silvius, P. M. Brown and T. J. O'Leary. *Biochemistry* **25**, 4249(1986).
**54.** P. M. Brown, J. Steers, S. W. Hui, P. L. Yeagle and J. R. Silvius. *Biochemistry* **25**, 4259(1986).
**55.** H. Chang and R. M. Epand. *Biochim. Biophys. Acta* **728**, 319(1983).
**56.** J. M. Seddon, K. Harlos and D. Marsh. *J. Biol. Chem.* **258**, 3850(1983).
**57.** D. A. Wilkinson and J. F. Nagle. *Biochemistry* **23**, 1538(1984).
**58.** S. Mulukutla and G. G. Shipley. *Biochemistry* **23**, 2514 (1984).
**59.** H. H. Mantsch, S. C. Hsi, K. W. Butler and D. G. Cameron. *Biochim. Biophys. Acta* **728**, 325(1983).

60. E. Rivas and V. Luzzati. *J. Mol. Biol.* **41**, 261(1969).
61. K. Gounaris, D. A. Mannock, A. Sen, A. P. R. Brain, W. P. Williams and P. J. Quinn. *Biochim. Biophys. Acta* **732**, 229(1983).
62. A. Wieslander, J. Ulmius, G. Lindblom and K. Fontell. *Biochim. Biophys. Acta* **512**, 241(1978).
63. G. G. Shipley, J. P. Green and B. W. Nichols. *Biochim. Biophys. Acta* **311**, 531(1983).
64. R. V. McDaniel. *Biophys. J.* **49**, 138a(1986).
65. H. C. Jarrell, J. B. Giziewicz and I. C. P. Smith. *Biochemistry* **25**, 3950(1986).
66. Y. Barenholz and T. E. Thompson. *Biochim. Biophys. Acta* **604**, 129(1980).
67. I. Pascher. *Biochim. Biophys. Acta* **455**, 433(1976).
68. E. Freire, D. Bach, M. Correa-Freire, I. Miller and Y. Barenholz. *Biochemistry* **19**, 3662(1980).
69. R. Skarjune and E. Oldfield. *Biochemistry* **21**, 3154(1982).
70. G. G. Shipley, L. S. Avecilla and D. M. Small. *J. Lipid Res.* **15**, 124(1974).
71. T. N. Estep, W. I. Calhoun, Y. Barenholz, R. L. Biltonen, G. G. Shipley and T. E. Thompson. *Biochemistry* **19**, 20(1980).
72. M. J. Ruocco, D. Atkinson, D. M. Small, R. P. Skarjune, E. Oldfield and G. G. Shipley. *Biochemistry* **20**, 5957(1981).
73. K.-A. Karlsson, B. E. Samuelsson and G. O. Steen. *Biochim. Biophys. Acta* **306**, 317(1973).
74. Y. Barenholz, J. Suurkuusk, D. Mountcastle, T. E. Thompson and R. L. Biltonen. *Biochemistry* **15**, 2441(1976).
75. R. P. Rand and V. A. Parsegian. *Ann. Rev. Physiol.* **48**, 201(1986).
76. J. Coorssen and R. P. Rand. *Studia Biophys.* **127**, 53(1988).
77. D. M. LeNeveu, R. P. Rand and V. A. Parsegian. *Nature* **259**, 601(1976).
78. E. A. Evans and V. A. Parsegian. *Proc. Nat. Acad. Sci., USA* **83**, 7132(1986).
79. E. A. Evans and D. Needham. *J. Phys. Chem.* **91**, 4219(1987).
80. T. J. McIntosh, A. D. Magid and S. A. Simon. *Biochemistry* **26**, 7325(1987).
81. D. C. Rau, B. K. Lee and V. A. Parsegian. *Proc. Nat. Acad. Sci., USA* **81**, 2621(1984).
82. J. N. Israelachvili and G. E. Adams. *J. Chem. Soc. Faraday Trans.* **74**, 975(1978).
83. J. N. Israelachvili and J. Marra. *Meth. Enzymol.* **127**, 353(1986)
84. A. C. Cowley, N. L. Fuller, R. P. Rand and V. A. Parsegian. *Biochemistry* **17**, 3163(1978).
85. A. Portis, C. Newton, W. Pangborn and D. Papahadjopoulos. *Biochemistry* **18**, 780(1979).
86. K. I. Florine and G. W. Feigenson. *Biochemistry* **26**, 1757(1987).
87. L. Ter-Minassian-Saraga and G. Madelmont. *FEBS Lett.* **137**, 137(1982).
88. S. A. Simon, T. J. McIntosh and R. Latorre. *Science* **185**, 65(1982).
89. L. Stamatatos and J. R. Silvius. *Biochim. Biophys. Acta* **905**, 81(1987).
90. T. J. McIntosh, A. D. Magid and S. A. Simon. *Biochemistry* **28**, 17(1989).
91. S. Nir. *Prog. Surf. Sci.* **8**, 1(1976).
92. P. Attard, D. J. Mitchell and B. W. Ninham. *Biophys. J.* **53**, 457(1988).
93. J. H. Crowe, F. A. Hoekstra, L. M. Crowe, T. Anchordoguy and E. Drobnis. *Cryobiology* **26**, 76(1989).

**94.** R. M. Gendreau (ed.). *Spectroscopy in the Biomedical Sciences.* CRC Press: Boca Raton, FL, 1987.
**95.** D. G. Cameron and R. A. Dluhy. In *Spectroscopy in the Biomedical Sciences (ed. R. M. Gendreau). CRC Press: Boca Raton, FL,* 1987, *pp.* 53–86.
**96.** D. G. Cameron, A. Martin and H. H. Mantsch. *Science* **219**, 180 (1983).
**97.** J. H. Crowe, F. A. Hoekstra and L. M. Crowe. *Proc. Nat. Acad. Sci., USA* **86**, 520(1989).
**98.** M. C. Phillips, B. D. Ladbrooke and D. Chapman. *Biochim. Biophys. Acta* **196**, 35(1970).
**99.** M. C. Phillips, H. Hauser and F. Paltauf. *Chem. Phys. Lipids* **8**, 127(1972).
**100.** D. Furtado, W. P. Williams, A. P. R. Brain and P. J.Quinn. *Biochim. Biophys. Acta* **555**, 352(1979).
**101.** A. Carruthers and D. L. Melchior. *Biochemistry* **22**, 5797(1983).
**102.** D. Papahadjopoulos, K. Jacobson, S. Nir and T. Isaac. *Biochim. Biophys. Acta* **311**, 330(1973).
**103.** M. C. Blok, E. C. M. van der Neut-Kok, L. L. M. van Deenen and J. de Gier. *Biochim. Biophys. Acta* **406**, 187(1975).
**104.** M. C. Blok, L. L. M. van Deenen and J. de Gier. *Biochim. Biophys. Acta* **433**, 1(1976).
**105.** D. Marsh, A. Watts and P. F. Knowles. *Biochemistry* **15**, 3570(1976).
**106.** R. I. MacDonald. *Biochemistry* **24**, 4058(1985).
**107.** J. Wilschut and D. Hoekstra. *Trends in Biochem. Sci.*, 479(November, 1984).
108. G. Strauss, P. Schurtenberger and H. Hauser. *Biochim. Biophys. Acta* **858**, 169(1986).
**109.** A. S. Rudolph, J. H. Crowe and L. M. Crowe. *Arch. Biochem. Biophys.* **245**, 134(1986).
**110.** A. S. Rudolph and J. H. Crowe. *Cryobiology* **22**, 367(1985).
**111.** T. D. Anchordoguy, J. F. Carpenter, S. H. Loomis and J. H. Crowe. *Biochim. Biophys. Acta* **946**, 299(1989).
**112.** J. Wilschut and D. Papahadjopoulos. *Nature* **281**, 690(1979).
**113.** N. Duzgunes, R. M. Straubinger, P. A. Baldwin, D. S. Friend and D. Papahadjopoulos. *Biochemistry* **24**, 3091(1985).
**114.** P. R. Cullis and M. J. Hope. *Nature* **271**, 672(1978).
**115.** L. T. Boni, T. P. Stewart and S. W. Hui. *J. Membr. Biol.* **80**, 91(1984).
**116.** S. W. Hui, T. P. Stewart, P. L. Yeagle and A. D. Albert. *Biochim. Biophys. Acta* **207**, 227(1981).
**117.** M. M. Hammoudah, S. Nir, J. Bentz, E. Mayhew, T. P. Stewart, S. W. Hui and B. J. Kurian. *Biochim. Biophys. Acta* **645**, 102(1981).
**118.** F. A. Hoekstra. *Plant Physiol.* **74**, 815(1984).

# Hydration of nucleic acids

E. WESTHOF

*Institut de Biologie Moléculaire et Cellulaire, Centre National de la Recherche Scientifique, 15, Rue R. Descartes, F-67084 Strasbourg-Cedex, France*

AND

D. L. BEVERIDGE

*Chemistry Department, Hall-Atwater Laboratories, Wesleyan University, Middletown, CT 06457, USA*

## 1. Introduction

It is common knowledge that living tissue contains around 70 % water, that water is certainly a unique liquid, and consequently that water plays a crucial role in both the structures and functions of biological macromolecules. The native conformations of biomolecules result from a balance between various intra- and intermolecular forces, e.g., interatomic repulsions, dispersion forces, hydrogen bonding forces, electrostatic forces, torsional forces about covalent bonds, and forces resulting from interactions with the solvent. The net result for proteins is a hydrophobic polypeptide core protected from the solvent creating regions with a microscopic environment controlled and adapted for specific functions. In nucleic acids, tertiary structure is a result of an equilibrium between electrostatic forces due to the negatively charged phosphates, stacking interactions between the bases due partially to hydrophobic and dispersion forces, hydrogen bonding interactions between the polar substituents of the bases, and the conformational energy of the sugar–phosphate backbone. In its preferred conformations, the polynucleotide backbone exposes the negatively charged phosphates to the dielectric screening by the solvent and promotes the stacked helical arrangement of adjacent bases. In this way, a hydrophobic core is created where hydrogen bond formation between bases as well as additional sugar–base and sugar–sugar interactions are favored. In such helical structures, only the internal atoms involved in hydrogen bonding between the bases are protected from solvent with most of the other atoms accessible to water. Water can influence conformation because it is a highly polarizable dielectric medium, because it competes with intramolecular interactions like hydrogen bonding, and because it gives rise to hydrophobic interactions. Thus, water molecules can contribute to the stability of helical and non-helical conformations of nucleic acids by screening the charges of the phosphates, by hydrogen bonding to the polar exocyclic atoms of the bases, and by influencing the conformations of residues with methyl groups via hydrophobic interactions. Besides, due to the periodicity of the helical conformations, water sites and bridges involving the nucleic acid polar atoms can lead to structured arrangements of water molecules.

### 1.1. *Water and DNA conformations*

The highly crystalline fibre diagram given by DNA fibres is obtained only in a certain humidity range, about 70% to 80%. The general characteristics of the diagram suggest that the DNA chains are in a helical form. [...] This diagram appears to correspond to scattering by individual helical units; [...]. That is, at high humidity a water sheath disrupts the spatial relationships between neighbouring helices, and only the parallelism of their axes is preserved. During the change 'crystalline → wet' a considerable increase in length of the fibres occurs. The helix in the wet state is therefore presumably not identical with that of the crystalline state.

R. Franklin, cited by A. Sayre in *Rosalind Franklin & DNA*. [1]

Interest in environmental effects on DNA conformations clearly started with the work of Franklin and Gosling [2] in the early 50s. Since those pioneering works, an enormous amount of data has been gathered by various physico-chemical techniques on the effects of humidity, salts, small organic molecules, etc. on nucleic acid conformations. An overview is given in the book by W. Saenger. [3] However, despite the accumulation of data, there is, as stated by Maddox, [4] 'only the foggiest understanding of how the environment determines the structure of A and/or B, or why particular nucleotide sequences should predispose towards Z instead'. This review is concerned mainly with precise structural and theoretical information on the modes of interaction between water molecules and nucleic acids. Besides, since nucleic acids are polyelectrolytes, the interactions with hydrated ions are also considered. In Section 5.2, the main suggestions made toward a general understanding of nucleic acid transitions will be presented and discussed. But before, with the hope of not behaving like the drunk looking for his lost key under a street-lamp because that's where the light is, [4] the arrangement of water molecules around nucleic acid fragments in crystals will be analysed in some detail.

In the appendix, most of the terms commonly used for describing nucleic acid conformations are defined and the main types of nucleic acid helices are delineated.

## 2. Hydration of bases, nucleosides, and nucleotides

In a 1969 review on water structure in organic hydrates, Jeffrey [5] remarked that surveys of crystal structures give a better view of crystallographers' interests than of water structure. Although the situation has greatly improved

Table 1. *Classification of crystal hydrates adapted from Jeffrey* [5] *with emphasis on nucleic acid components.*

| Water framework | Crystal hydrates |
|---|---|
| *Infinite three-dimensional hydrogen-bonded networks* | Ices |
| *Pentagonal dodecahedron clathrates* | Rare gas hydrates |
| *Liquid-like water trapped between the packed crystallized macromolecules*: First hydration shell of the macromolecule is ordered. The structure is determined by intermolecular contacts between the macromolecules. | 'Sea ice', *t*RNAs |
| *Two- and one-dimensional hydrogen bonded networks*: Water molecules form nets, sheets, columns, ribbons, or chains. The structure is strongly influenced by the functional groups of the solute molecules. | Polynucleotides, nucleotides |
| *Isolated water molecules*: Water molecules have mainly a space-filling function in the crystal. The structure is determined principally by the hydrogen bonding functions of the solute molecules. | Bases, nucleosides |

since then, especially in the small size molecule field, this remark is still valid today for the larger size molecules since, as before, the environment of the water molecules is only infrequently described in detail. With large structures, the problem is compounded by the limitations of X-ray diffraction methods: not only is it impossible to locate the hydrogen atoms but also the precise location and the unambiguous identification of the oxygen atoms are difficult, subjective sometimes, and controversial. The problems inherent to the determination of water structure around biomolecules have been discussed in recent reviews and will not be detailed here. [6–8]

Following Jeffrey, [5] crystal hydrates can be classified according to the type of framework built by the water molecules with the high hydrates displaying three-dimensional networks and the low hydrates using water molecules mainly as a filler of the crystal lattice (Table 1). Crystals of large nucleic acids have a high solvent content, as high as 75% in some *t*RNA crystals, [9] making them akin to sea ice in which tiny pockets of brine are entrapped between interconnected ice crystals. In crystals of such large nucleic acids, most of the solvent molecules occupy the channels and voids between the packed crystallized molecules of the solute and can be considered in a state identical to bulk-phase solution. Those solvent molecules lead to featureless electron density. A small part of the crystal solvent is crystallographically ordered and forms the first hydration shell or the bound water of the macromolecule. However, the identification and analysis of this hydration shell is important for understanding the role of water in the structure and stability of nucleic acids. On the other hand, in the small

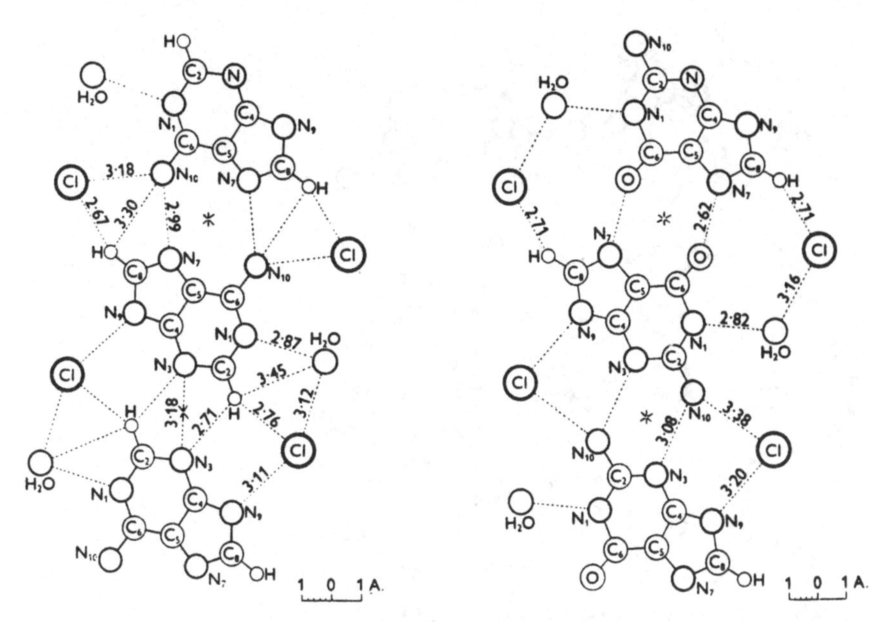


Figure 1. Views of the hydrogen bonding environment in the crystal structures of adenine hydrochloride hemihydrate **[10]** (left) and of guanine hydrochloride monohydrate **[11]** (right). The adenine molecules are protonated at the N1 position and the guanine molecules at the N7 position. In its normal tautomeric form, there is a hydrogen atom bound at position N1 of the guanine. Notice that in both instances the self-pairs are formed via Hoogsteen-type hydrogen bonding and not Watson–Crick (N10 should be N6 in adenine and N2 in guanine): N6—H ··· N7 in adenine pairs and N7—H⁺ ··· O6 in guanine pairs. In addition, guanine molecules hydrogen bond via N3 and the exocyclic amino group N2. In each crystal, the unique water molecule, with an approximate tetrahedral arrangement, donates its protons to two chloride anions (at 3.12 Å) and accepts protons from two imino nitrogens (N1 ··· Ow ··· N1 in adenine, 2.81 Å, and N1 ··· Ow ··· N1, 2.82 Å, N1 ··· Ow ··· N9, 3.27 Å, in guanine). Reprinted with permission from refs 10 and 11.

hydrates, the crystal structure is dominated by intermolecular hydrogen bonding between the bases (Figure 1), by ionic coordination between phosphate groups and bases (Figure 2), by stacking between the bases (Figure 3), or by the strong hydrogen bonding tendency of the hydroxyl groups of sugar moieties (Figure 4). As a consequence, in the small hydrates, the structural role of water molecules is secondary and restricted to filling the interstitial spaces in the crystal lattice. In fact, in several instances, two water molecules symmetry-related by a two-fold screw axis are twice hydrogen bonded to each other with the other hydrogen atoms interacting with the solute molecules (Figures 4 and 5). In the latter two structures, the symmetry generates an infinite two-fold helix or zigzag chain of water molecules along a crystallographic axis. Such an arrangement implies a chain of three-centred hydrogen bonds. In the structure shown in Figure 3, the water molecules form instead hydrogen bonded ribbons between nucleoside molecules. In-

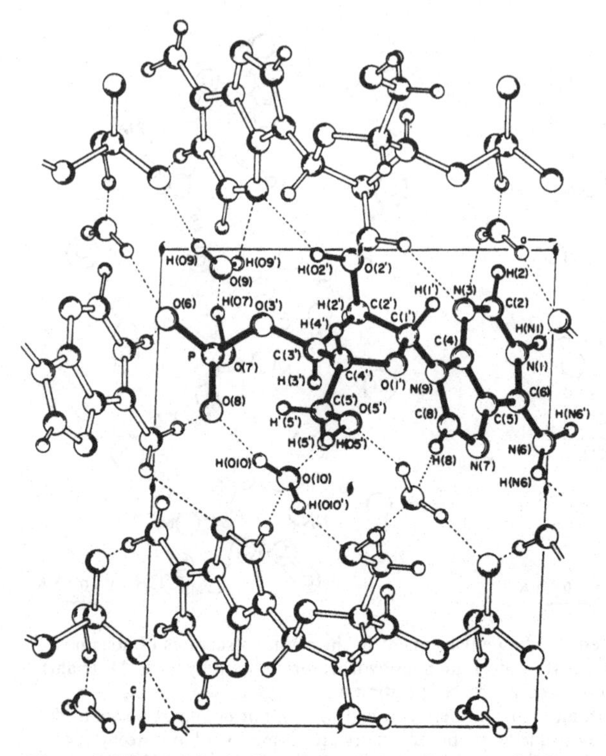


Figure 2. View of the hydrogen bonding environment in the crystal structure of adenosine 3′-phosphate dihydrate. [12] One water has a close to tetrahedral arrangement with a long distance to the C8—H of an adenine (3.26 Å and 2.24 Å for C8 ··· Ow and H(C8) ··· Ow, respectively). The second molecule has only three neighbours, with one short contact with the protonated phosphate oxygen, O7 ··· Ow 2.61 Å and H(O7) ··· Ow 1.56 Å, and a longish one with the ring nitrogen N3 of the adenine, N3 ··· Ow 3.13 Å and H(Ow) ··· N3 2.40 Å. Reprinted with permission from ref. 12.

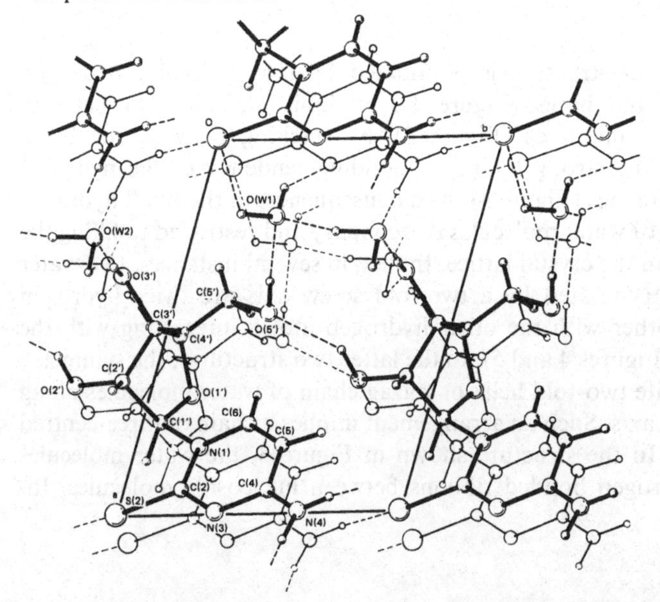

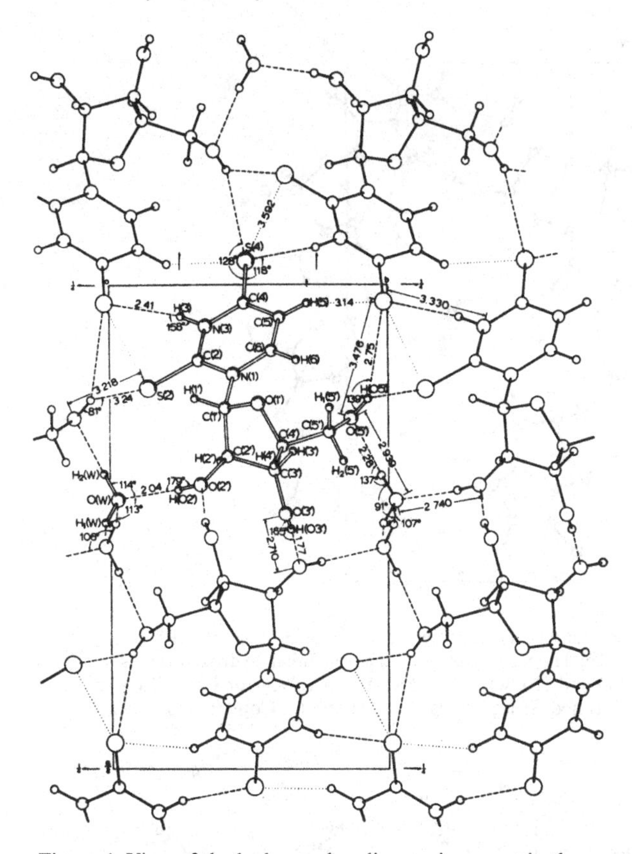


Figure 4. View of the hydrogen bonding environment in the crystal structure of 2,4-dithiouridine monohydrate. **[14]** Two-fold screw related water molecules form hydrogen bonds with one another in an approximately tetrahedral arrangement. The symmetry generates an infinite two-fold helix in the direction perpendicular to the view. Reprinted with permission from ref. 14.

between those two extremes, crystals of nucleotides or polynucleotides display various one- and two-dimensional hydrogen bonded networks of water molecules hydrating the nucleic acids. It was noticed by Clark **[16]** that, although the environment of the water molecules is usually close to tetrahedral (whatever the number of neighbours), the water molecule is also quite often surrounded by a planar three-fold array of ligands. This early

---

Figure 3. View of the hydrogen bonding environment in the crystal structure of 2-thiocytidine dihydrate. **[13]** The two water molecules are hydrogen bonded to each other (2.74 Å and 1.87 Å for Ow···Ow and H(Ow)···Ow, respectively, with an angle Ow—H···Ow of 141°) with one water molecule making a weak hydrogen bond to the sulfur atom (3.28 Å and 2.74 Å for S2···Ow and S2···H(Ow), respectively). Reprinted with permission from ref. 13. Copyright © (1971) American Chemical Society.

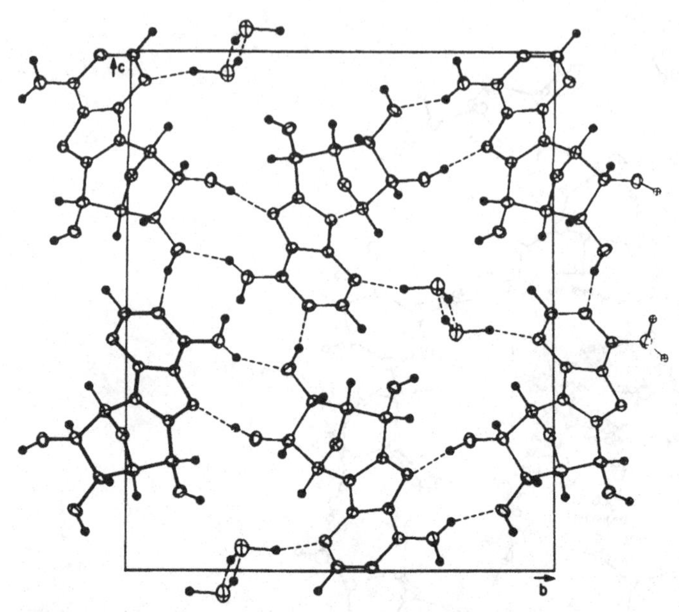


Figure 5. View of the hydrogen bonding environment in the crystal structure of 8,5′-cycloadenosine monohydrate. [15] The water molecule makes similar hydrogen bonds: 2.86 Å and 1.74 Å for Ow ··· N3 and H(Ow) ··· N3; 2.87 Å and 1.99 Å for Ow ··· Ow and H(Ow) ··· Ow, respectively. Reprinted with permission from ref. 15. Copyright © 1980 American Chemical Society.

Figure 6. The hydrogen bonding environment in some neutral adenine derivatives: (*a*) 3′-O-acetyladenosine, [18] (*b*) 2′-amino-2′-deoxy-α-D-adenosine, [19] (*c*) deoxyadenosine. [20] The same for some protonated adenine derivatives: (*d*) adenosine-3′-phosphate, [12] (*e*) adenosine-2′,5′-uridine phosphate. [21] The adenine protonated at the N1 position exhibits a tendency for C8—H ··· Ow interaction. Reprinted with permission from ref. 18. Copyright © 1971 American Chemical Society.

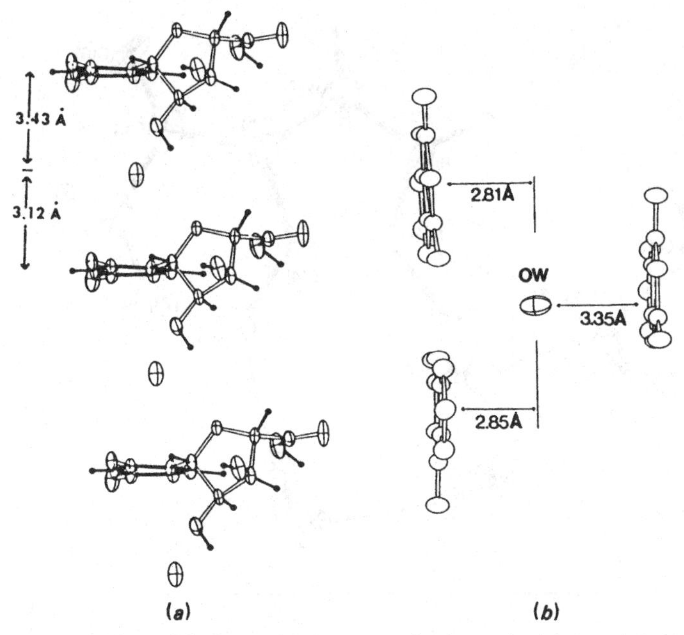


Figure 7. View of the intercalated water molecules (*a*) in the crystal structure of 5-nitro-1-(β-D-ribosyluronic acid)-uracil monohydrate (the 5-nitro group is omitted): sandwich structure; and (*b*) in the crystal structure of 8-azaguanine hydrochloride monohydrate: half-sandwich. Reprinted with permission from ref. 22. Copyright © 1981 Adenine Press.

observation is upheld by the most recent review of hydrogen bonding in crystal structures of nucleic acid components **[17]** in which about one-third of the observed three-coordinated water molecules have a planar environment.

## 2.1. *Small hydrates*

Despite limitations inherent to the method, **[6–8]** important information can be gathered from analysis of crystal structures of small hydrates. For example, a comparison of crystal structures of adenine derivatives led Rao and Sundaralingam **[18]** to suggest that protonated adenines have a tendency for C8—H···O hydrogen bonding, apparently through an increase in the polarizability of the C8—H bond (Figure 6). Parthasarathy and coworkers **[22]** discovered the intercalation of water molecules between nucleic acid bases (Figure 7). In such 'sandwich structures', there is no stacking of the bases since they are separated by the water molecules. Also, the water molecules participate only in three instead of four hydrogen bonds without hydrogen bonding to the bases above or below.

Characteristic patterns of hydration seen in crystal structures of nucleic acid constituents were later observed in oligomeric structures. For example,

(*a*)

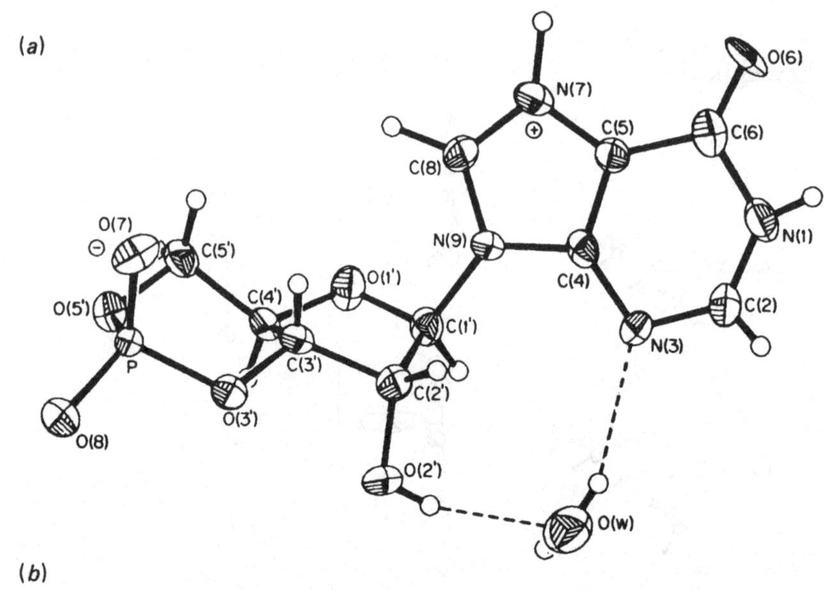


(*b*)

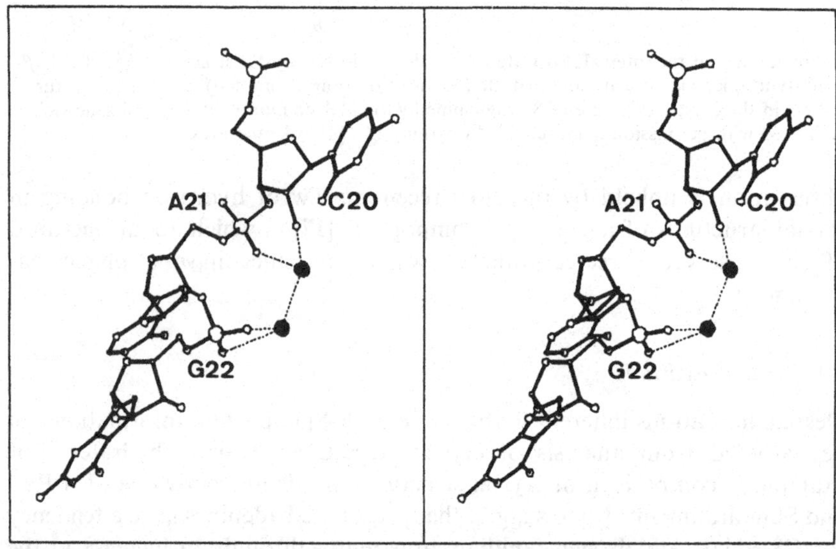


Figure 8(*a*). View of the water bridge linking the sugar hydroxyl group O2′ with the adenine ring N3 atom in the crystal structure of 3′,5′-cIMP. [23] The distances and angles are as follows: O2′ ··· Ow and H(O2′) ··· Ow, 2.74 Å and 1.79 Å, with the angle O2′—H—Ow of 166°; Ow ··· N3 and H(Ow) ··· N3, 2.99 Å and 2.01 Å, with the angle Ow—H—N3 of 166° also. The water bond angle is 111°. The water molecule makes a fourth hydrogen bond to an anionic phosphate oxygen with distances 2.93 Å and 2.14 Å, respectively for Ow ··· O1P and H(Ow) ··· O1P. This type of water bridge requires the *anti* orientation of the base with respect to the sugar and has only been observed with the sugar in the C3′-*endo* pucker domain. An identical water bridge is frequently observed in ribopyrimidines between O2′ and the exocylic O2 atom of the pyrimidine base in the *anti* orientation. Reprinted with permission from ref. 23. (*b*) 3′-Phosphate–water–sugar bridge in *yeast t*RNA-asp in a sharp turn region. Notice the water bridge between phosphates 21 and 22.

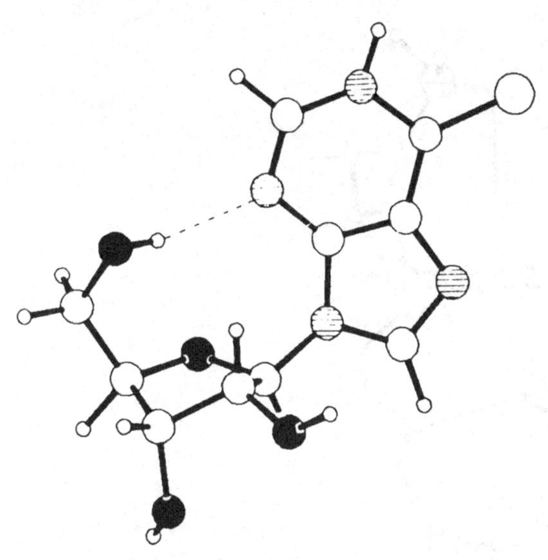

Figure 9. View of the intramolecular hydrogen bond between the sugar hydroxyl group O5′ and the adenine ring N3 atom. This folded conformation of purinenucleosides requires the *syn* orientation of the base with respect to the sugar and occurs preferentially with the sugar in the C2′-*endo* domain and the *gauche*$^+$ conformation of the exocyclic group O5′. No naturally occurring pyrimidinenucleoside has been observed in the *syn* orientation. From ref. 30.

RNA structures frequently display a water bridge between the O2′ hydroxyl group and the base atom N3 or O2 (Figure 8(*a*)). Such a water bridge has been noticed in small RNA helices, [**24, 25**] in RNA-drug intercalation complexes, [**26, 27**] and in *t*RNA crystals. [**28**] However, water bridges between the ribose hydroxyl group O2′ and an anionic phosphate oxygen of the 3′-phosphate, as suggested by NMR investigations on several polynucleotides, [**29**] were never observed in helical conformations of nucleic acids. Nevertheless, when the sugar–phosphate backbone makes sharp turns and adopts non-helical conformations as in *t*RNA structures, 3′-phosphate–water–sugar bridges are observed (Figure 8(*b*)). This observation stresses the stabilizing roles and structural importance of water bridges in non-standard conformations. Purinenucleosides can easily adopt the *syn* orientation about the glycosyl bond [**30**] (Figure 9). In such a conformation, the occurrence of the water bridge between O2′ and N3 is prevented and there is commonly a hydrogen bond between the hydroxyl O5′ and the base atom N3. When in the *syn* orientation, 5′-purinenucleotides, in which the O5′—H···N3 intramolecular hydrogen bond cannot take place, are stabilized instead by a water bridge between the guanine amino group and an anionic oxygen atom of the 5′-phosphate (Figure 10). This situation is observed in the Z-DNA forms. [**33**] The N7 sites of purine derivatives are frequently acceptors of hydrogen bonds (Figures 10 and 11). In some instances, the amino group in position

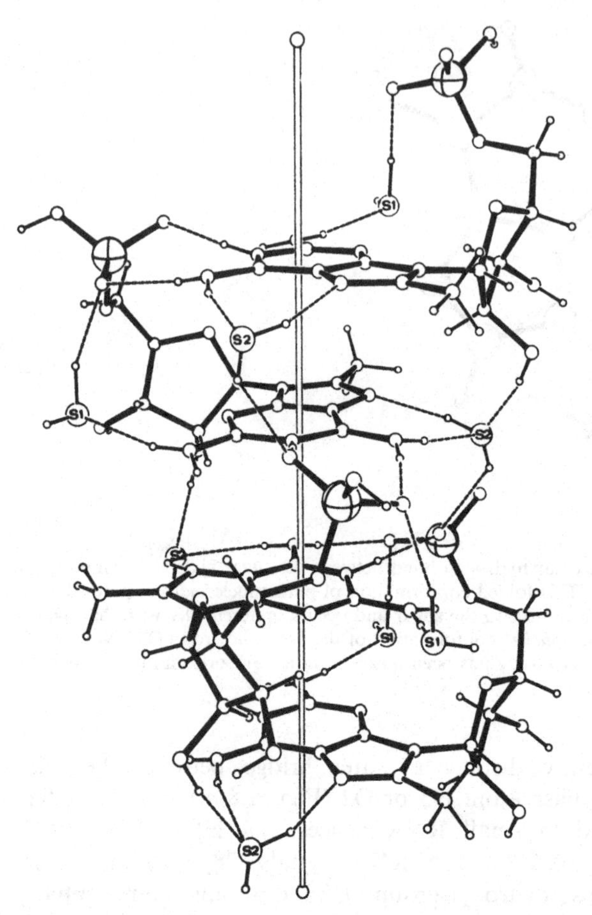


Figure 10. View of the hydrogen bonding environment in the crystal structure of 2-amino-8-methyladenosine-5′-monophosphate dihydrate. [31] Water molecule S1 bridges the guanine amino group N2 to an anionic oxygen of the 3′-phosphate with distances for N2···Ow and H(N2)···Ow of 2.88 Å and 2.03 Å; for Ow···O1P and H(Ow)···O1P of 2.95 Å and 1.90 Å, respectively. Water molecule S1 makes a third hydrogen bond to itself symmetrically-related (2.97 and 2.23 Å). Water molecule S2 has four ligands: N7 of the adenine (2.81 and 1.93 Å), N6 of the same adenine (3.12 and 2.33 Å), O2′ of an adjacent ribose (2.69 and 1.86 Å), O2P of another anionic phosphate oxygen (2.80 and 2.19 Å). The adenine moiety is protonated at N1 and the N6 and N1 nitrogens are linked to two anionic phosphate oxygens as in Figure 2. Interestingly, the interplanar spacing between bases is 3.7 Å, instead of the usual 3.3–3.4 Å, [32] as is observed in the Z-DNA forms. [33] Reprinted with permission from ref. 31. Copyright © 1982 American Chemical Society.

6 can donate a hydrogen bond simultaneously to the water molecule bound at N7, although this is not always clear-cut as shown in Figure 12. Quite often, the two sites are individually hydrated without water bridges. The structure of sodium inosine 5′-phosphate [36] presents the situation in which one water molecule bridges the N7 atoms of two symmetrically-related

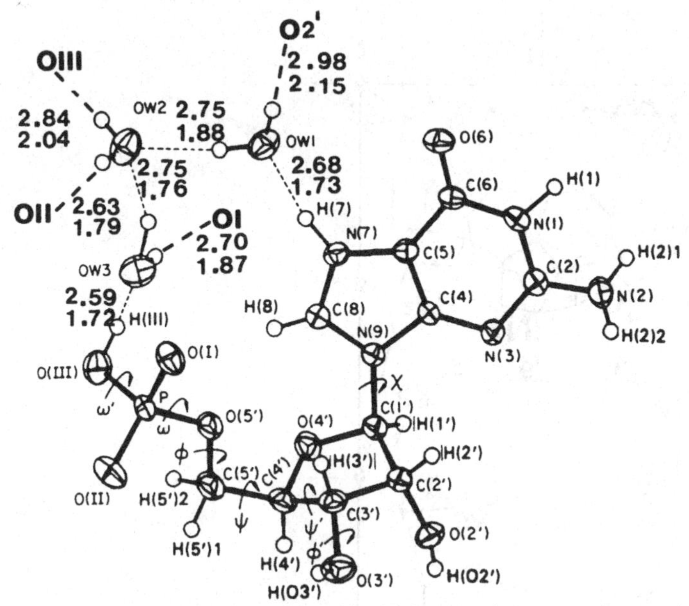


Figure 11. View of the hydration bridge in the crystal structure of guanosine-5′-monophosphate trihydrate. [34] Three water molecules link the guanine protonated at the N7 position to the phosphate oxygen OIII. Only Ow2 makes four hydrogen bonds. Adapted from ref. 34.

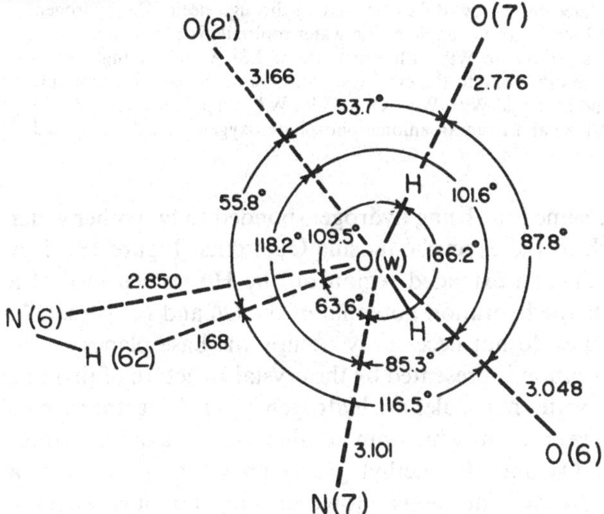


Figure 12. View of the hydrogen bonding environment around the water molecule in the crystal structure of 5′-methyleneadenosine 3′,5′-cyclic monophosphonate monohydrate. [35] The hydrogen atoms were not determined. It is clear, however, that N6—H donates a hydrogen bond to the water and, on the basis of the distances, that the anionic oxygen atom O7 accepts a hydrogen bond from the water. The second hydrogen atom of the water appears to make a three-centered hydrogen bond to the ring nitrogen N7 of the same adenine and to the other anionic oxygen atom O6 of a symmetrically-related molecule. Reprinted with permission from ref. 35. Copyright © 1972 American Chemical Society.

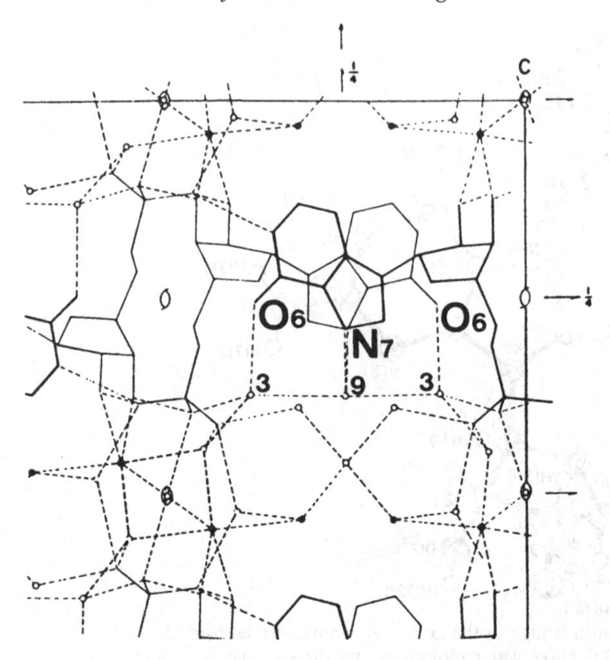


Figure 13. View of two hydration bridges in the crystal structure of a hydrated monosodium inosine-5′-phosphate. [36] Water molecule 9 donates its two protons to the ring nitrogen N7 atoms of two nucleotides related by the two-fold axis passing through itself. The hydrogen bonding distances are both 2.59 Å and the angle at the water molecule is 84°. Water molecule 3 donates one of its protons to W9, with a distance of 2.81 Å and an angle N7—W9—W3 of 90°, and the other one to the exocyclic atom O6 of the guanine molecule, with a distance of 2.73 Å and an angle W9—W3—O6 of 91°. W3 accepts protons from another water molecule (2.91 Å) and from an anionic phosphate oxygen (2.72 Å). Adapted from ref. 36.

molecules, while at the same time being hydrogen-bonded to two other water molecules binding each to the exocyclic inosine O6 atoms (Figure 13). It is clear from Figure 14 that an extended region on the Hoogsteen face of a guanine base represents the hydration potential of the O6 and N7 sites. [37]. Also, the water molecules do not necessarily occupy the base planes.

A very interesting situation is presented by the crystal structure of thymine monohydrate. [38] The water molecules are hydrogen bonded together across glide planes forming zigzag chains which run parallel to the *c*-axis. The water molecules sit between O4 and the methyl group on C5 and also give a hydrogen bond to the exocyclic atom O4. However, the amount of water of crystallization is only 83 % of the theoretical value. Moreover, for the water molecule, the thermal parameter in the *c*-axis direction is extremely large and such that a root-mean square vibrational amplitude of 0.45 Å would be expected for the water oxygen. It was presumed that the *z*-coordinate of the water oxygen represents an average value of a distribution of parameters rather than an actual position. A configuration of indefinite sequences of

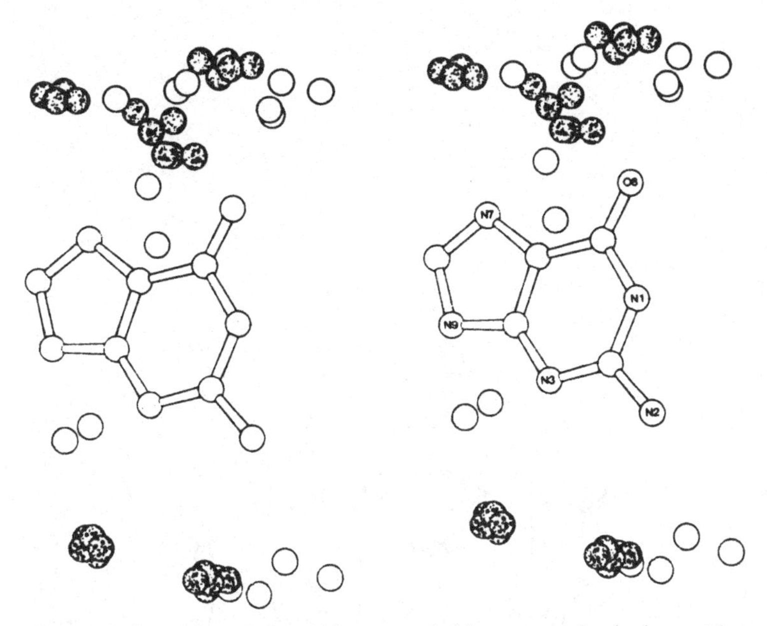


Figure 14. Stereoview of the positions occupied by water molecules in crystal structures of various guanine derivatives. Stippled circles correspond to positions occupied by water molecules in GpC salts. Reprinted with permission from ref. 37. Copyright © 1981 Adenine Press.

short chains of hydrogen-bonded water molecules resulted in which five water molecules were distributed over six consecutive unit cells (Figure 15(*a*)). The crystal structure of guanine monohydrate [**39**] presents an analogous situation (Figure 15(*b*)). In that structure, pairs of guanine molecules are hydrogen bonded, via N1—H ··· O6 between the Watson–Crick and the Hoogsteen sides as well as on the other side via N9—H ··· N3 hydrogen bonds across a center of symmetry. This results in sheets of hexagonal, graphite-like arrays. Through the centers of these hexagonal arrays, water molecules hydrogen bond to each other, to O6, and to the amino group N2. The distances between water oxygen atoms across the centers of symmetry are 2.53 and 2.54 Å, appreciably shorter than usual. Again, the thermal or Debye–Waller parameters for the oxygen water molecule are extremely large with the major axis of the thermal ellipsoid nearly parallel to the axis of the column. Those observations were interpreted in the same way as in thymine monohydrate: the water molecules form short chains within which the *z*-coordinate of the water oxygen varies so that the O ··· O distance is about 2.8 Å with a vacant site once the water molecules lose register with the guanine bases. A very similar situation has been encountered in the crystal structure of the C5-brominated $d(C—G)_3$ in the Z-form [**40**] in which there is a chain of water molecules linked to successive O2 atoms of cytosine molecules in the minor groove. It was noticed that the

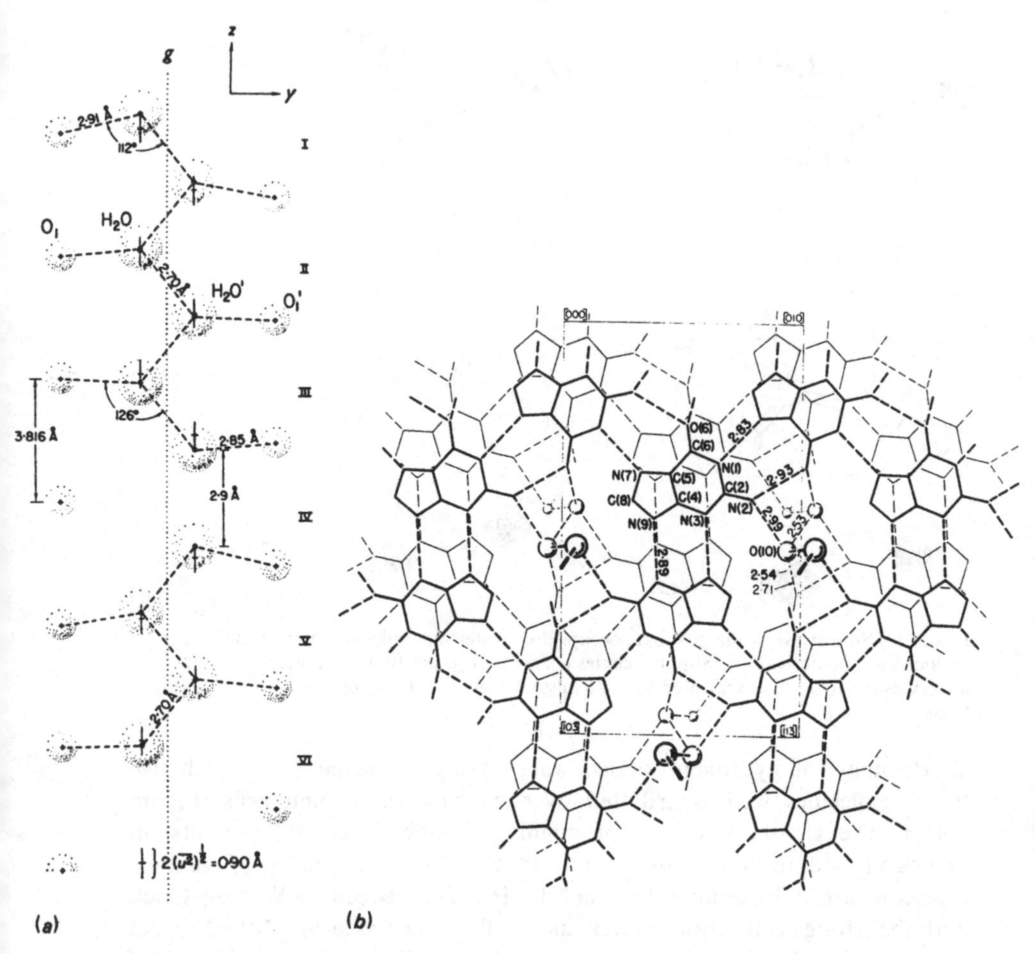


Figure 15. (*a*) Model showing how to distribute five water molecules on each side of a glide plane over six consecutive unit cells so as to avoid the small distance of 2.55 Å between water molecules in the crystal structure of thymine monohydrate. The five water molecules form two short zigzag chains of six and four elements. The horizontal strokes denote the average positions related by the glide plane. Reprinted with permission from ref. 38. (*b*) View down the *c*-axis of the crystal structure of guanine monohydrate displaying the chain of water molecules which presents the same disorder as in (*a*). Reprinted with permission from ref. 39.

occupancies of those water molecules were especially variable and a localized dynamics was inferred.

Crystals of nucleotides, although not necessarily more hydrated than those of bases or nucleosides, display cyclic arrangements of hydrogen bonds in which water molecules participate [41] (Figure 16). At least one of the anionic oxygen atoms of the phosphate group is usually involved in these cycles of

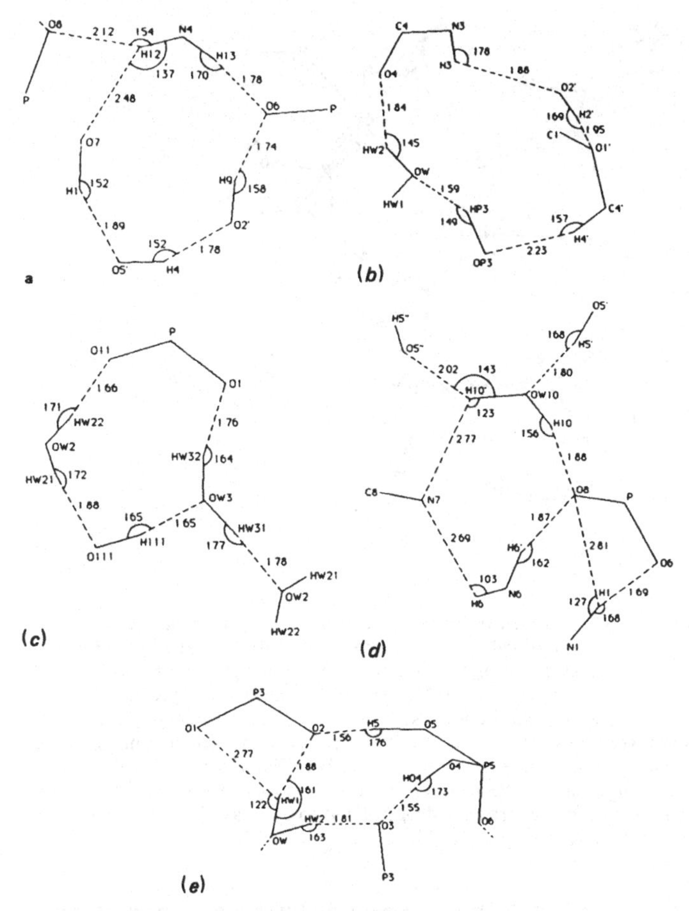


Figure 16. Closed cycles of hydrogen bonds occurring in the structures of (*a*) cytidine-3′-monophosphate, monoclinic form; **[42]** (*b*) uridine-3′-monophosphate monohydrate; **[43]** (*c*) guanosine-5′-monophosphate trihydrate; **[34]** (*d*) adenosine-3′-monophosphate dihydrate; **[12]** (*e*) 2,2′-anhydro-1-β-D-arabinofuranosyl cytosine 3′,5′-diphosphate monohydrate. **[44]** The A···H distances (in Å) and the XH···A angles (in °) are given. The covalent distances OH and NH were normalized to 0.97 and 1.03 Å, respectively. Reprinted with permission from ref. 41. Copyright © 1985 Butterworth and Co. (Publishers) Ltd.

hydrogen bonds. Figure 17 shows several of these cyclic hydrogen bonds in the crystal structure of calcium thymidylate. **[45]**.

The dinucleoside monophosphate dCpG presents the very interesting situation of crystallizing in two different conformations in the presence of various cations: a parallel mini-helix with protonated cytosine self-pairs, $(C—C)^+$ where a proton is shared between two N3 atoms, and guanine self-pairs, G—G, (with 4 water molecules and either a sodium **[46]** or an ammonium **[47]** ion); and a Z-type mini-helix with alternating *anti-syn*

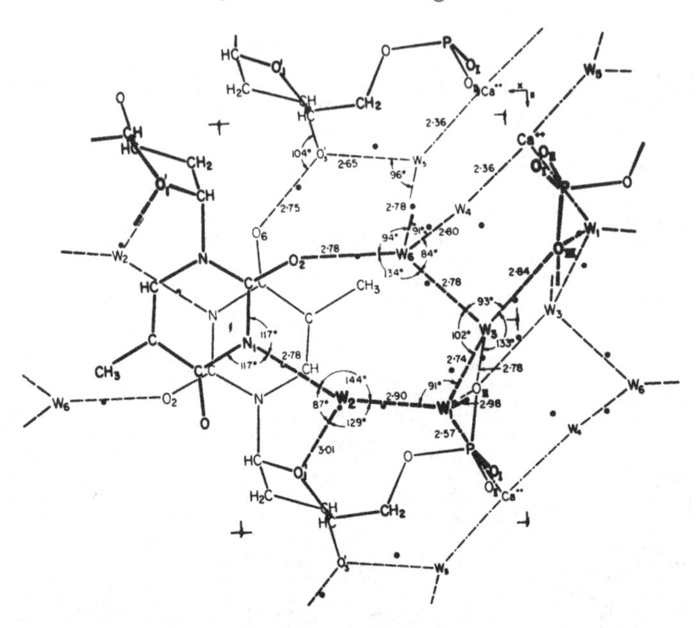

Figure 17. Suggested scheme of hydrogen bonding and calcium ion–water molecule coordination in the crystal structure of calcium 5′-phosphate thymidine hexahydrate **[45]** displaying several several closed cycles of hydrogen bonds. Only one apparent hydrogen bond does not involve a water molecule (O3′ ··· O6; the latter is called O4 in present day nomenclature). There are five water–water contacts (2.74 ··· 2.90 Å). The calcium ion has seven-fold coordination (average Ca—O distance of 2.42 Å): four phosphate anionic oxygens (from three different molecules) and three water molecules, W1, W4, and W5. The apparent thermal motions of these three water molecules are lower than those of the other three water molecules. W2, which makes only one strong hydrogen bond to N1, has the highest thermal amplitude. Reprinted with permission from ref. 45.

orientation for C and G and Watson–Crick pairing (with 6.5 water molecules and an ammonium ion). **[48]** In the first structure, the guanine N7 is hydrogen bonded to the ammonium group which, via a water molecule, interacts with an anionic phosphate oxygen of the 5′-phosphate (see also Figure 18). The second structure presents the first case of a Z helix at the dimer level. In the major groove, there are two intrahelical bridges involving two water molecules each: one interstrand bridge links the N4 amino groups of the two cytosine bases and one intrastrand bridge links the N4 of C1 to the O6 of G1. The first bridge has been noticed in the structure of d(CGCGCG) while, for the second one, an interstrand bridge was observed instead of an intrastrand bridge. **[49]** In the minor groove, a water molecule links the exocyclic O2 atoms of the cytosine bases, simultaneously hydrogen bonding to a water molecule linking the guanine amino group N2 to another water molecule which binds to O1P. This type of network has been systematically observed in larger Z-DNA structures. **[40, 50, 51]** One ammonium ion also links the water molecule bridging the O2s to the O2P.

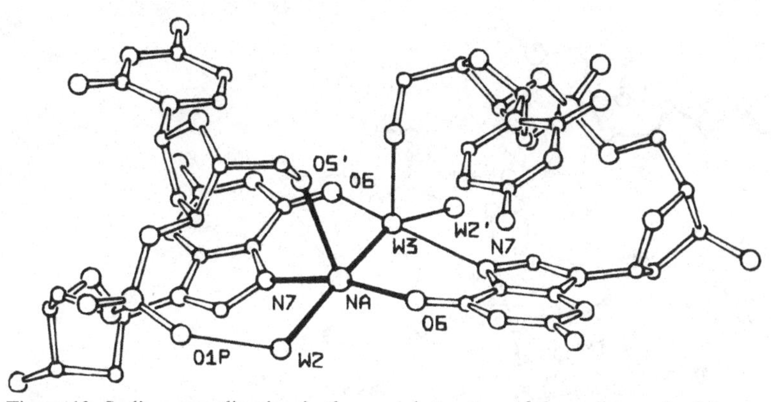


Figure 18. Sodium coordination in the crystal structure of the sodium salt of the dinucleoside dCpG. [46] The ammonium ion adopts a similar mode of intermolecular binding in the ammonium salt of dCpG. Reprinted with permission from ref. 46. Copyright © 1987 Adenine Press.

The other ammonium interacts in the major groove and appears particularly important for the crystal stability. The same Z-DNA fragment was also cocrystallized with the drug mitoxantrone [52] (see Section 2.3). In that structure, there are no intramolecular bridges in the major groove since the side chains of the drug interact with the guanine N7 and O6 atoms. The minor groove is also perturbed and the cytosine O2s are linked via a chain of 4 water molecules, the middle two of that chain binding to the guanine N2 and to the 3′-phosphate O2P.

## 2.2. *Ion binding*

Pelotonné sur lui-même, Sydney [BRENNER] exhibait un masque fermé de bouledogue. De temps à autre, l'un de nous égrenait à nouveau la litanie des manipulations ratées pour y trouver la faille. [...] Soudain, Sydney pousse un cri. Bondit comme un diable. Hurle: 'Le magnésium! C'est le magnésium!' [...] Sydney avait un juste. C'était bien le magnésium. C'était lui qui donnait aux ribosomes leur cohésion.

François Jacob, in *La Statue Intérieure*. [53]

Nucleic acids are polyelectrolytes and, consequently, the importance of ions for the stabilization of deoxy and ribo nucleic acids has long been recognized. In the presence of ions, nucleic acid constituents crystallize as higher hydrates than in their absence and the observed interaction schemes between the nucleic acid and the hydrated ions are thus probably better snapshots of the situation in the bulk. When discussing metal ion binding to nucleotide ligands two limiting cases are usually considered: [54] (*a*) site binding with a direct coordination of the metal ion to ligands of the nucleotide, leading to the formation of inner-sphere complexes after partial dehydration of the metal ion; (*b*) ion atmosphere binding wherein the metal ions with their intact hydration shells interact indirectly through water

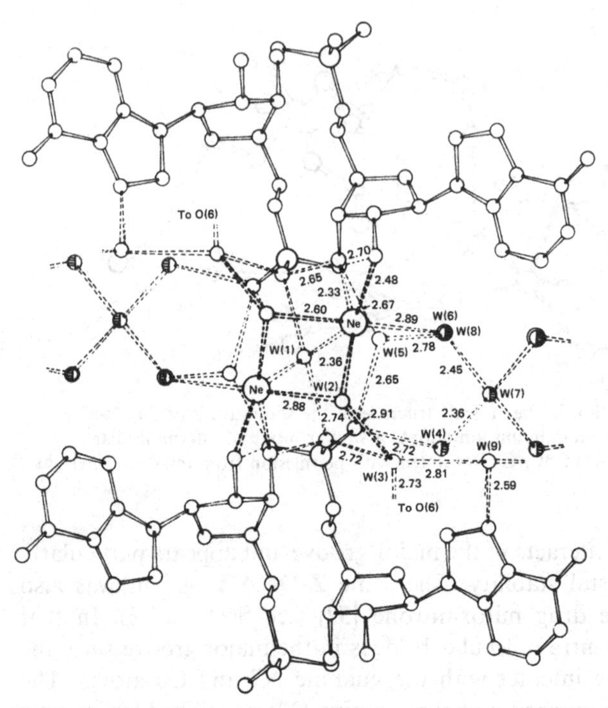


Figure 19. View of the surroundings of the phosphate group and the sodium ion in the crystal structure of hydrated monosodium inosine-5′-phosphate. [36] The shaded circles represent disordered water molecules. Reprinted with permission from ref. 36. Copyright © 1969 American Chemical Society.

molecules with the nucleotides, forming outer-sphere complexes. The kinetics of magnesium binding to polynucleotides have been well studied and follow the former structural description. [55] First, an outer-sphere complex is formed with a rate close to the limit of diffusion control. In a second step, one or more water molecules of the ion hydration shell exchange with polar atoms of the polynucleotide leading to inner-sphere complexation with a rate determined by the process of water dissociation, around 100000 per second for magnesium and 1000 times faster for calcium ions. The second step was observed only with short oligoriboadenylates, indicating that inner-sphere complexation requires particular conformations of the nucleotide chains and the presence of the adenine base and of the ribose hydroxyl group O2′.

The crystal structures of metal complexes of nucleic acids and their constituents have been reviewed. [56, 57] Here, only the complexes with alkali and alkali earth metals will be considered owing to their importance for nucleic acid stability in biological systems. In the literature, no crystal structure of complexes between nucleotides and magnesium ions can be found. However, there are several structures of complexes between purine and pyrimidine 5′-nucleotides and sodium or barium ions [57] (Figure 19).

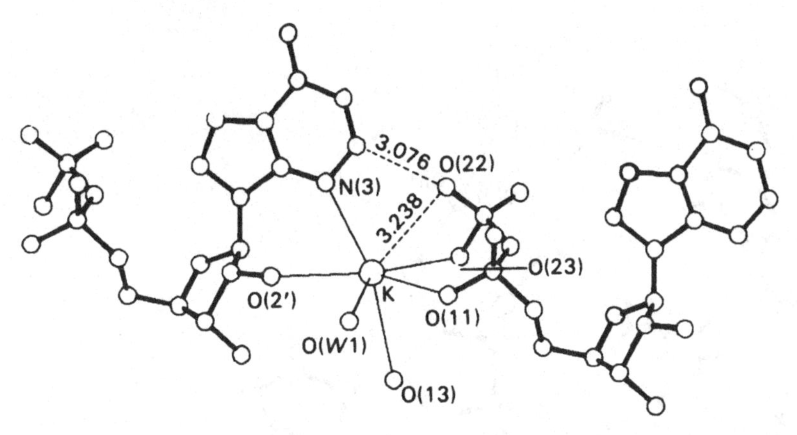


Figure 20. View of the surroundings of the potassium ion in the structure of potassium ADP dihydrate. **[60]** Note how the potassium sits in a position often occupied by a water molecule in ribo derivatives. The sugar pucker is C2′-*endo*. The distances K–O2′ and K–N3 are 2.86 and 3.16 Å, respectively, with an angle at the ion of 70.4°. Reprinted with permission from ref. 60.

The structures display similar crystal packings where layers of stacked nucleotides alternate with water channels containing the metal ions. Except in one case, the metal ion is directly coordinated to both the O2′ and O3′ ribose hydroxyls with the other ligands occupied by water molecules. Another interaction scheme involves direct contact to the purine ring N7 or to the pyrimidine exocyclic O2. In one complex between barium and 5′-AMP, **[59]** the binding of the metal ion to the nucleotide ligands are exclusively via water molecules. In the structure of potassium ADP, **[60]** the potassium ion is highly desolvated and interacts directly with three anionic phosphate oxygen atoms, making at the same time a bridge between the hydroxyl O2′ atom and N3 of the adenine base (Figure 20). As mentioned above, the latter position is commonly occupied by a water molecule.

Three features appear interesting. Firstly, the ion is rarely bound directly to the phosphate anionic oxygen atoms in ribonucleotide complexes, but more often in deoxyribonucleotide complexes. Secondly, often one or two water molecules of the ion hydration shell are common to two ions so that they share an apex or an edge of the coordination spheres (two such sodium ions are separated by 3.3–3.8 Å). And finally, the crystal packing is characterized by alternate columns of stacked bases and of hydrated ions.

Again, in structures of complexes with dinucleotides there are no magnesium complexes, only sodium and calcium ion complexes. No example of outer-sphere complexation can be found in such complexes. The nucleotide ligands are either phosphate anionic oxygen atoms or base atoms (O2 in pyrimidine, N7 in purine with O6 in guanine). The direct binding to base atoms can be exquisitely specific as in the sodium binding to O2s in the minor groove of ApU helical dimers, **[25]** since a similar binding would not be

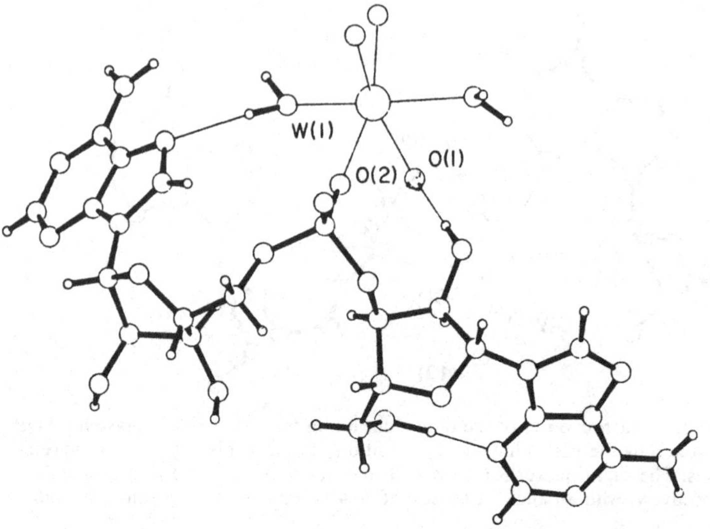

Figure 21. View of the surroundings of the calcium ion in the crystal structure of the hydrated calcium salt of adenylyl-3′,5′-adenosine. [**62**] The calcium ion is positioned on two-fold crystallographic axes and, consequently, only three kinds of ligands are used for calcium binding. Atom O(1) is an anionic phosphate oxygen of a symmetrically related molecule. Reprinted with permission from ref. 62. Copyright © 1980 American Chemical Society.

geometrically possible in the case of UpA helical dimers where the distance between the two O2s is around 8 Å instead of the required 4 Å (because of the right-handed helical sense). In UpA helical dimers, an ion could geometrically bind only the N3 atoms of the two adenine bases (see Section 5.1). Another example of a specific complex involving the N7 and O6 of two guanine bases, but this time involving symmetrically-related bases instead of being intramolecular, is observed in the sodium and ammonium salts of dCpG (Figure 18).

Ion binding to the phosphate oxygen atoms always involve symmetrically-related helical dimers and at least one anionic phosphate oxygen. In the sodium GpC structure, [**24**] two sodium coordination octahedra have one face in common, and consequently share three ligands: two anionic oxygen atoms and one water molecule which sits on the two-fold axis relating the two octahedra by symmetry. In two structures the dimers do not form base-paired mini-helices but adopt stretched single-strand-like conformations (the sodium salt of pTpT [**61**] and the calcium salt of ApA [**62**]). In the ApA structure, the calcium ion is bound by a direct interaction with the phosphate group and by outer-sphere water-mediated interaction with the N7 of the 3′-terminal adenosine (Figure 21). The pTpT complex involves the thymine O2s and one anionic oxygen atom of three independent nucleotides.

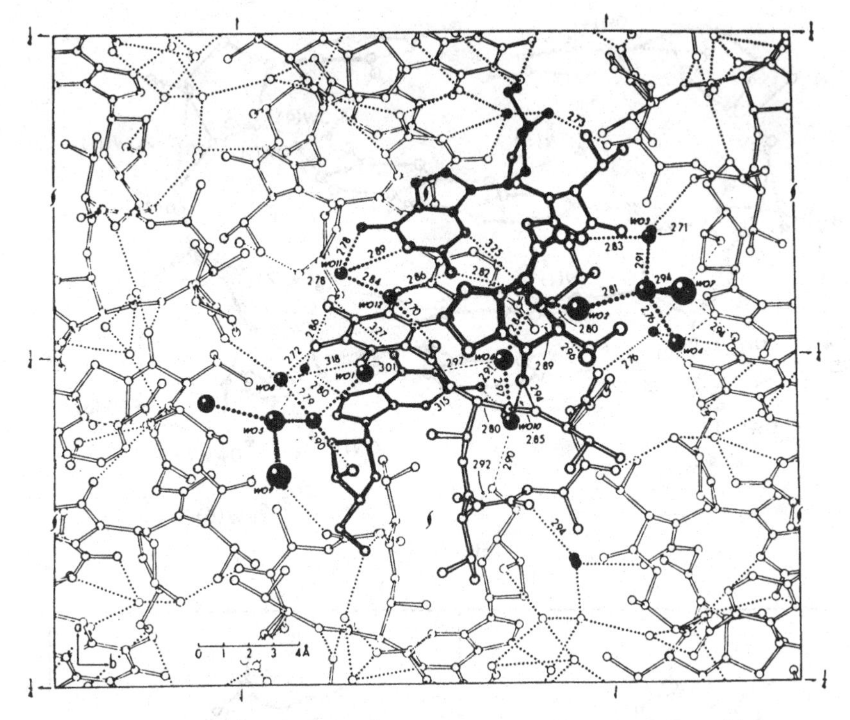

Figure 22. View down the *c*-axis of the actinomycin-deoxyguanosine [**63**] crystal. Deoxyguanosine molecules are striped and the actinomycin molecule is indicated with dark open lines and circles. Black circles indicate water molecules. Reprinted with permission from ref. 63. Copyright © 1972 Macmillan Magazines Ltd.

## 2.3. *Hydration in nucleic acid complexes*

Contacts between nucleic acids and ligands, like drugs or proteins, are not restricted to direct hydrogen bonding. Indeed, the non-bonded van der Waals' interactions appear of overwhelming importance, together with contacts mediated by water molecules and ions. The importance of water molecules was already apparent in the deoxyguanosine-actinomycin D complex in which, despite the presence of only 12 water molecules among the 140 atoms of the asymmetric unit, 80% of the 152 hydrogen bonds in each unit cell involve interactions with water molecules [**63**] (see Figure 22). Incidentally, in that structure, there is no base–base hydrogen bonding and hydrogen bonding of the Watson–Crick sites to water molecules can be observed. Thus, in one deoxyguanosine molecule, there is a water bridge between O6 and N1 and via another water molecule to N2; while, in the other, there is a water bridge between N1 and N2 only. Besides the direct hydrogen bonds mentioned below, there are water bridges between the drug and the nucleic acid. For example, a water molecule links one guanine amino

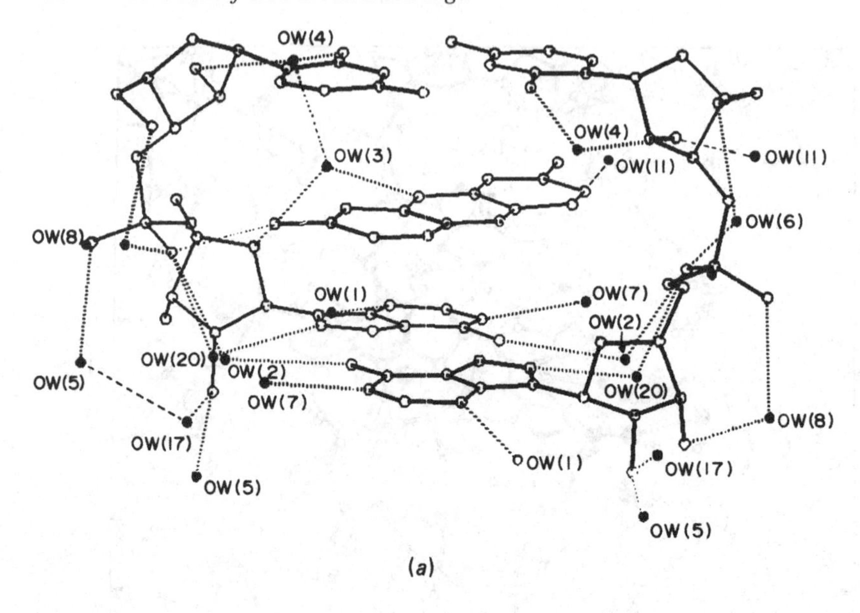


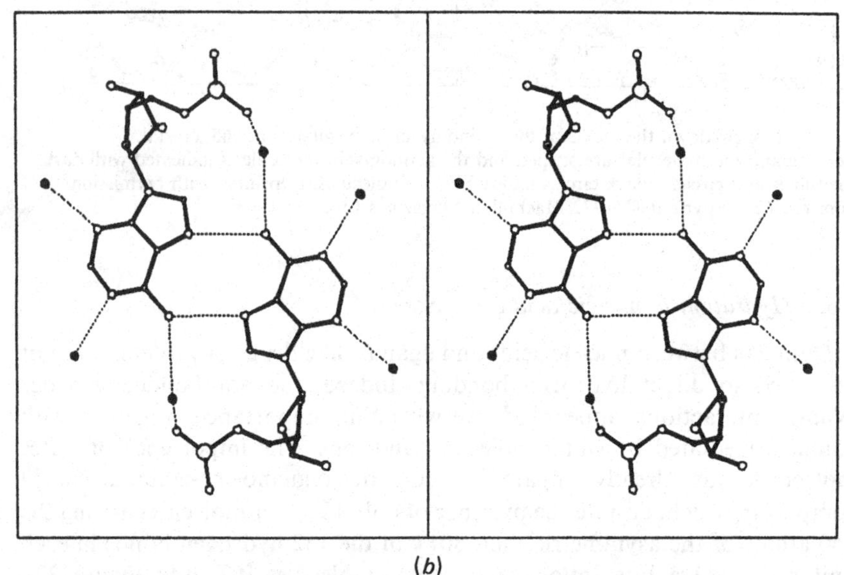


Figure 23. View of the 1:1 complex between *r*CpA and proflavine sulfate. [27] In this complex, the sulfate group did not cocrystallize. The cytosine bases form protonated $(C—C)^+$ pairs and the adenine bases form N6—H···N7 hydrogen bonds. In this structure, the helical axis of the mini-helix is parallel to the (crystallographic) two-fold symmetry axis relating the two strands. Consequently, the proflavine molecules are crystallographically disordered. (*b*) Stereo view of the hydration pattern around the A–A base pair of the CpA-proflavine drug complex. [27] Notice the water bridge linking the phosphate group of one nucleotide to the amino group N6 of the opposite adenine base.

group to the first carbonyl group of the cyclic peptide. Also, two water bridges link the terminal O3′ atom to the carbonyl and amino groups of the actinomycin D chromophore.

Crystallographic studies on drug binding to nucleic acid fragments have been reviewed. **[64–6]** In ribooligonucleotide intercalation complexes, the hydroxyl O2′ and the sugar ring O4′ atoms are involved in water bridges similar to those seen in helices without the increased separation between base pairs. For example, in the rCpG–proflavine **[26]** and rCpA–proflavine **[27]** (Figure 23(*a*)) complexes, there is a water molecule bridging the O2 of cytosine and its O2′ atom, which via another water molecule is linked to the O4′ atom of the purine residue. In the rCpA–proflavine complex, that second water molecule is itself hydrogen bonded to the quaternary nitrogen of the drug. In addition to the hydrogen bonds between self-pairs of adenine, there is another hydrogen bond between two-fold related rCpA molecules that is mediated by a water molecule between the exocyclic amino nitrogen of one adenine ring and one anionic oxygen of the opposite phosphodiester group, N6 ··· Ow ··· O1P (Figure 23(*b*)). In the structure of ApApA, **[67]** where such self-pairs between adenine rings occur, the exocyclic amino group of adenine is linked directly to an anionic phosphate oxygen of the symmetry-related backbone with a water molecule bridging N1 and the same anionic phosphate oxygen. A very similar hydrogen-bonding pattern (with the phosphate group replaced by a sulfate group) exists in the crystal structure of the adenosine–proflavine complex. **[78]** In both the rCpA–proflavine and the dCpG–terpyridinium platinum **[68]** complexes, there is a water bridge between the terminal O5′ of the cytosine and the anionic phosphodiester oxygen O1P in the 3′ direction. A water molecule, making a short contact to the C8 of the adenine, is also hydrogen bonded to the anionic phosphate oxygen O1P of the same rCpA molecule. **[27]**

In the fascinating crystal structure of the complex between the dimer dCpG and the drug proflavine, **[69]** water molecules hydrogen bond so as to form four edge-linked pentagons in the major groove. The O6 and N7 atoms of the two guanine bases are individually hydrated by two water molecules linked to each other. Also, the water bound to O6 of one dCpG is linked to a water molecule interacting with the quaternary nitrogen atom of the intercalated proflavin molecule.

Another example where water molecules contribute to the stabilization of drug–nucleic acid complexes is found in the beautiful work on the structure of daunomycin complexed to d(CpGpTpApCpG). **[70, 71]** In that structure, two water molecules with low thermal parameters bridge the antibiotic to the DNA fragment. It is interesting that there is only one direct hydrogen-bond interaction between the antibiotic and the nucleic acid (hydroxyl O9 of daunomycin to N2 and N3 of G2). The other interactions involve non-bonded van der Waals' contacts as well as a hydrated sodium ion which bridges N7 of G12 to two carbonyl oxygen atoms of the drug via its

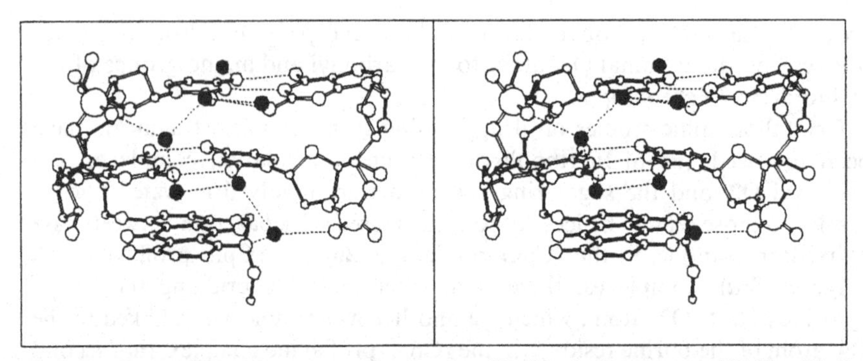

Figure 24. Stereo view of the 1:1 complex between dCpG and mitoxantrone. [52] The dCpG molecules form mini-helices in the Z-form and the drug stacks between the nucleic acid fragments. Only one mitoxantrone molecule is shown. Drawing courtesy M. A. Viswamitra and B. Ramakrishman.

hydration water molecules. In the structure of the open complex between 1,10-phenanthroline-platinum (II)-ethylenediamine and 5′-phosphoryl-thymidylyl-3′,5′-deoxyadenosine, [72] in which reversed Hoogsteen base pairing between A and T of adjacent molecules occurs, ammonium ions link the amine nitrogens of the sandwiched drug to the exocyclic O4 atoms of the thymine bases. Drug–DNA water-mediated bridges are also present in the complex between dCpG and mitoxantrone, [52] a modified anthraquinone with two side chains. The structure consists of dCpG dimers in the Z-conformation sandwiching the drug whose side chain atoms interact with the N7 and O6 atoms of guanine bases above and below. In one case, a protonated amino group is hydrogen bonded to N7 with a terminal hydroxyl group hydrogen bonding to O6. The second side chain interacts with its protonated amino group directly to O6 and via a water molecule to N7. In the minor groove, two water molecules link the cytosine O2 to O2P of the 5′-phosphate and one of these bind to an exocyclic atom oxygen of the chromophore (Figure 24).

The exocyclic amino group N2 and the ring nitrogen N3 of guanine are involved in three complexes between nucleic acid and a drug as different as actinomycin D, [63, 73] triostin A, [74] or daunomycin. [70, 71] In the first two cases, the interactions involve the carbonyl oxygen and the amide group of a threonine and an alanine residue, respectively. In the crystal structure of the B-dodecamer, d(CGCGAATTCGCG), [75] guanines of symmetrically-related molecules are also hydrogen bonded via the N2 and N3 atoms. In short, it does therefore appear that direct hydrogen bonding between drugs and nucleic acids is infrequent, that water or ion mediated bridges between the drug and the nucleic acid are important, and that a major contribution to the stability of the drug–nucleic acid complexes comes from non-bonded van der Waals' interactions. The spreading of charges through hydrogen-bonded networks of water molecules [76] is conspicuous in several complexes

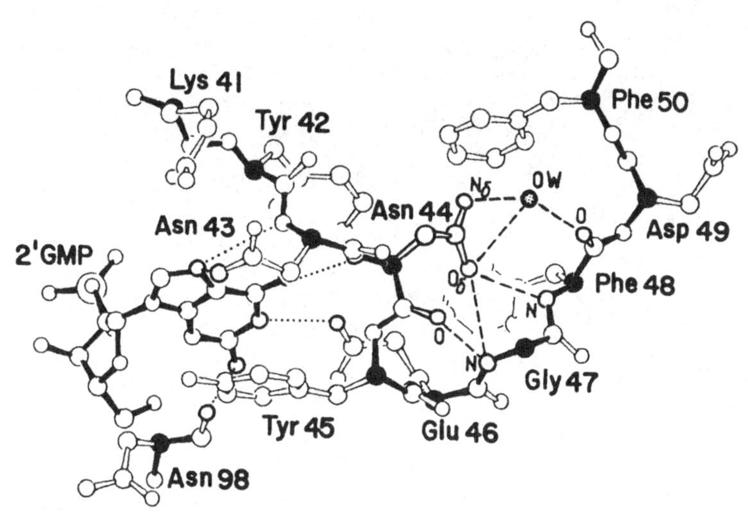


Figure 25. View of the guanine-binding site in the complex between RNase T1 and 2′-guanosine-monophosphate. [79] The bonds of the 2′-GMP and of the main chain atoms of the protein are solid. Hydrogen bonds between the guanine and the protein are dotted and those between protein atoms are represented by broken lines. Note the participation of a bridging water molecule in the hydrogen-bond network holding the peptide loop. Reprinted with permission from ref. 79. Copyright © 1988 IRL Press Ltd.

between nucleic acids and cationic drugs in which the charged species interact with one another *via* bridges of water molecules. **[77, 78]**

In complexes between bases, nucleosides, or nucleotides and proteins, the hydration sites are often occupied by protein atoms. For example, in the crystal structure of the complex between ribonuclease T1 and 2′-GMP **[79]** (Figure 25), there are five hydrogen bonds between N1, N2, O6, and N7 of the guanine base, three of which are to main chain atoms of the protein, Asn(43)N—H···N7, Asn(44)N—H···O6, and Asn(98)C=O···H—N2, while two hydrogen bonds involve side chain atoms, Asn(43)N$\delta$···N7 and Glu(46)O$\epsilon$···N1. Moreover, the guanine base is sandwiched between the phenolic rings of two tyrosine residues. Similarly, in the complex between catabolite gene activator protein and 3′,5′-cAMP **[80]** (Figure 26), the hydroxyl group of Ser128 bridges the N7 and N6 atoms of the adenine. However, the involvement of water molecules in such complexes should not be neglected. For example, in the well-described structure of the complex between NADPH and *L. casei* dihydrofolate reductase **[81]** (Figure 27), the only direct contacts between the adenine ring and the protein are hydrophobic and all hydrogen bonds are made via water molecules: Gln(101)=O···Ow···N7 and His(77)C=O···Ow···N1. Several other water molecules participate in the binding of the ribophosphate groups to the protein. In the structure of the complex between a nicked octanucleotide and DNaseI, **[82]** although one water molecule bridges a terminal O3′ to a

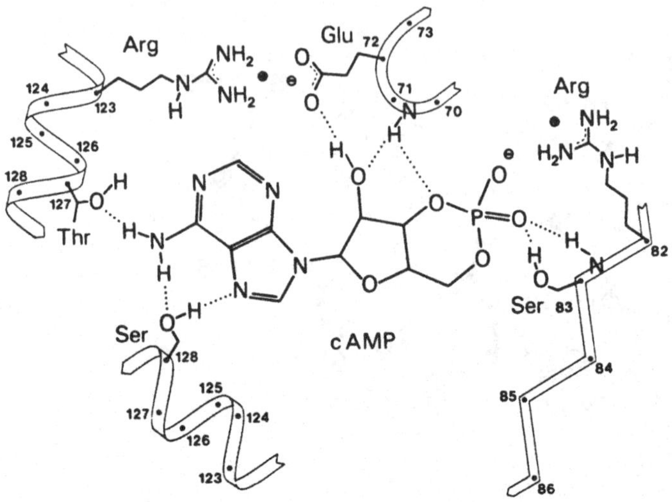

Figure 26. Schematic view of the binding site of cyclic-adenosine-3′,5′-monophosphate (cAMP) in its complex with the catabolic gene activator protein (CAP). Reprinted with permission from ref. 80. Copyright © 1987 Academic Press Inc. (London) Ltd.

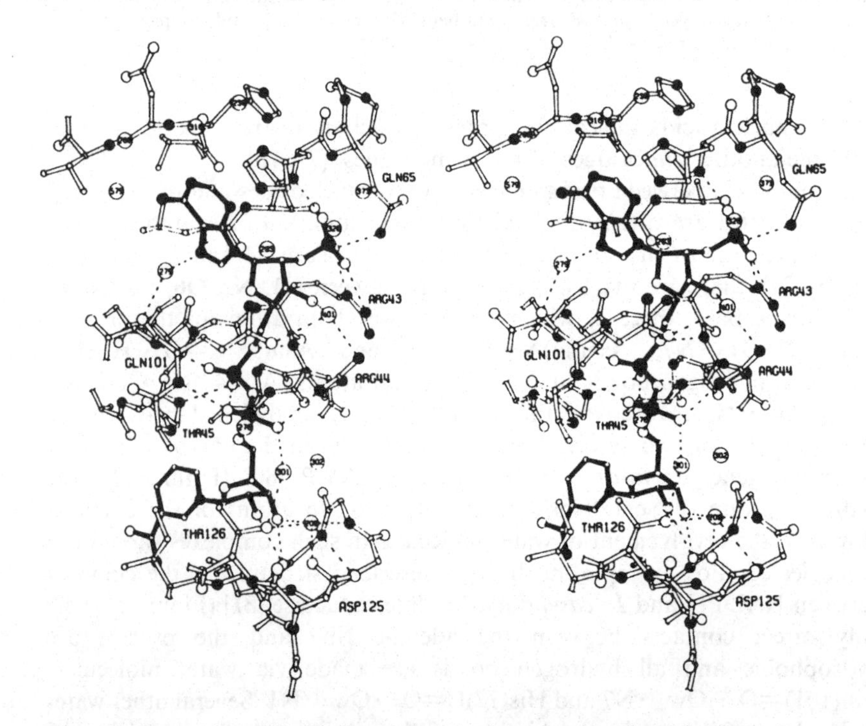


Figure 27. Stereo view of the binding site of nicotinamide adenosine diphosphate 2′-monophosphate (NADPH) in its complex with *L. casei* dihydrofolate reductase. [81] Water molecules are represented by numbered circles. Reprinted with permission from the author.

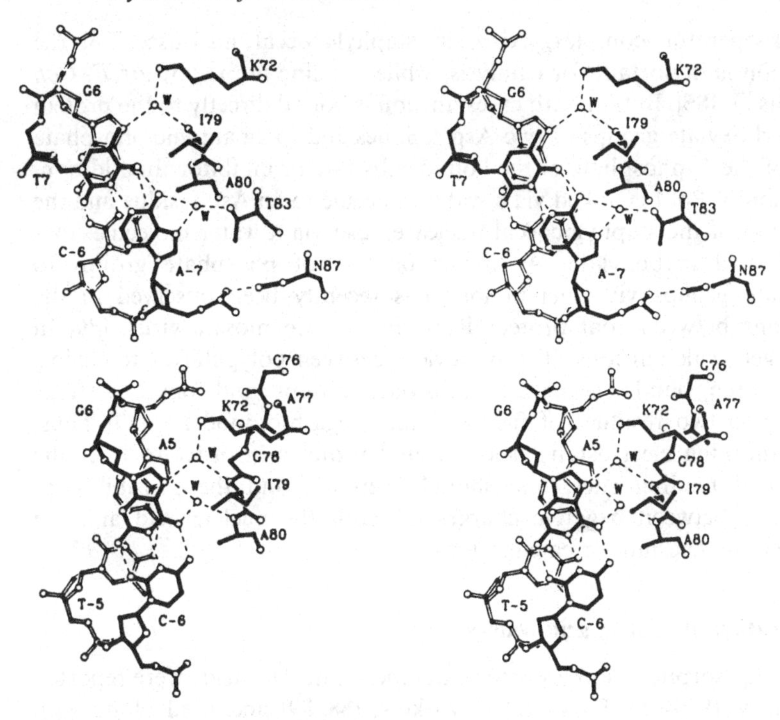


Figure 28. Stereo views of some water-mediated contacts between the DNA operator and the *trp* repressor protein. Reprinted with permission from ref. 83. Copyright © 1988 Macmillan Magazines Ltd.

nitrogen atom of His249, several water molecules link protein atoms in the environment of the cutting site as if to hold the peptide conformations in the proper orientation for binding to the nucleic acid. The recent crystal structure of the *trp* repressor/operator complex **[83]** (Figure 28) clearly shows that water-mediated polar contacts between the protein and the DNA fragment contribute to the specificity of binding. While there are only fourteen direct hydrogen bonds between operator and repressor, there are six solvent mediated contacts, three to the phosphate groups and three to the base pairs. Water molecules participate in the binding of Asp46 to phosphate 1 via a calcium ion and in the binding of phosphate 3 to the hydroxyl groups of the side chains of Thr53 and Thr81. Of the three water molecules interacting between the bases and the protein, one water molecule bridges the N7 and N6 atoms of A7 to the $\gamma$-hydroxyl of Thr83. A second water molecule links O6 of G6, N7 of A5, and the peptide nitrogen of Ala80; while a third water molecule links N7 of G6, the peptide nitrogen of Ile79, and the N terminal of Lys72. All these contacts are at the N-terminal part of the second helix of the helix-turn-helix motif which interacts in the major groove.

As is the case with drug binding to nucleic acids, metal ions can help the binding of substrate to proteins. Such an example occurs in the *trp* repressor–operator complex. Also, in staphylococcal nuclease, [**84**] the calcium ion is important for catalysis, while the zinc ion is not for *E. Coli* polymerase I. [**85**]. But, in both cases, the ion is bound directly to the protein by the carboxylate groups of two Asp residues and to an anionic phosphate oxygen of the 5′-phosphate of the bound substrate or inhibitor. In addition, the zinc ion of Pol I is linked via a water molecule to an Asp residue and the calcium ion of the staphylococcal nuclease, also via a water molecule, to a glutamic acid carboxylate. A similar binding of phosphate groups to carboxylate groups via calcium ions has recently been observed in the interactions between coat protein RNA in tobacco mosaic virus. [**86**] In alginate gels, calcium ions fit into crevices between polyguluronate chains, forming ionic bonds between carboxylate groups and oxygen atoms belonging to two residues of facing chains ('egg-box model'). [**87**] Thus, when computing electrostatic fields around proteins in order to infer the nucleic acid binding sites, one should keep in mind the possibility of interactions between negative charges on both the nucleic acid and the protein via magnesium or calcium ions.

## 3. Hydration of nucleic acid helices

Water vapor sorption isotherms for a number of nucleic acids were reported in the early 1960s by Falk and coworkers, [**88, 89**] and used along with measurements of the distribution of OH vibrational frequencies to describe the hydration of DNA. They found that around 20 waters/nucleotide were required to maintain the B-form, with 11–12 spectrally perturbed with respect to liquid water. Some six of these were assigned, in order of expected decreasing affinity, to the anionic oxygens on phosphates, the ester O3′ and O5′ oxygens of the phosphates and the furanose sugar O4′. The remaining 5–6 of these waters were assigned to the hydrogen-bond acceptor and donor groups on the base pair edges lining the major and minor grooves, with acceptor sites ranked higher in affinity than donor sites. Two of the waters were found to be bound to phosphate with especially high affinity. [**89**] This inner layer of the first hydration shell, termed Class 1 by Wolf and Hanlon, [**90**] is incapable of freezing and is thought to be impermeable to electrolyte, or altered at least in that ions cannot diffuse in this layer with the same activation energy characteristic of liquid water. Kopka, Fratini, Drew, and Dickerson [**91**] later proposed a revised list based on relative crystallographic occupancies of hydration sites, and placed the minor groove spine first, followed by the anionic phosphate oxygens, the sugar O4′ atoms, the base edge Ns and Os, and lastly the esterified phosphate oxygens. Various other proposals for relative affinities of nucleic acid hydration sites have emerged from theoretical studies, as described in Section 4 below.

The normalized relative affinities for hydration around the various hydrophilic atoms of DNA in one crystal structure of each of the three main

Table 2. *Average number of water contacts (within* 3.4 Å*) made by hydrophilic atoms in the three main helical forms. The values are normalized so that the largest relative affinity in each form has a value of* 1.0 *(in each case it is an anionic phosphate oxygen).*

| | | Number of contacts | | |
|---|---|---|---|---|
| | Atom type | A-form | B-form | Z-form |
| Minor groove | O2 (pyr) | 0.5 | 0.4 | 1.0 |
| | N2 (Gua) | 0.3 | 0.1 | 0.5 |
| | N3 (pur) | 0.5 | 0.4 | 0.0 |
| Major groove | N4 (Cyt) | 0.4 | 0.8 | 0.4 |
| | O4 (Thy) | 0.4 | 1.7 | — |
| | O6 (Gua) | 0.8 | 0.4 | 0.7 |
| | N6 (Ade) | 0.3 | 0.6 | — |
| | N7 (pur) | 0.6 | 0.4 | 0.9 |
| Backbone | O5′ chain | 0.1 | 0.3 | 0.0 |
| | O3′ chain | 0.0 | 0.0 | 0.1 |
| | O4′ | 0.3 | 0.3 | 0.1 |
| Anionic oxygens | O1P | 1.0 | 1.0 | 1.0 |
| | O2P | 0.7 | 0.7 | 1.0 |
| | O5′ terminal | 0.5 | 1.7 | 0.4 |
| | O3′ terminal | 1.3 | 0.6 | 0.4 |

The sequences used were d(GG5BrUA5BrUACC) for the A-form, d(CGCGAATTCGCG) for the B-form and d(5BrCG5BrCG) for the Z-form. Adapted from ref 92.

helical forms are given in Table 2. The anionic phosphate oxygen atoms are the most hydrated, the sugar ring oxygen atom O4′ is intermediate, and the esterified O3′ and O5′ backbone atoms are the least hydrated. The hydrophilic atoms in the two grooves of the helix are about equally well hydrated, at half the level of the phosphate oxygen atoms. The relative order of hydration affinities is thus: anionic phosphate oxygens, polar base atoms, and sugar oxygen atoms. This sequence of relative affinities of nucleic acid atoms for hydration departs from the one proposed by Falk and coworkers [**88, 89**] in which the base atoms had the smallest affinities. The involvement of a nucleic acid atom in hydrogen bonding depends on its accessibility, and modulation of the water affinities of the hydrophilic atoms with the nucleic acid conformation, helical form, or crystal packing contacts is observed. Table 3 contains the atomic accessibilities of the polar atoms considered in Table 2 calculated with the coordinates determined in the monocrystal and with the coordinates determined from fiber diffraction work (except for the Z-form). Several observations contained in Table 2 can be rationalized after consideration of the values presented in Table 3. There are, however, exceptions. For example, the low hydration of N4 of cytosine bases despite its high accessibility or the high hydration of O4 of thymine bases despite its low accessibility. The comparison between accessibilities determined from

Table 3. *Atomic accessibilities (in Å²) for a water molecule* (r = 1.4 Å) *in the three main helical forms of DNA based on monocrystal coordinates* [**112*a***, 40, 51] *and on fiber diffraction* [**326**].

| | Atom type | Atomic accessibilities A-form crystal/fiber | B-form crystal/fiber | Z-form crystal[a] |
|---|---|---|---|---|
| Minor groove | O2 (pyr) | 6.4/8.5 | 3.0/2.3 | 5.9 |
| | N2 (gua) | 8.8/17.1 | 11.7/6.6 | 7.7 |
| | N3 (pur) | 4.7/6.0 | 2.8/0.6 | 0.0 |
| Major groove | N4 (cyt) | 19.1/14.1 | 10.1/14.2 | 12.6 |
| | O4 (thy) | 5.1/5.7 | 5.6/6.0 | — |
| | O6 (gua) | 8.7/5.2 | 4.5/6.0 | 8.0 |
| | N6 (ade) | 8.4/8.0 | 9.9/11.5 | — |
| | N7 (pur) | 8.0/5.6 | 9.9/10.1 | 21.6 |
| Backbone | O5′ chain | 0.2/0.0 | 0.8/0.0 | 0.5/4.8 |
| | O3′ chain | 5.0/5.0 | 3.2/5.4 | 2.7/4.7 |
| | O4′ | 2.3/3.2 | 1.7/3.0 | 0.0/2.8 |
| Anionic oxygens | O1P | 37.3/36.7 | 37.5/37.8 | 19.8/32.6 |
| | O2P | 26.0/23.5 | 31.2/26.3 | 37.833.6 |
| | O5′ terminal | 23.8/22.1 | 38.4/21.1 | 23.9 |
| | O3′ terminal | 36.2/30.7 | 36.8/35.2 | 34.8 |

The accessibilities were calculated with the program of Richmond [*327*] and the following values for the van der Waals' radii, including implicitly hydrogen atoms: C = 1.9 Å; N = 1.7 Å; O = 1.4 Å; P = 1.9 Å.

[a] For the Z-form crystal, the first number corresponds to the cytosine nucleotide and the second one to the guanine one in the cases of the backbone and anionic oxygens.

crystal coordinates and from fiber coordinates displays unexpected discrepancies. For instance, the guanine amino group N2 is less accessible in the 'crystal' than in the 'fiber' in case of the A-form but not in case of the B-form, and the opposite is true for the amino group N4 of cytosine bases. Those two groups are the most exposed groups in a G–C base pair of B-DNA. In the crystal, these oligomers are not straight helices and these differences could reflect a closing of the minor groove in the A-form and an opening of the minor groove in the B-form at G–C base pairs.

In the A-form, the anionic phosphate oxygen pointing into the major groove is the least hydrated. As mentioned by Kopka *et al.* [**91**], the esterified O3′ and O5′ phosphate oxygen atoms do not nucleate ordered water molecules. In crystal structures of nucleic acid fragments, the O4′ sugar ring oxygen is often prevented from hydrogen bonding because it is shielded by the attached base or by stacking on a neighbouring base. [**93**] Also, in Z-DNA, only the guanine sugar O4′ atoms are accessible for hydrogen bonding, since the cytosine sugar O4′ atoms point toward the guanine ring in the 3′-direction. [**33**] In A-DNA, the O4′ atom is, however, frequently involved in water binding through a water bridge with the O2 of a preceding

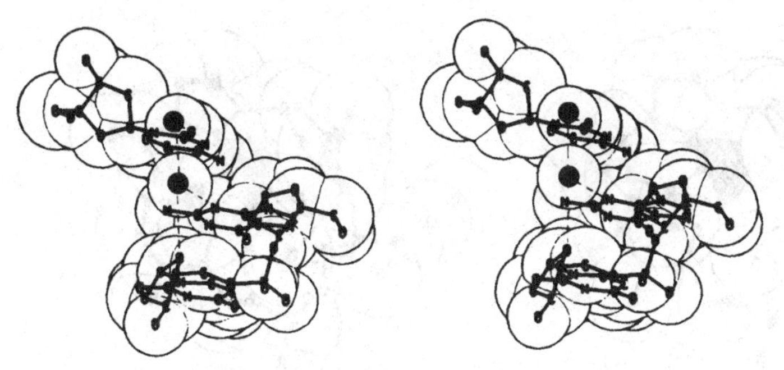

Figure 29. Stereo view showing two water molecules interacting in the minor groove of the A-form octamer d(CCA5BrUA5BrUGG): **[94]** one water molecule bridges the N3 of an adenine to the O4′ of the next sugar and is itself linked to the O2 of an uracil base on the other strand.

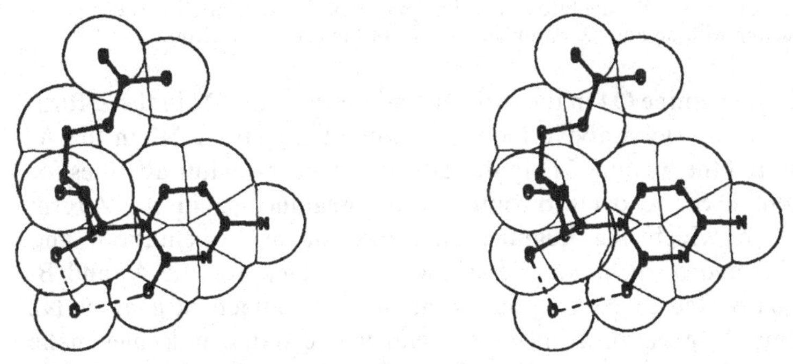

Figure 30. Stereo view showing a water molecule bridging the ribose hydroxyl O2′ to the O2 exocyclic atom of the attached base.

pyrimidine or the N3 of a preceding purine **[94]** (Figure 29). This tendency is still more prevalent in RNA helices where a molecule of water frequently bridges the base atoms O2 or N3 with the hydroxyl atom O2′ of the same residue and, at times, with also the O4′ of the next residue (Figure 30). In the B-dodecamer, **[40, 75]** the distance between O4′ and the water molecule bridging hydrophilic atoms in the minor groove is greater than 3.2 Å, which can be a close non-bonded O···O contact as seen in many hydrate crystals. **[6]** However, in two recent B-DNA structures, **[95, 96]** water molecules bridging O2 or N3 of the base to the O4′ atom of the next sugar have been systematically observed. In short, the backbone atoms O3′ and O5′, and to a smaller and more variable extent, the furanose ring oxygen atom O4′ seem to have the lowest affinity for water molecules. In contrast, the terminal O3′ and O5′ hydroxyls have a high affinity for solvent molecules and often participate in water bridges between symmetrically-related molecules in the three helical forms.

Among the hydrophilic atoms of the bases, there is a modulation of the relative water affinities with the nucleic acid form. Thus, the guanine O6 in

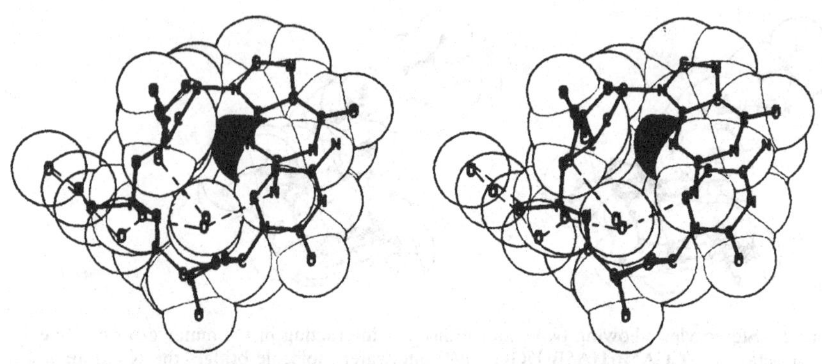

Figure 31. Stereo view of a 5′-GpC-3′ fragment of the Z-form hexamer d(5BrCG5BrCG5BrCG) **[51]** showing the inaccessibility of the guanine N3 atom, the water bridge between the N2 exocyclic nitrogen and an anionic phosphate oxygen of the cytosine nucleotide. The water molecule involved in that bridge is itself linked to another water molecule interacting with an anionic phosphate oxygen of the guanine nucleotide.

the A-form, the thymine O4 in the B-form, and the cytosine O2 in the Z-form have relative occupancies above the mean, while the guanine N2 in the A- and B-forms and the guanine N3 in the Z-form present very low affinities for water molecules. Steric overcrowding for the guanine N3 in the Z-form (Figure 31), in which the guanine base has the *syn* orientation, and intermolecular hydrogen bonding and packing contacts for the A- and B-forms can be invoked to explain the low number of contacts of guanine N2 and N3 atoms. In general, as in the proteins where water molecules make more hydrogen-bond contacts with the carbonyl oxygens than with the peptide N—H, exocyclic carbonyl groups of nucleic acids tend to have a higher affinity for making hydrogen bonds with water molecules than do exocyclic amino groups. **[97]**

### 3.1. *Hydration motifs*: *bridges and networks*

A passage from one of Heinrich von Kleist's letters (November 16, 1800) comes to mind. He had been passing through an arched gateway: 'Why, I thought, does the vault not collapse, though entirely without support? It stands, I replied, because all the stones want to fall down at the same time'.

Erwin Chargaff, in *Heraclitean Fire* **[98]**

The geometries of nucleic acid double helices are such that water molecules are able to bridge hydration sites of the same residue, of adjacent residues on the same strand, or of distant residues on either of the two helical strands. In crystals, hydration bridges between symmetrically-related residues are frequent and, although important for understanding the physico-chemistry of nucleic acid crystallization and overall lattice stability, are of less interest in the present context. Such water bridges between symmetrically-related molecules involve chiefly the terminal O3′ and O5′ hydroxyls in the three

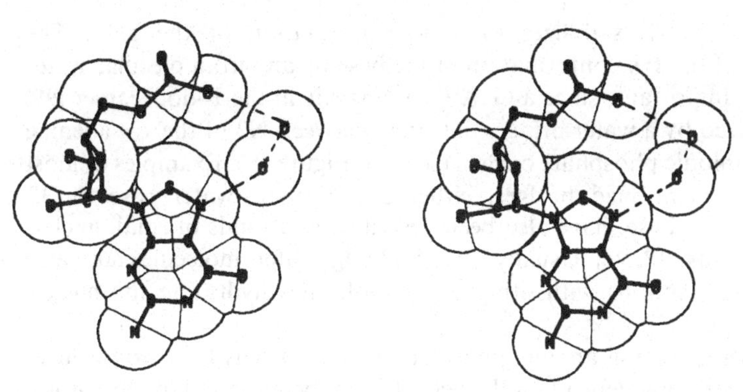

Figure 32. Stereo view of a two-water bridge between N7 of a guanine ring and an anionic phosphate oxygen of its 5′-phosphate. From ref. 28.

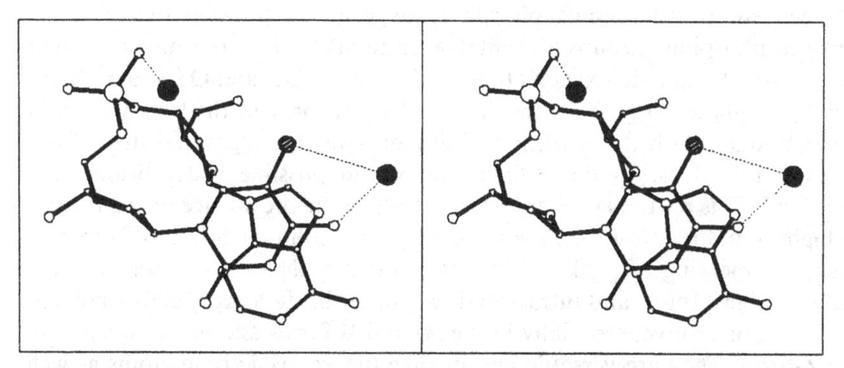

Figure 33. Stereo view showing two water molecules around T19 in the B-dodecamer: [**40**] one water molecule is lodged between the methyl group and the 5′-phosphate and the other sits between the methyl group and the exocyclic O4 atom of the thymine ring. Dotted line represents hydrogen bonding distances.

forms, as well as the O2P anionic phosphate oxygen atom in the B- and Z-forms. Interestingly, the sugar ring oxygen atom O4′ is involved in the A-form (with O5′ and O3′ atoms) and in the Z-form (with O2P atoms) in water bridges connecting symmetrically-related molecules.

Intranucleotide water bridges are structurally important, especially in periodic structures. Several of those intraresidue bridges observed in crystals of nucleic acid fragments, e.g., the water bridge between N7 and O6 (or N6) of purine residues, are observed also in crystals of the three helical forms.

Water bridges between a phosphate anionic oxygen atom and a base atom are typical of each form: in the A-form, between N7 (of purines) or N4 (of pyrimidines) and O1P generally via two water molecules (Figure 32); in the B-form, water molecules have been seen either trapped between the methyl and O4 of thymine residues or between the methyl and the 5′-phosphate (Figure 33); while, in Z-form, the water bridge between the guanine amino group N2 and O1P stabilizes the *syn* conformation of the base. The stabilization of the *syn* conformation of the base by an intramolecular water bridge seems important, since a G–A (*syn*) mispair in the B-dodecamer **[99]** is also stabilized by a water bridge, this time, between N3 of the *syn* adenine and its O1P anionic phosphate oxygen (see also Figure 10). Examples of most of the above mentioned bridges, either 5′-phosphate–water–base or 3′-phosphate–water–base, have also been described in crystals of small nucleic acid constituents. **[8, 92]** Usually, the bridging water molecules are also involved in interactions with other water molecules hydrating the nucleic acid.

Water bridges linking anionic phosphate oxygen atoms from consecutive residues also strongly depend on the helical form, being found in the A-form and the Z-form while absent in the B-form as remarked by Saenger, Hunter, and Kennard. **[100]** Not all A-form crystals **[101]** present one-water bridges between successive anionic phosphate oxygens, even when the distances between phosphate groups are suitable. In the B-form, water molecules link the phosphate anionic oxygen atoms and the ester O3′ and O5′ atoms of the same phosphate. In general, the A- and Z-forms present the least variety in water bridges, with the A-form disliking one-water bridges and the Z-form two-water bridges. In the B-form, almost all possible water bridges are exploited. This particularly might play a role in the prevalence of the B-form at high humidity. However, the relatively low resolution of the B-form may lead to a merging of peaks so that two-water bridges may appear as one-water bridges. Inter- and intrastrand two-water bridges are mainly observed in the major groove, especially in the A- and B-forms and less frequently in the Z-form. They are versatile and involve the exocyclic base atoms as well as the N7 of purines and no dependence on sequence is yet clearly apparent.

The situation is different in the minor groove where water molecules, involved mainly in one-water bridges are present in all three forms, and appear to contribute significantly to the structural stability and integrity of the nucleic acid helices. Quigley **[102]** has remarked that biological macromolecules are deficient in hydrogen-bond donors and this is particularly true in the minor groove where the only donors are the guanine amino group and, in RNA, the hydroxyl O2′. In the minor groove, water molecules bind to N3 of purines, O2 of pyrimidines, O4′ of the sugar rings, and O2′ of ribose rings. By bridging those hydrogen-bond acceptors, they satisfy the tendency of polar atoms to form hydrogen bonds. Nucleic acid structures in which such dangling polar groups could not participate in the hydrogen-bond network would loose stabilization free energy with respect to those structures with hydrogen-bond bridges. Those water molecules bridging polar atoms of the minor groove are similar to the internal water molecules found in protein structures, **[97]**, albeit structurally more important owing to their greater number per residue and to the periodicity of helical structures. The most common water bridges found in the minor groove of all

Table 4. *Structurally important water bridges between polar groups in the minor groove of the three helical DNA forms and in helical RNA fragments or tRNAs. From ref.* 8.

| | Type of bridge | Helical form |
|---|---|---|
| Intrastrand | N2 (*i*)···Ow···N3 (*i*) | RNA, A-DNA |
| | N3 (*i*)···Ow···O4′ (*i*+1) | A-DNA, B-DNA |
| | O2 (*i*)···Ow···O4′ (*i*+1) | A-DNA, B-DNA |
| | O2′ (*i*)···Ow···N3 (*i*) | RNA |
| | O2′ (*i*)···Ow···O2 (*i*) | RNA |
| | O2′ (*i*)···Ow···O4′ (*i*+1) | RNA |
| Interstrand | O2 (*i*)···Ow···O2 (*j*+1) | B-DNA |
| | N3 (*i*)···Ow···O2 (*j*+1) | B-DNA |
| | O2 (*i*)···Ow···O2 (*j*±1) | Z-DNA |

Letter in parenthesis refers to residue number with positive sign for the 5′- to 3′-OH direction. (*j*) refers to a residue on the other strand with positive sign for the 5′- to 3′-OH direction.

three DNA forms, in RNA fragments, and in *t*RNA crystals are shown in Table 4.

Because of the helical periodicity in nucleic acid structures, some of those water bridges can lead to regular and striking hydration networks. Thus, in the B-form (Figure 34), the water molecules bridging the N3 and O2 atoms and, in the Z-form (Figure 35), the water molecules bridging the O2 atoms are themselves connected by water molecules and form a 'spine of hydration' running down the minor groove. [**40, 50, 51, 103**] In B-DNA, the presence of the amino group of a guanine residue leads to a discontinuity in the floor of the minor groove with an apparent disruption of the spine while, in Z-DNA, an adenine residue breaks the continuity of the spine locally. [**104**] In two recent crystal structures of B-oligomers, [**95, 96**] the spine of hydration is 'zipped open' so that there is a string of hydration on each strand with a concomitant widening of the minor groove (Figure 36(*a*)). Because of the right-handed helical sense of B-DNA and the *anti* orientation of the bases, water bridges making up the 'spine of hydration' across the minor groove can only occur from one strand to the other in the 3′-direction, i.e., when looking down the minor groove from left to right and down one base. At a 5′-TpA-3′ step, on pure geometric grounds, one could envisage a water bridging the opposing N3 nitrogen atoms of the adenine bases. However, such a bridge has not, as yet, been observed in crystal structures (see Section 2.2). In crystal structures, the bases making up the base pairs are characterized by a negative propeller-twist angle between them. It has been suggested that, a negative propeller-twist between the bases of the A–T base pairs leads to clashes between the adenine bases in the minor groove of a 5′-TpA-3′ step. [**105**] In order to relieve such bad contacts, base pairs could either open toward the minor groove, decrease the twist angle between base

(a)

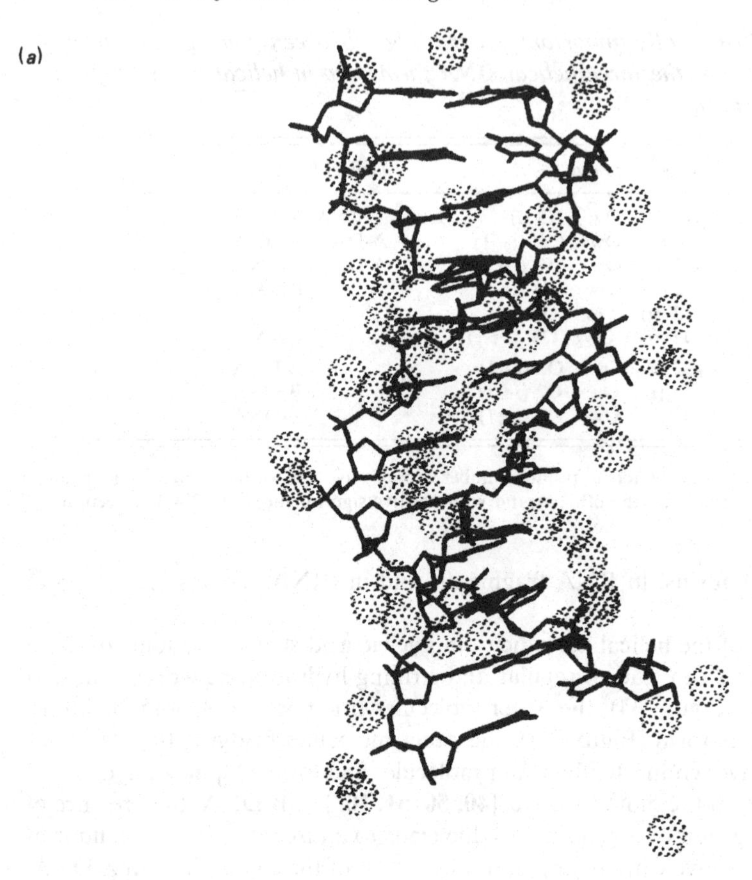

(b)

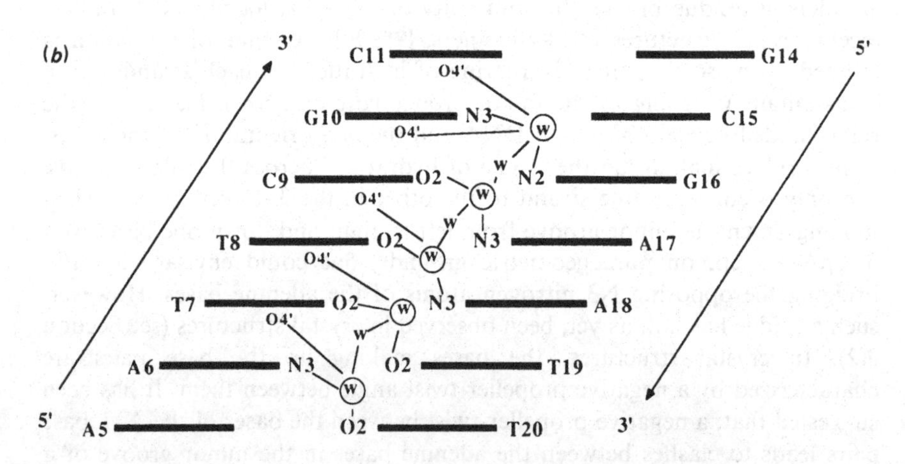


Figure 34. (*a*) View down the minor groove showing the spine of hydration in the B-dodecamer after the rerefinement of ref. 40. (*b*) A schematic drawing of the interactions.

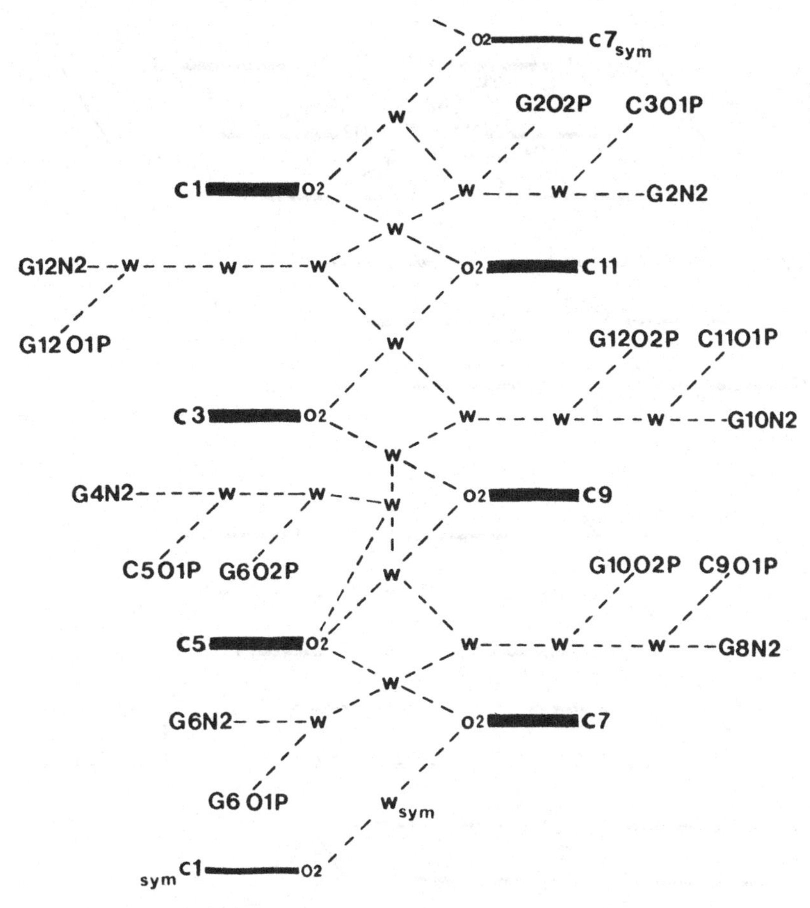


Figure 35. Schematic view of the interactions made by the water molecules in the minor groove of the Z-DNA hexamer d(5BrCG5BrCG5BrCG). **[51]**

pairs, or reduce the propeller-twist angle. But, in any case, the distance between the N3 nitrogen atoms across the minor groove increases, preventing the occurrence of a water bridge. Energy calculations **[106]** showed that the presence of a spine of hydration in polyd(A–T), in which every other step is a 5′-TpA-3′ step, is not favoured, while it is favoured in polyd(A)–polyd(T). In short, in a mixed sequence DNA, the spine of hydration could exist in runs of As or Ts and should be disrupted at 5′-TpA-3′ steps (Figure 36(*b*)).

Very recently, a neutron high-angle fibre diffraction study, **[339]** using Fourier difference synthesis after $H_2O/D_2O$ exchange clearly showed that water is ordered in the minor groove of the D conformation of polyd(A–T) (see Table A1). The water peaks were located close to the dyads relating the adenine bases (and thus at TpA steps). According to the above mentioned

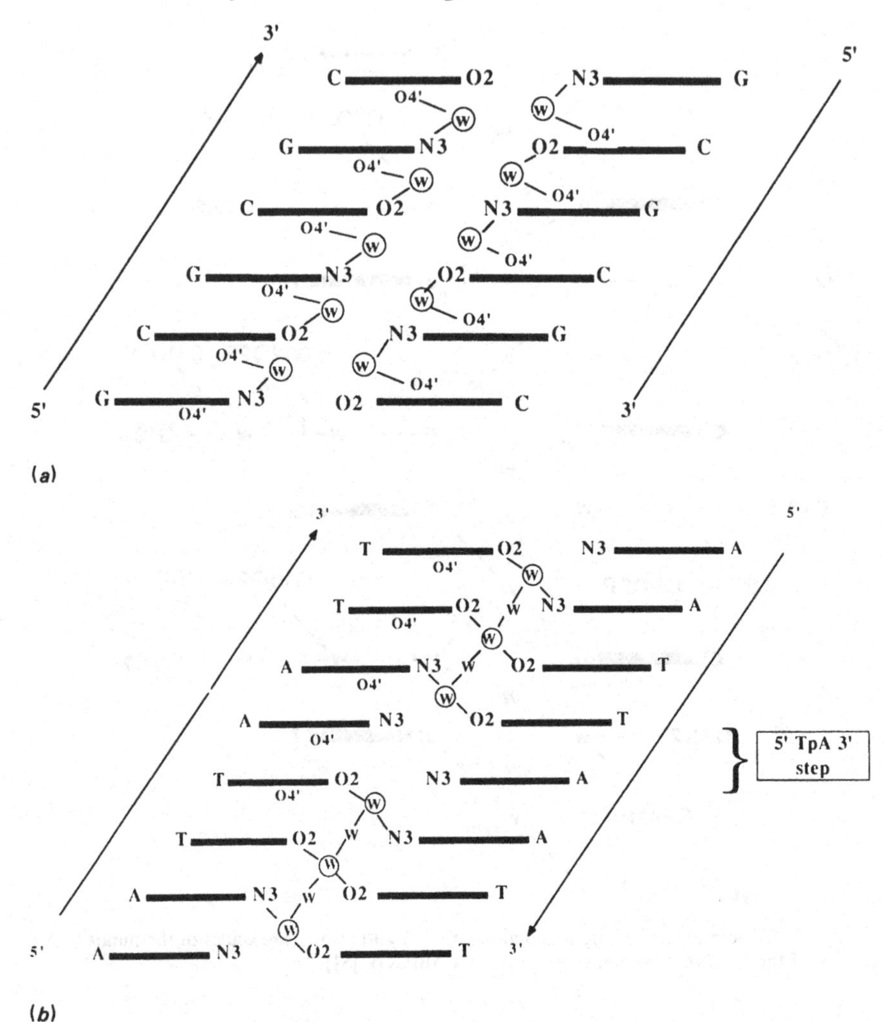


Figure 36. Schematic drawing of the observed interactions of water molecules in the minor groove of B-DNA type helices: (*a*) the two strings of water molecules along each strand; (*b*) the spine of hydration, broken at a TpA step.

calculations, **[106]** such a location is not favourable; but, those calculations were done on DNA conformations close to the B-form and not on D-form DNA.

Water molecules bridging the anionic phosphate oxygen atoms can also lead to apparent strings of water along the backbone of the A- and Z-forms (see above). In two crystals of an A-octamer **[94]** with alternating A–T (or A–5BrU) sequences, two-water bridges linking polar atoms in the major groove form beautiful edge-linked pentagonal arrangements of water

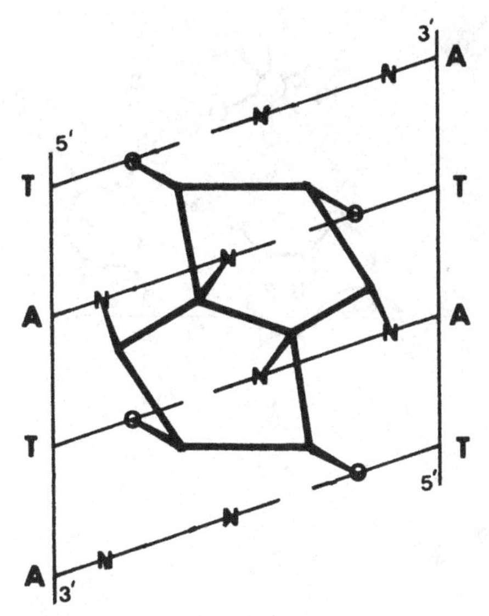


Figure 37. Schematic drawing of the water molecules forming pentagons and covering the major groove of the octamer d(GG5BrUA5BrUACC). [**94**] Adapted from a drawing by R. E. Dickerson.

molecules in the major groove (Figure 37). In the crystal structure [**26**] of the complex between the dimer d(CG) and the drug proflavine, water molecules hydrogen bond so as to form four edge-linked pentagons. Also, in the A-form tetramer d(5ICCGG), [**107**] there are two pentagonal arrangements of water molecules, one on the major groove of the C–G portion and one interacting with an anionic phosphate oxygen. Similarly, there is a group of pentagon-linked water molecules interacting with a water molecule of the 'spine of hydration' of the B-form dodecamer. [**40**] Although it has been suggested [**69**] that pentagonal water networks are well adapted to the hydration of nucleic acids, pentagonal clusters have also been found in the protein crambin. [**108**]

### 3.2. *Hydration around unusual base pairs*

Crystal structures of oligomers containing non-Watson–Crick type base pairs (e.g., G–T instead of G–C) have shown that water molecules rearrange and tend to cluster around unusual base pairs, as if compensating for the loss in hydrogen bonds which is often typical of such pairs [**109–14**] (Figure 38). A similar observation was made in an analysis of the hydration of *t*RNA molecules which contain several non-Watson–Crick base pairs involving either common or rare bases [**28, 115**] (Figure 39(*a*)). Despite the lower resolution of the *t*RNA analysis, some of the best-defined water molecules involve unusual base pairs. Furthermore, the same observation was made in

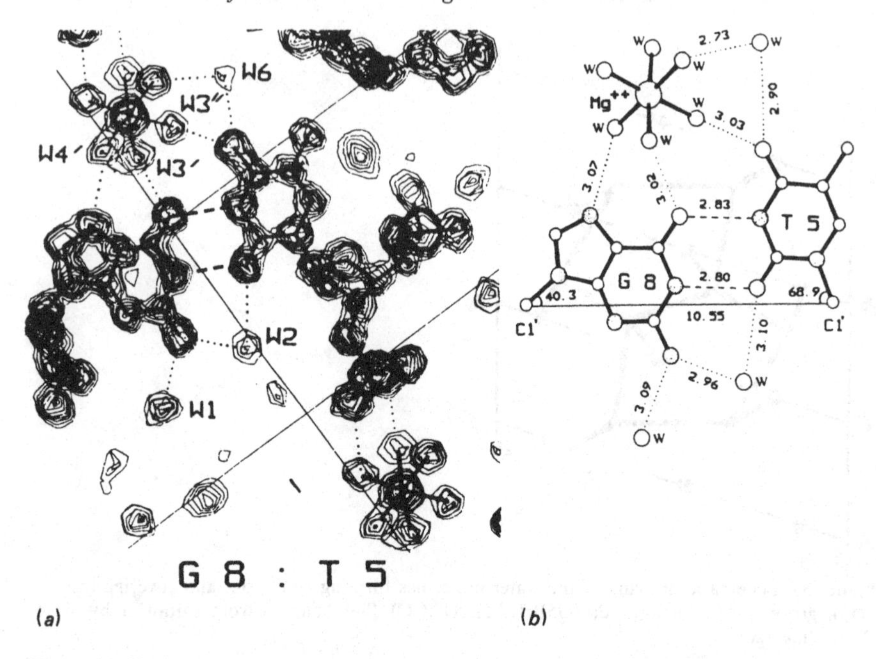


Figure 38. Environment around a G–T base pair in the structure of the Z-DNA hexamer (dCGCGTG): **[110]** (*a*) the electron density and (*b*) the interpretation. Reprinted with permission from ref. 110. Copyright © 1985 IRL Press Ltd.

three different kinds of *t*RNA crystals. In unusual base pairs, the reorientation of the bases with respect to their orientation in Watson–Crick pairing shape either internal notches or protrusions around which water molecules can fit. Apparently, irregularities around the observed unusual base pairs lead to locally attractive hydration sites whereby water molecules can help stabilize the base pairing. It has been suggested **[116]** that an important component in fidelity control of replication mechanisms originates in a local exclusion of water around the nucleic acid base pairs inside the active site of the enzyme machinery. Such a desolvation by the enzyme should destabilize mismatched bases with respect to standard Watson–Crick pairings, thereby contributing to fidelity. In addition, the less regular base pairs should not fit as well in the active site pocket of the enzyme.

In large molecules with complex tertiary structure as observed in *t*RNAs, non-canonical base pairing occurs between bases far apart in the primary sequence. Such pairs present new phosphate–water–base bridges, similar to those seen in crystal structures containing A–A pairs (see Section 2.3). Two examples of corresponding water bridges in an A–A pair and an A–G pair in *yeast* *t*RNA–asp are shown in Figure 39(*b*).

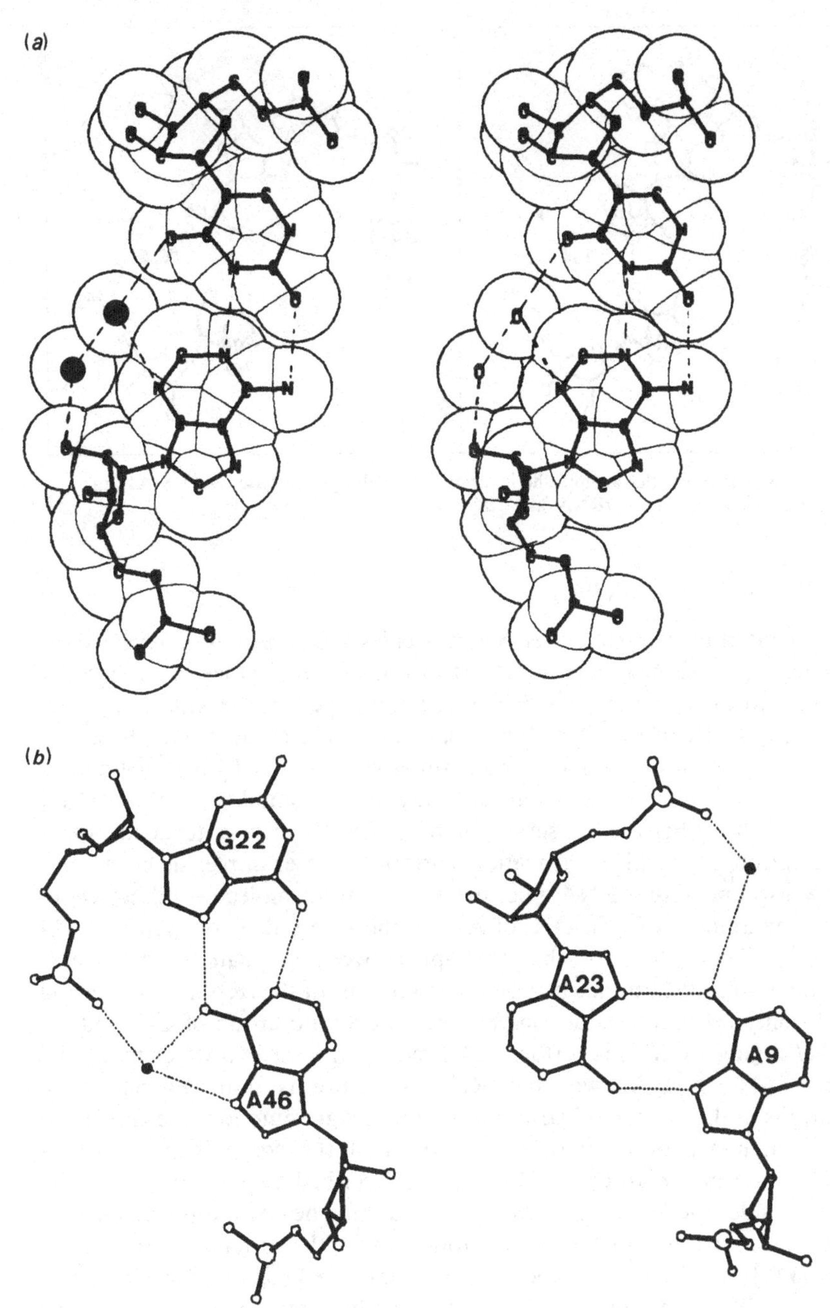


Figure 39. (*a*) Two water molecules interacting in the minor groove of the unusual base pair pseudouridine–adenosine in the orthorhombic form of *yeast t*RNA–phe. [**28**] (*b*) Two examples of phosphate-water-base in *yeast t*RNA–asp in purine–purine pairs of base triples. Compare these bridges with Figure 23(*b*).

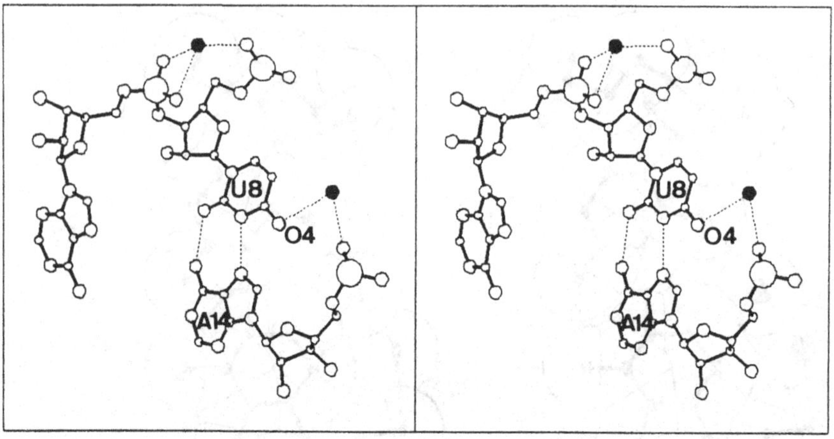

Figure 40. Water molecules around the invariant *trans* Hoogsteen base pair U8–A14 in the crystal structure of *yeast* *t*RNA–asp. **[28]**

### 3.3. *Internal water in* tRNAs

Refinement of four crystals of *t*RNA molecules at 3 Å resolution revealed the presence of water molecules localized in notches or cavities of the tertiary structure formed by the polynucleotide folding. **[28]** These water molecules bridge ribose hydroxyl O2′ atoms, base exocyclic atoms, and phosphate anionic oxygen atoms which would otherwise not be hydrogen bonded, thereby contributing in a significant way to the stability of the tertiary folding. When there is conservation of bases, there is a tendency for a conservation of localized hydration patterns. For example, in both *yeast* *t*RNA–asp and *yeast* *t*RNA–phe, there is a solvent molecule linking O4 of U8 to the anionic oxygen O1P of A14 in the reversed Hoogsteen U8–A14 base pair (Figure 40). Also, the base triple between base pair 10–25 and base 45, which is at the interface between the two arms of the *t*RNA molecule and is held only through interactions between the amino group of G45 and the O6/N7 region of G10, is further stabilized by at least two water molecules (Figure 41). It is known that RNA structure is quite insensitive to variations in the degree of hydration but that magnesium ions are crucial for the maintenance of the tertiary structures of the complex structures of *t*RNAs and ribosomal RNAs. However, as described above, water molecules bound inside pockets formed by loops and stretches of the polynucleotide backbone contribute further to the stabilization of the structure. A theoretical study **[117]** has also emphasized the stabilizing effect of bound water to loop structures. These internal water molecules can be regarded as an integral part of the *t*RNA structure itself.

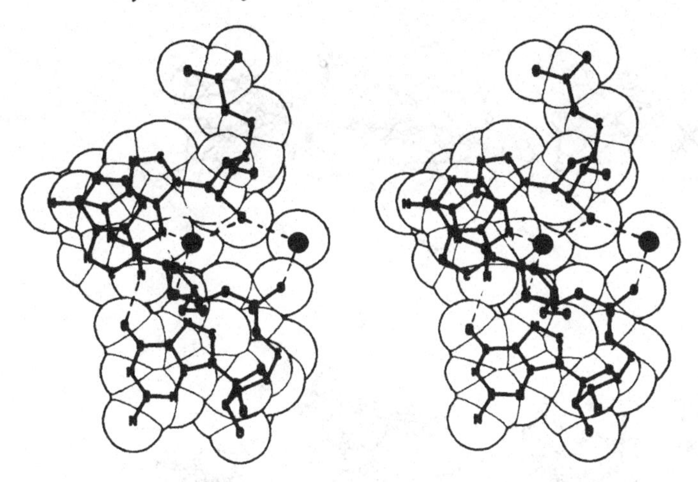

Figure 41. Stereo view of a water contact observed in *yeast* *t*RNA–phe and *yeast* *t*RNA–asp: a water molecule links O2′ of A9 to N3 and O2′ of G45, thereby filling a cavity formed by the tertiary structure. The O2′ of G45 is itself linked to an anionic oxygen atom of G10 by another water molecule. A9 is stacked under G45 and G10 is below. **[28]**

### 3.4. *Ionic hydration*

Magnesium binding to nucleic acids was first observed crystallographically in *t*RNA crystals. **[56, 118]** Later, magnesium ions were observed in Z-DNA crystals, **[119, 120]** and very recently in a B-DNA crystal. **[96]** In Z-DNA crystals, the binding is overwhelmingly intermolecular involving two helices and in the great majority of cases it is of the outer-sphere type (Figure 42). The concave shape of the Z-DNA major groove is probably the geometrical underlying cause for the presence of intermolecular contacts within the crystal packing environment. Moreover, the base ligands in the Z-DNA narrow and deep minor groove are quite inaccessible to ligands of metal complexes. The hydrated metal complexes usually interact via water molecules on one side with the guanine O6 and N7 and on the other side with one anionic phosphate oxygen atom. **[121]** Very interesting sodium and magnesium ion clusters with shared edges and face were observed in one crystal. **[50]** A bridging sodium ion with direct contact to N7 and an anionic phosphate oxygen of symmetrically-related molecules was also observed **[51]** (Figure 43). With magnesium and sodium ions, only partial dehydration occurs (1–2 water molecules are removed) and direct contact to nucleic acid ligands occur mainly at N7 of purines and at anionic phosphate oxygen atoms. Fully hydrated ions interact as often with the bases as with the sugar–phosphate backbone.

Such intermolecular interactions are important for understanding nucleic acid condensation and crystallization. In protein crystals, electrostatic and hydrogen bonding interactions between oppositely charged residues and

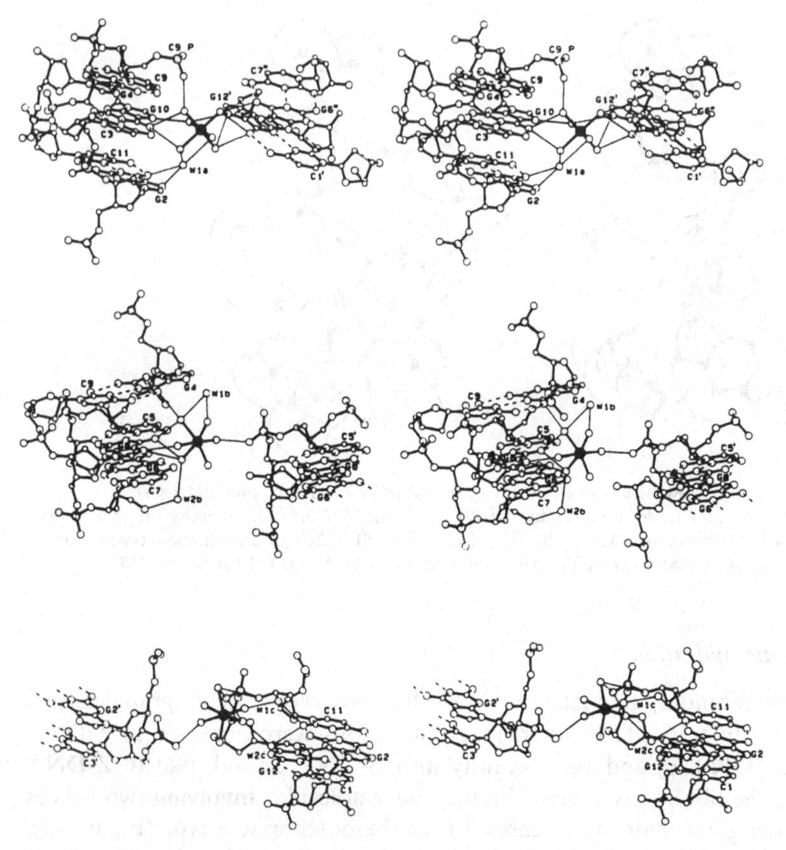


Figure 42. The three main ion binding sites in crystals of Z-type DNA oligomers. Reprinted with permission from ref. 120. Copyright © 1987 Adenine Press.

between polar residues can promote the association between protein molecules to form crystal structures. In contrast, nucleic acids have strong repulsions between their anionic phosphates and intermolecular hydrogen bonding can only occur either between amino groups of bases and the sugar–phosphate backbone or between ring nitrogen or oxygen of the bases and terminal O3′ or O5′ atoms (in RNA structures, the sugar O2′ atom is frequently involved). Therefore, the stabilization of nucleic acid crystal structures should strongly depend on the presence of counterions to neutralize and eventually bridge phosphate groups. Some of the counterions are fully hydrated and, by necessity, located in empty pockets between oligomers. The crystallographic identification of these is particularly difficult because of their mobility and disorder. Others are partially hydrated, bound very specifically in a tight pocket to two symmetrically-related oligomers. For the crystallization process to occur, there must be an equilibrium between ionic delocalization of fully hydrated ions without specific binding

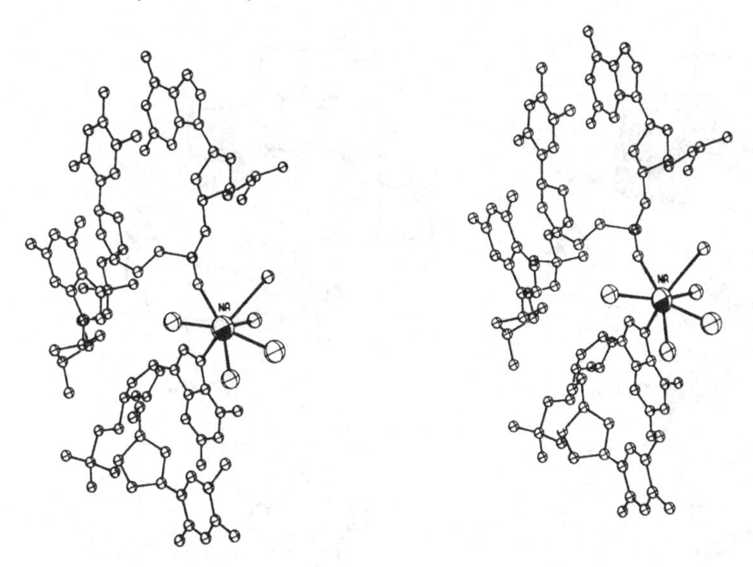


Figure 43. Binding of a hydrated sodium ion to two molecules of the hexamer d(5BrCG5BrCG5BrCG) [**51**] with direct contacts to the N7 of a guanine ring and to an anionic phosphate oxygen atom of a neighbouring molecule. This position corresponds roughly to the middle site shown in Figure 42.

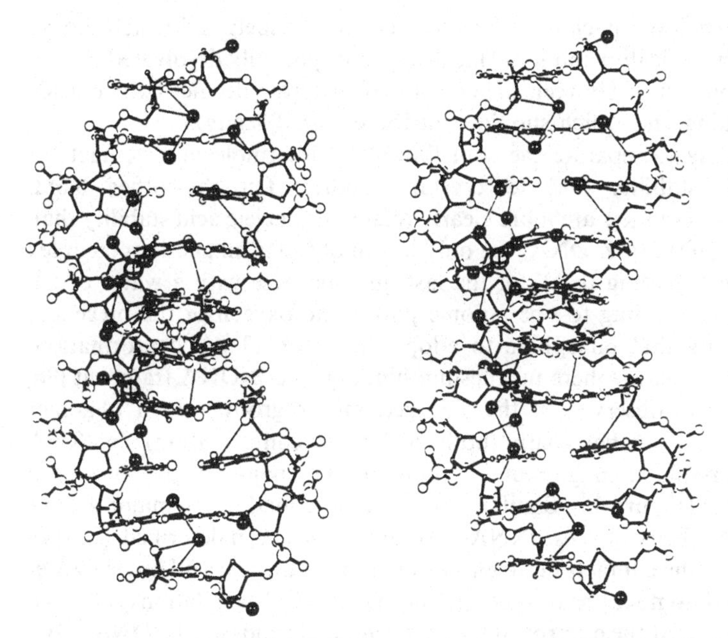

Figure 44. Hydration of the minor groove of the decamer d(CCAAGATTCC). [**96**] There is a crystallographic two-fold axis horizontally in the plane of the paper through the center of the helix. Notice the hydrated magnesium ions. Adapted from ref. 96.

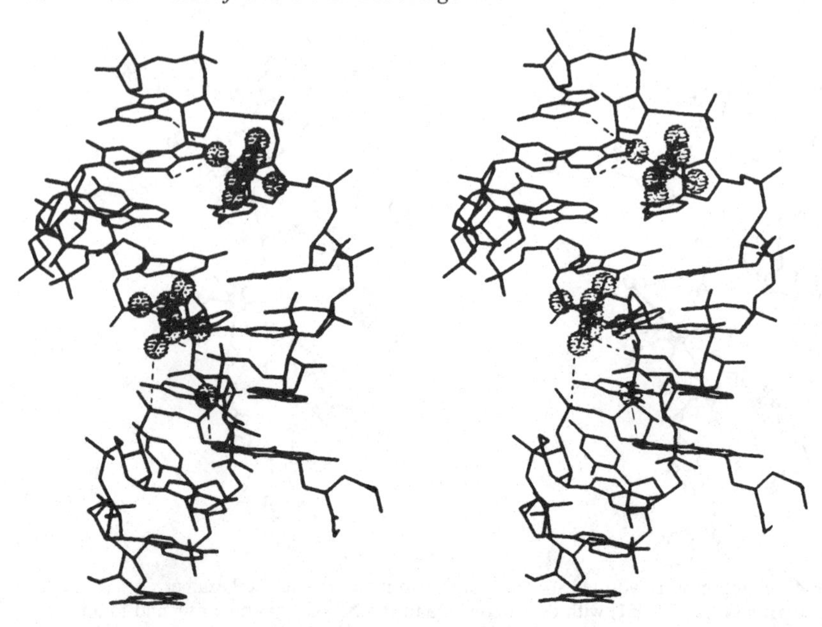

Figure 45. Two hydrated magnesium ions binding in the minor groove of the anticodon helix of *yeast* *t*RNA–phe in the monoclinic form. [**115**]

(which requires loose packing or the presence of channels and which should contribute entropically) and ionic localization of partially desolvated ions in specific regions on or between the crystallized macromolecules (which leads to tight packing and which should contribute enthalpically).

It is not easy to separate the contribution of intermolecular contacts to nucleic acid stability and to crystal stability. On the other hand, intramolecular contacts are more clearly related to nucleic acid stability than to crystal stability. In Z-DNA, the only intramolecular magnesium complex involves direct binding to N7 of the last guanine base with a water of the hydration shell binding to one anionic phosphate oxygen of the preceding residue, forcing that phosphate to adopt the so-called ZII conformation. [**122**] Recently, outer-sphere magnesium binding to a B-DNA fragment [**96**] (Figure 44) was observed in the enlarged minor groove of an oligomer containing two G–A base pairs (both in the *anti* conformation). In *t*RNA helices, outer-sphere magnesium binding in the major groove has been observed [**115**] (Figure 45). Similarly, a spermine bound in the major groove of the acceptor helix of *yeast* *t*RNA–asp form A has been described. [**123**] In all these cases, the contacts between the ion and the nucleic acid are via water molecules and do not appear to be specific. Theoretical calculations [**124**] had already shown that the electrostatic field strength depends on the DNA form, being strongest in the minor groove of the B-form and in the major groove of the A-form.

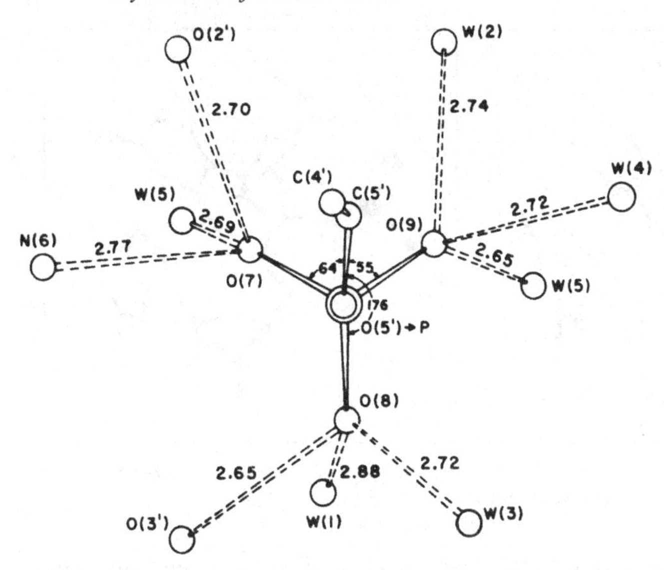


Figure 46. The hydrogen-bonding environment around the phosphate group in the crystal structure of 5′-IMP. [36] The view is along the ester bond O5′—P. The disposition of the hydrogen bonds around each P—O bond is roughly tetrahedral. Reprinted with permission from ref. 36. Copyright © 1969 American Chemical Society.

While binding to bases in helices appears to be preferentially of the outer-sphere type, binding to phosphate groups occurs in loops and bends of the sugar–phosphate backbone and is often of the inner-sphere type. Several examples of such complexes can be seen in the various crystal forms of *t*RNA molecules. [**56, 118**] Such complexes, without being highly sequence specific, require a precise geometrical and stereochemical environment beside the strong electrostatic field created by several phosphate groups forming the loop.

Fiber diffraction studies [**125**] have shown that cesium ions bind in the minor groove of DNA at positions similar to those occupied by the water molecules of the spine of hydration and are thus directly coordinated to pyrimidine O2s and purine N3s. On the other hand, in the major groove, the cesium ions bind close to the phosphate groups but are fully hydrated.

In some instances, three water molecules are seen forming hydration cones around phosphate anionic oxygen atoms. Figure 46 shows examples from the crystal structure of 5′-IMP, [**36**] and Figure 47 shows an example found in the crystal structure of the B-dodecamer. [**40, 75**] Other cases can be found in the Z-form of the d(5BrCG5BrCG5BrCG) hexamer. [**51, 345**] This kind of hydration pattern around anionic phosphate oxygen atoms is also quite common in crystals of nucleotides containing alkali and alkaline-earth metal cations. Such arrangements appear to be typical of the electrostriction effect, whereby the interactions of the water molecules with the anionic charge,

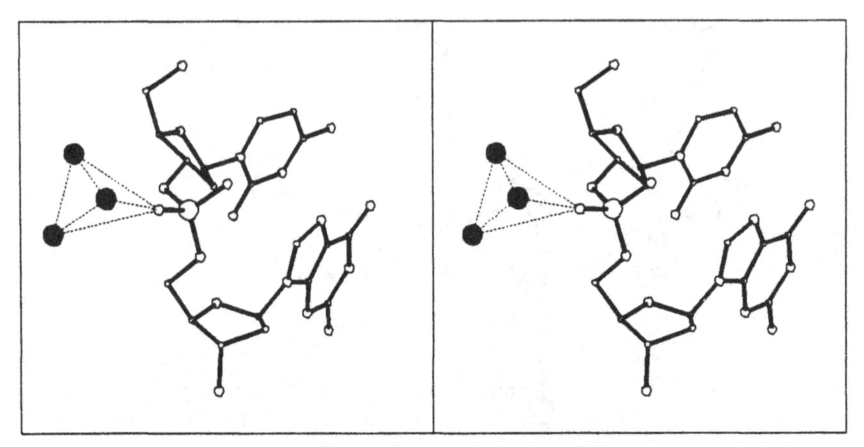

Figure 47. Cone of hydration centered on an anionic phosphate oxygen in the B-dodecamer d(CGCGAATTCGCG). From the coordinates of ref. 40.

rather than the intermolecular hydrogen bonding between the water molecules dominate the local geometrical disposition of the water molecules.

Mention should be made to the way the sulfate ion is bound to the sulfate-binding protein involved in active transport in *Salmonella typhimurium*: **[126]** there are no charged residues, cations, or water molecules within van der Waals' distance to the sulfate dianion. The ion is held in place by seven hydrogen bonds: five main-chain peptide NH groups, one serine hydroxyl, and one indole NH of a tryptophan side chain.

## 4. Theoretical and computational studies

Water, water, everywhere...

Samuel Coleridge, in *The Rhyme of the Ancient Mariner.*

As detailed in the preceding sections, considerable important and interesting information has been acquired from the crystallographic vantage point about solvation sites, water bridges and, in some cases, elaborate water networks. However, except in small fragments, only a small fraction of the total water, sometimes less than 25%, is found to be crystallographically ordered. Thus the description of nucleic acid hydration provided by this approach is fragmentary. In this section we attempt to fill in some of the details, first conceptually and then with the aid of theoretical calculations.

Sinanoglu and coworkers in a series of papers beginning in the mid-1960s **[127]** constructed an approach called 'solvent effect theory' for estimating the contribution of environmental effects to DNA structure. They included terms due to polarization of the water treated as a dielectric continuum via Onsager reaction field theory, van der Waals' effects, and surface forces in the water. The latter tended to dominate in this theory, and the idea that the stability of the DNA double helix might be due to the surface tension of water proved to be quite provocative at the time and stimulated further appreciation of the role of water in the system. Subsequently, solvent effect theory has met with some stiff criticism, [**128**] and the surface tension idea is considered oversimplistic in light of the intricate picture of the hydration structure of nucleic acids now emerging from crystallography. In retrospect, these studies were at least thermodynamic in concept, and certainly polarization energy due to solvent remains an aspect which has not yet been definitively treated. Continuum theories have formed the basis for several subsequent computational studies of tautomeric equilibria [**129**] and complexation [**130, 131**] of nucleic acid bases.

Lewin, [**132**] in a highly speculative but influential paper in 1967, drew attention to diverse possibilities for water bridges to form and to stabilize certain structural features of the DNA double helix. The counterion distribution was predicted to follow a helical pattern across the grooves stabilized by hydration. However, many of Lewin's specific proposals have not generally held up well against crystallographic data, but the idea that water bridges are an important element of nucleic acid hydration was firmly planted by this paper.

### 4.1. *Quantum mechanical calculations*

Pullman, Pullman, and workers reported a series of studies based on quantum mechanical (QM) calculations [**133–47**] aimed at enumerating the solvation sites of the ribose sugar, [**133**] phosphodiester linkage, [**134–7**] and the purine and pyrimidine bases. [**138–47**] These calculations were each a partial exploration of the water–solute pairwise interaction energy surface, with solvation sites associated with energy minima. The calculated sites for the Watson–Crick nucleotide base pairs are shown in Figure 48. The relevance of these results to condensed phase hydration, where N-body interactions and consideration of statistical weight enter the picture, is just now being assessed *via* molecular simulation. The problem is that a strongly bound Pullman solvation site may have a low statistical weight factor, depending on the shape of the surface. One prevailing concept which emerged from these studies is the idea of 'cones of hydration' around anionic phosphates (Figure 49), discussed in chapter 3.4 and further below. In ionic hydration, the strength of the pairwise energetics are a dominant force.

Perahia, Jhon, and Pullman, [**148**] developed empirical energy functions to enumerate calculated solvation sites for larger systems, which were used with minimization procedures to delineate energetically favorable binding sites on B-DNA. Their results support the association of 4–6 waters/nucleotide with phosphate hydration. Base and sugar hydration sites were enumerated at polar sites as expected. The minimizations predict water bridges between N7 and X6 (X = O in G, $NH_2$ in A) and base–sugar bridges to the furanose O4′

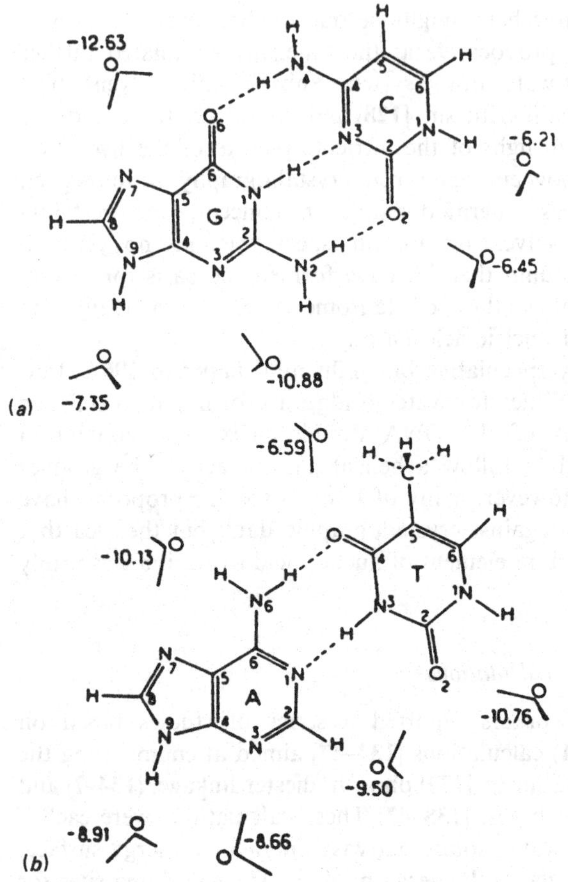

Figure 48. (*a*) Pullman solvation sites for a G–C base pair. Energies in kcal $mol^{-1}$. (*b*) Pullman solvation sites for an A–T base pair. Energies in kcal $mol^{-1}$. From ref. 145.

from purine N3 and pyrimidine O2 sites. However, in crystal structures of B-form oligomers, the latter bridges are observed between consecutive nucleotides (see Table 4) and only rarely within the same nucleotide (an example can be seen in the crystal structure of d(CCAAGATTGG) [**96**]).

Following extensive work in the Pullman Laboratory on QM electrostatic potentials, electrostatic mapping of the DNA environment was carried out by Lavery and coworkers [**149–52**] based on empirical energy functions (Figure 50). The electrostatic potential measures the interaction energy of a positive test charge (prototype electrophile) and the corresponding field indicates the interaction energy of a point dipole (hydrophilic species). These studies led to a recognition of the electronegative qualities of the grooves of DNA due the superposition of anionic phosphate potentials, and the idea that the grooves may not only be hydration sites but favorable positions for

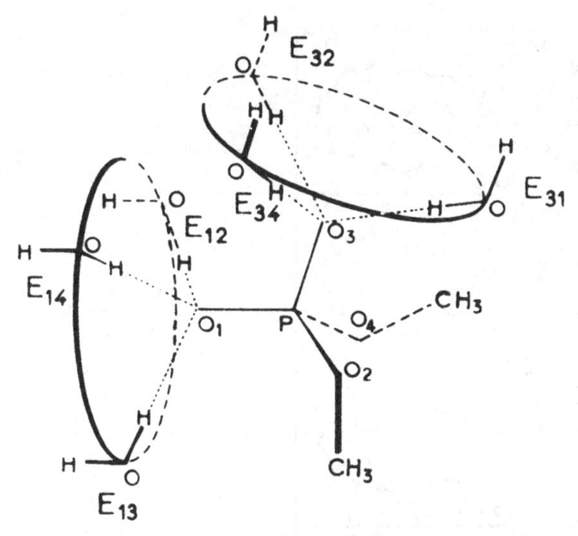


Figure 49. Structure proposed by Pullman, Pullman, and Berthod **[137]** for the hexahydrate of H2PO4–, illustrating the 'cones of hydration' around anionic oxygens. Reprinted with permission from ref. 137. Copyright © 1978 Springer-Verlag New York Inc.

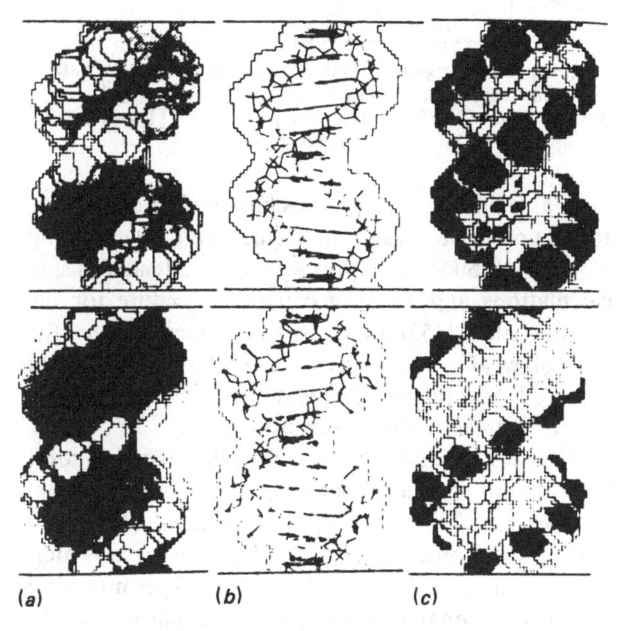


Figure 50. (top) Electrostatics of poly(dG)·poly(dC) reported by Lavery, Pullman and Pullman. **[150]** (bottom) Electrostatics of poly(dG)·poly(dC) screened by $Na^+$ counterions. **[150]** (*a*) Potential on the molecular surface envelope, (*b*) diagrammatic representation of a helix turn, and (*c*) field on the molecular surface envelope. Reprinted with permission from ref. 150. Copyright © 1982 Springer–Verlag New York Inc.

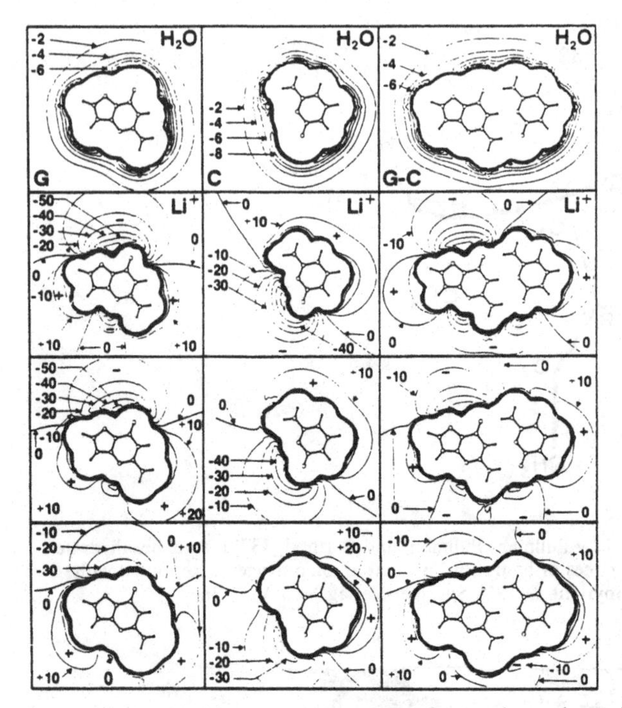


Figure 51. Calculated energy contour maps for guanine and cytosine interacting with water (top line), $Li^+$ (second line), $Na^+$ (third line), and $K^+$ (bottom line), based on energy functions determined from QM calculations by Clementi and coworkers. **[197]** Reprinted with permission from ref 197. 

the residence of counterions as well. The electric fields provided an alternative means of predicting interactions with water molecules and enumerating hydration sites on DNA. **[151]** Such an approach was extended using accurate electrostatic calculations and an approximate procedure for the rapid detection of hydration sites. **[153]** Large spatial zones around the optimum binding position of weakly bound as well as of strongly bound water molecules were observed. Several bridging water molecules, observed crystallographically, were reproduced by the calculations.

Clementi, Corongiu and coworkers **[154–63]** used QM calculations to form a discrete data base of interaction energies from which analytical potential functions were derived for base–water, **[154]** sugar–water, **[155]** and phosphate–water **[156, 157]** interactions. Combining these with water–water, **[158, 159]** ion–water, **[160]** and nucleic acid–ion potentials **[161]** permits rapid estimates of N-particle configurational energies (assuming pairwise additivity) for use in molecular simulations. Extensive characterization of these potentials via energy contour maps has been described by Clementi and Corongiu for water and ions interacting with nucleotide bases and base pairs, **[162]** and later reviewed in expanded and updated form by Clementi. **[163]**

An example for G–C base pairs is shown in Figure 51, which illustrates the extent to which potentials vary for the different alkali metal cations. Due to the size of the systems treated, the level of OM basis set that could be employed was relatively small, and subject to superposition errors that could lead to overestimates of the calculated interaction energies. In some cases, counterpoise corrections were included. Other groups developing potentials and force fields for use in simulation studies have taken a more semiempirical approach, calibrating against experimental data on liquids and aqueous solutions. However, the experimental data available is not necessarily comprehensive, and ultimately this is a 'seat of the pants' exercise. The extensive comparisons and characterizations necessary to provide a basis for unequivocally assessing potentials has not yet been feasible due to the inordinate amount of computer time required, but should now be forthcoming.

Detailed aspects of the hydrogen bonding of water to nucleotide bases has been the subject of a number of *ab initio* QM investigations by Del Bene. [**164**] QM studies of tautomeric equilbria in water relevant to nucleotide base chemistry in aqueous solution have been reported in a series of papers by Kwiatkowski and coworkers, [**165–8**] and reviewed in a recent informative summary of theoretical studies of biomolecular hydration by Kwiatkowski and Berndt. [**169**] A molecular orbital study [**170**] on the effect of water binding to base pairs indicated also that the binding of water to adenine yields larger solvation energy for the complementary pair A–U than for the non-complementary A–C pair.

## 4.2. *Monte Carlo simulations*

Monte Carlo (MC) simulation on the water structure around constituents of nucleic acids based on the analytical potentials derived from QM calculations were also described by Clementi and coworkers. [**157, 162**] MC studies based on alternative choices of potentials for nucleic acid bases and water have been reported by Danilov and coworkers, [**171–4**] who subsequently extended their work to a consideration of base associations. The preference of bases to stack rather than pair in water was reproduced in this simulation, and linked to an energetically more favorable water structure around the stack. MC calculations were carried out by Pohorille and coworkers [**175, 176**] on stacked dimers of nucleic acid bases in $CCl_4$ and water.

The analysis of liquid state computer simulation results is based at the most fundamental level of consideration on the calculated molecular distribution functions. The analysis of nucleic acid hydration presents special difficulties, since the calculated distribution functions contain admixtures from ionic, hydrophobic, and hydrophilic modes of solvation, which are distinctly different in character. A general procedure for the analysis of the local solution environment of a dissolved solute applicable to nucleic acids

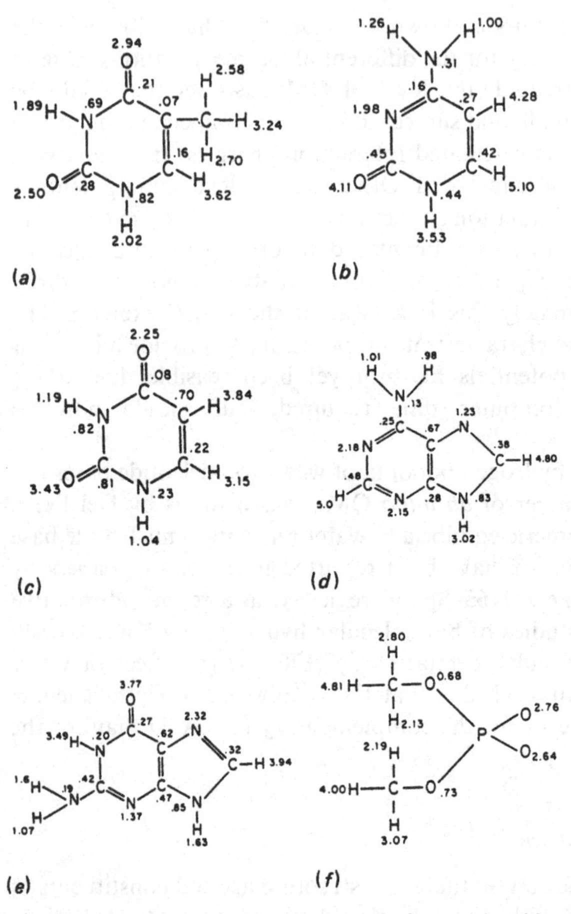


Figure 52. First shell coordination numbers for nucleic acid constituents calculated by Beveridge and coworkers [179] from MC simulation and proximity analysis: (*a*) thymine; (*c*) uracil; (*d*) adenine; (*e*) guanine; (*f*) DMP in *gg* conformation.

has recently been devised, **[177, 178]** whereby solvent molecules in the various N-molecule configurations are assigned to the closest solute atom (the proximity criterion), a procedure formally equivalent to partitioning configuration space in the simulation cell on the basis of the Voronoi polyhedra of the isolated solute. The subsequent analysis is organized in the general framework of the theory of quasicomponent distribution functions. **[177]** Using this procedure, the spatial extent of the first coordination shell of each solute atom can be well defined and the analysis can proceed on a solute atom, functional group, or subunit basis. The general idea can be extended to partition analysis of the hydration of a DNA double helix into contributions from major groove, minor groove, and backbone.

Recent computer simulation studies reported by Beveridge and coworkers

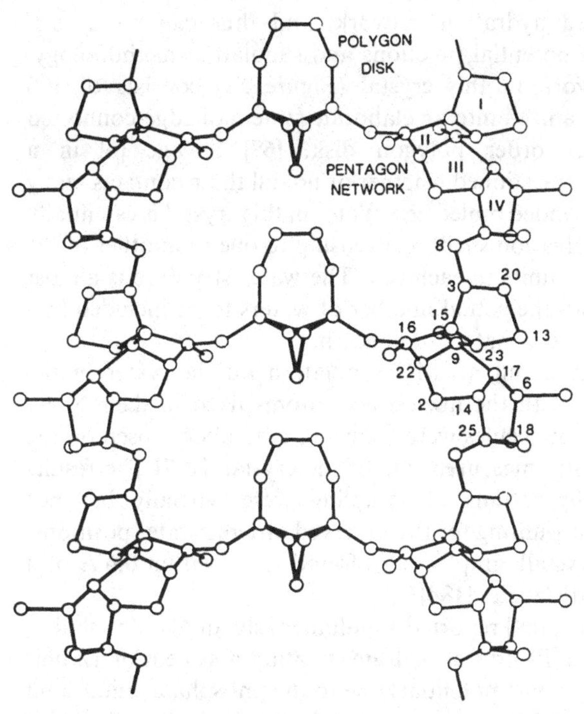


Figure 53. Ordered water network determined crystallographically for the 2:2 dcpG–proflavin crystal hydrate. Reprinted with permission from ref 69. Copyright © 1980 Macmillan Magazines Ltd.

[**179–83**] have dealt with various aspects of nucleic acid solvation via proximity analysis. The principle quantities obtained from proximity analysis are solute coordination numbers and solute–solvent pair interaction and binding energies. The hydration of constituent base, sugar, and phosphate groups, [**179–81**] and the effect of hydration on the conformational stability of the phosphodiester group in water [**182, 183**] were treated. The first shell coordination numbers for hydrated nucleic acid constituents calculated from proximity analysis are shown in Figure 52. The folded, *gauche-minus*, *gauche-minus* form of the phosphodiester group was found by free energy simulations to be preferentially stabilized by hydration over the *trans* extended form of the molecule. Preliminary studies of the hydration of d(CpG)2 in canonical A- and B-forms using proximity analysis has been described by Howell. [**184**]

The small crystal hydrates of nucleic acid constituents, nucleosides, and nucleotides discussed in the preceding sections which have well-defined solvent serve as a testing ground for the performance of molecular simulations, and present as well specific research problems accessible to theoretical study. The dinucleoside–proflavine (dCpG–Pf) crystal hydrate [**185**] has received considerable theoretical attention since it features an extensive and well-defined hydration network, and thus can serve as a convenient critical test of potential functions and simulation methodology. The ordered water network in this crystal (Figure 53) consists of 100 molecules per unit cell organized into an elaborate system of edge-connected pentagons and a higher order polygon disk. [**69**] Savage [**6**] in a comprehensive review has questioned whether or not all these contacts really correspond to hydrogen-bonded molecules. Water in this crystal is essentially all in a fairly close (first solvation shell) relationship to one or another of the four dCpG–Pf asymmetric units in each cell. The water structure is almost completely determined, but the actual number of waters to be included in a simulation has proved to be a matter of concern.

Mezei *et al.* [**186**] carried out an MC simulation on the water in the crystallographic unit cell, with the nucleic acid atoms fixed in their X-ray positions. Here 108 water molecules were included, a number chosen on the basis of independent density measurements of the crystal. [**187**] The results after some 2000 K configurations of sampling were partially but not completely successful in accounting for the observed ordered water positions, with some 64% of the crystallographic sites found to be within 0.6 Å of a calculated 'generic' solvation site. [**188**]

Clementi and coworkers [**189**] reported simultaneously an MC simulation on the water of the dCpG–Pf crystal hydrate, treating a system of 12 unit cells so that the long range ionic potentials due to the phosphate anions and proflavin cations would not have to be truncated unrealistically. They also varied the number of waters in the crystal and, on the basis of energetics and positional analysis [**190**] as well as solvation sites, [**191**] predicted that the optimum number of waters in the crystal should be 122–132 per unit cell. With this number, extremely good agreement on the subset of 100 waters closest to the crystallographic positions was obtained. However, the number is considerably larger than that obtained from density mesurements. The agreement with observed water positions is bound to improve at higher densities, since the space is more tightly packed with waters and some will obviously end up close to the observed positions. The authors discuss possible sources of errors in their calculation in terms of potential functions and the neglect of cooperative effects. Recent molecular dynamics calculations on this crystal are described in the following section. Additional MC simulations on nucleic acid crystal hydrates have been reported by Goodfellow and coworkers. [**192, 193**]

The molecular assembly of waters around DNA was studied by MC simulation as reported in a series of papers by Clementi and Corongiu. [**194–201**] Particularly, simulations on 447 water molecules surrounding the dodecamer d(ATGACTGCATCC) in canonical A [**194(*b*), (*c*)**] and B [**194(*a*), (*d*), 195(*a*)**] forms were carried out and analyzed extensively on the basis of an 'average' hydration structure determined from maxima in the solvent probability distribution function for water oxygens and hydrogens.

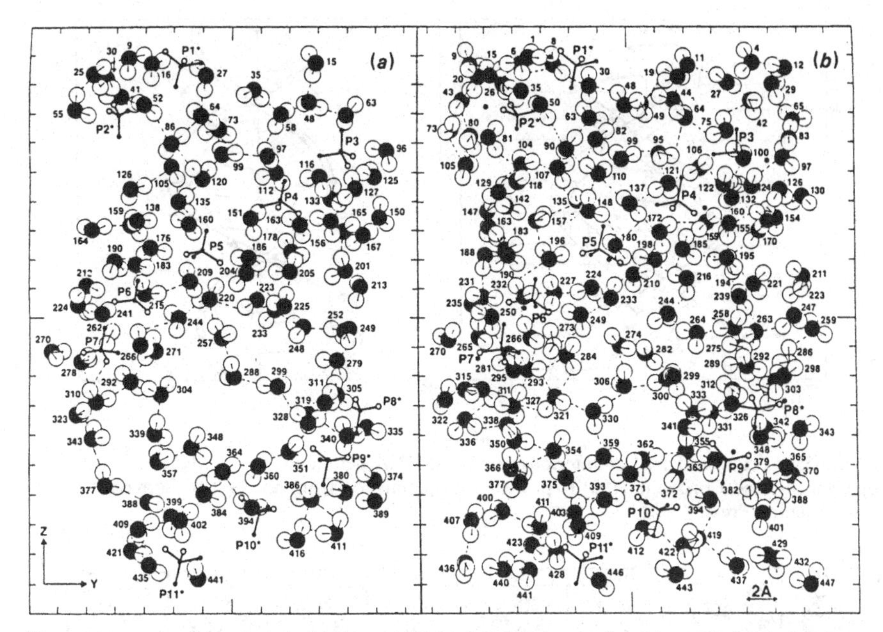


Figure 54. Hydration of (*a*) B-DNA, and (*b*) $Na^+$ B-DNA calculated by MC simulation by Clementi and Corongiu [**197**] and showing their filaments (dotted line). Reprinted with permission from ref. 197(*a*). Copyright © 1981 Adenine Press.

In the average structure, extensive occurrence of water bridges especially between base-pair steps was noted. The existence of 'filaments' of waters (Figure 54) spanning the major groove was reported from the simulation. However, in solvent water at this density, one expects all water molecules to be hydrogen bonded and a network will thus inevitably be established, so that the sequence of waters in a Clementi filament may be somewhat arbitrary. Water bridges and filaments connecting successive phosphates were also seen. The analysis of energetics supported the relative affinities phosphates > sugars > bases suggested as noted earlier by Falk, Hartmann, and Lord [**88**].

Clementi and Corongiu subsequently extended their B-form calculations to include sodium and other alkali metal counterions, [**195(*b*)–201**] first treated as fixed at the phosphate anions and subsequently allowed to move. The water structure in both cases turned out to be somewhat electrostricted compared with the waters-only simulation described above (see also Figure 54). This led to increased structuration of the nucleic acid hydration complex via cross-linking and local cyclic structures. In the calculation in which counterion motions were sampled, the $Na^+$ were displaced from the phosphates so that each moiety could be independently hydrated. The ions were found to reside in two helical envelopes, asymmetrically penetrating

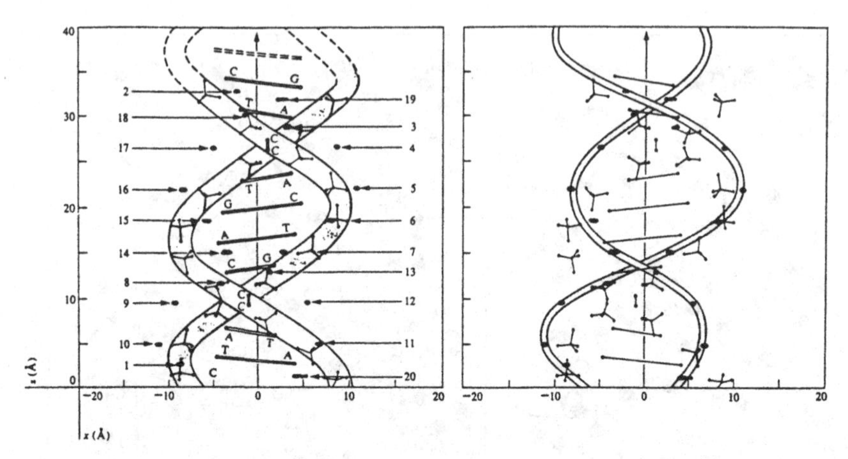


Figure 55. Location of $Na^+$ counterions around B-DNA determined from MC calculations by Clementi and Corongiu. [190] Reprinted with permission from ref. 197(*a*). Copyright © 1981 Adenine Press.

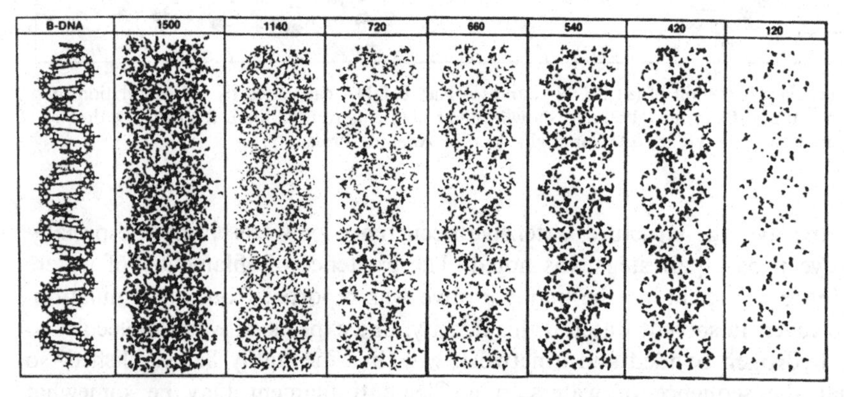


Figure 56. Water and $Na^+$ solvation of B-DNA from MC calculations by Clementi and Corongiu [**197*(a)***] based on various numbers of water from 120 to 1500. Reprinted with permission from ref 197(*a*). Copyright © 1981 Adenine Press.

into the major groove and situated just outside of the minor groove (Figure 55). The ion positions were notably sequence dependent, and thus suggested to play a role in sequence dependent interactions and helix unwindings. Simulations on DNA, water, and ions together are subject to considerable convergence difficulties characteristic of a liquid mixture, and the length of runs necessary to produce results free of quasiergodic problems has not yet been unequivocally established.

A series of MC simulations with various numbers of water molecules ranging from 22 to 447, subsequently extended to 1500, [**196**] were carried out by Clementi and Corongiu to simulate the experimental conditions established in adsorption/desorption experiments (Figure 56). At $Nw = 22$,

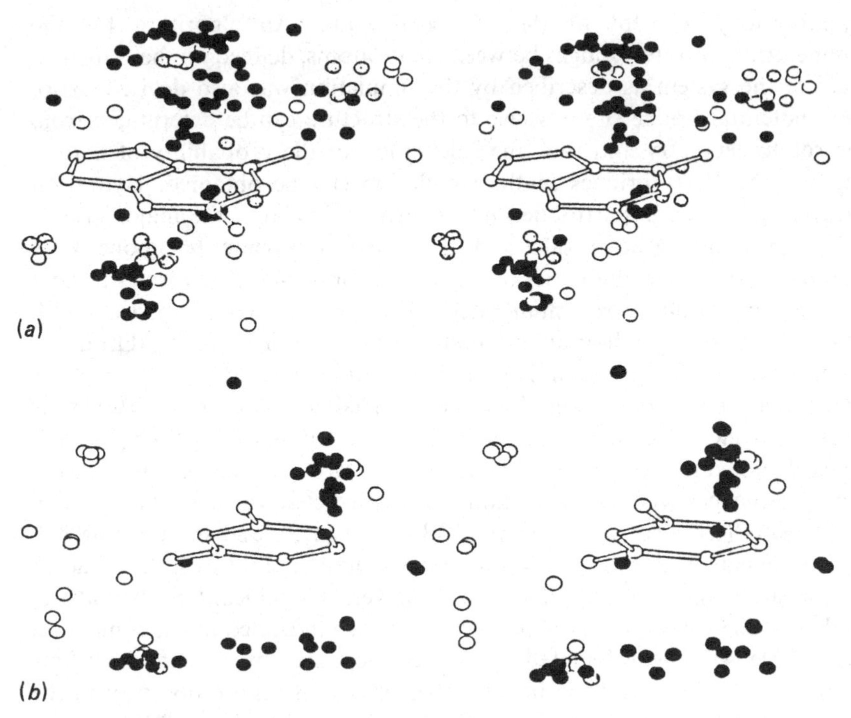


Figure 57. Comparison in stereo of calculated hydration density **[203]** (filled circles) with observed ordered water position (open circles) **[202]** in various crystals containing nucleotide bases: (*a*) guanine, (*b*) cytosine. Reprinted with permission from ref. 202. Copyright © 1988 Adenine Press.

the waters were found to associate with the phosphates. The grooves began to fill at a level of hydration corresponding to a slight excess of 6 waters/nucleotide. At lower water content, the hydration energy of the A-form of d(ATGATCGCATCC) was found to be greater than that of the B-form, probably due (but not stated explicitly) to the formation of a $PO_2^- \cdots W \cdots {}^-O_2P$ 'suspension' bridge arrangement in the water structure. At more extensive levels of hydration, the hydration energy of the B-form becomes larger than that of A-form.

Most recently, a comparison of the calculated hydration density in [rGpC]aq with observed ordered water positions in G, C, and rGpC crystal structures was reported by Berman, Sowri, Ginell, and Beveridge **[202]** and Subramanian, Pitchumani, Beveridge, and Berman. **[203]** In addition to proximity analysis, the entire solvent distribution local to the solute was examined. The calculated hydration density for the nucleic acid solute was depicted for presentation here as the linear superposition of points representing the centers-of-mass of solvent waters from a series of individual configurations taken at equally spaced intervals along the production (post-equilibration) segment of the MC realization. An algorithm for the enumeration of water bridges between solute atoms, defined on the statistical state of the system as described by the simulation was also devised. Also, each potential bridge site intrinsic to the structure can be determined from the solute geometry and thus the calculation of the probability of finding each of the water bridges in the simulation can be obtained. Hydration density distributions partitioned into contributions from the major groove side, the minor groove side and the sugar–phosphate backbone were examined, and the probabilities of occurrence for one- and two-water bridges in the simulation were enumerated. The results were compared with observations of crystallographic ordered water sites from X-ray diffraction studies on G and C containing small molecules and in crystal structure determinations of the sodium, [**24**] calcium, [**204**] and ammonium [**205**] salts of rGpC, which all display A-form double-stranded mini-helices (Figure 57). The calculated results are generally consistent with the observed sites, except for cytosine N4 where a hydration site is predicted yet none observed in rGpC salts (see also Section 3 and Tables 2 and 3), and for guanine N3, which appears in this calculation to compete unfavorably with the adjacent donor site at guanine N2. There is, however, a significant probability of finding a one-water N3(G)···W···N2(G) bridge indicated in the simulation (see Table 4). The calculated one- and two-water bridges in the rGpC hydration complex coincide in a number of cases to those observed in the ordered water structure of the sodium rGpC crystal hydrate. [**24**]

Subramanian and coworkers [**206, 207**] have recently reported MC simulation studies of hydration of a fully screened charge model of DNA, considering the dodecamer d(CGCGAATTCGCG) together with 1777 water molecules in a hexagonal cubic cell under periodic boundary conditions. The calculated total first coordination number of 10.4 waters/nucleotide compares closely to the critical level of hydration indicated by Falk *et al.* [**88**] to be necessary for DNA stability, and further supports the idea that the experimental number of 10 waters/nucleotide corresponds to an intact first hydration shell. The addition of the second shell coordination of the minor and major groove gives a total of 17.4 waters/nucleotide, close to the value of 20 required in Falk's experiments for the stability of the B-form. The calculated results on the number of waters required to form an intact first shell and to fill the grooves of B-DNA completely corresponds closely to the estimates from solvent accessibility calculations by Alden and Kim [**269**] (see Section 5.2 and Table 6).

In the minor groove (Figure 58), the localization of hydration density occurs at N3(A) and N2(G) sites and describes essentially monodentate hydrogen bonding. The crystallographic sites in the minor groove are essentially limited to the A–T tract, and indicate the existence of N3(A)···W···O2(T) bridges which comprise the first shell of the 'spine of hydration'. The reason for the discrepancy between calculated and observed

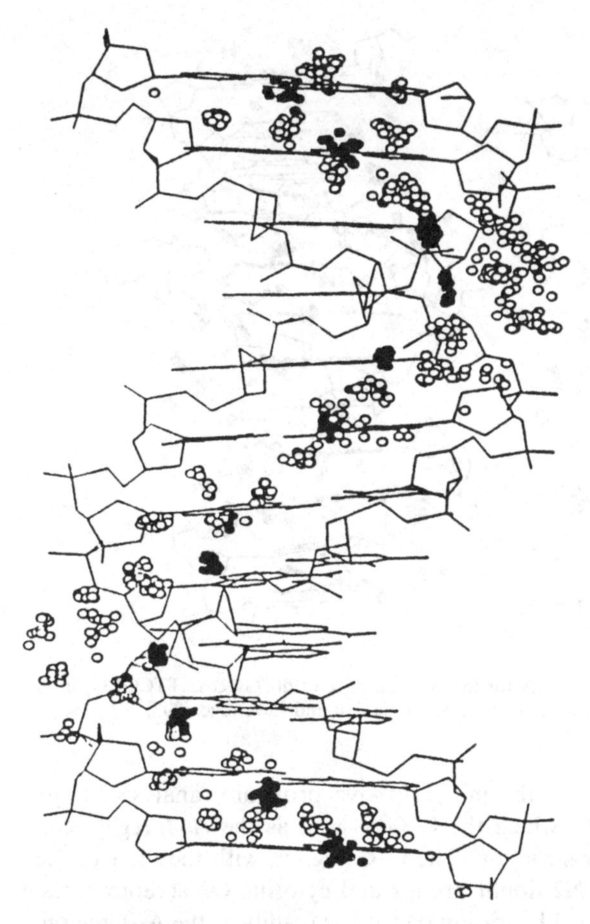

Figure 58. Hydration density in the minor groove of d(CGCGAATTCGCG) in the canonical B-form as determined by MC simulation by Subramanian and Beveridge **[207]** presented as a superposition of water oxygen positions obtained from a sequence of 16 configurations extracted from equally spaced intervals along the simulation. Solid circles indicate first shell waters of hydration with respect to nucleotide bases, and open circles indicate the remaining minor groove waters.

results lies to a considerable extent in the structural differences between the canonical and crystallographic forms (see Section 3 and Table 3). The former was assumed in the calculations. In the latter, the inherently greater propeller twist in A–T versus G–C base pairs orients the bases such that N3(A) and O2(T) are 3.7 Å apart, a propitious distance for bridging by a single water molecule. In the canonical form, the corresponding distance is 4.4 Å, and in our calculation the O2(A)···W···N3(A) bridge is seen only about 18% of the time. The calculated hydration furthermore does not show much density at O2(T) sites, preferring instead successive N3(A) sites in the first coordination shell.

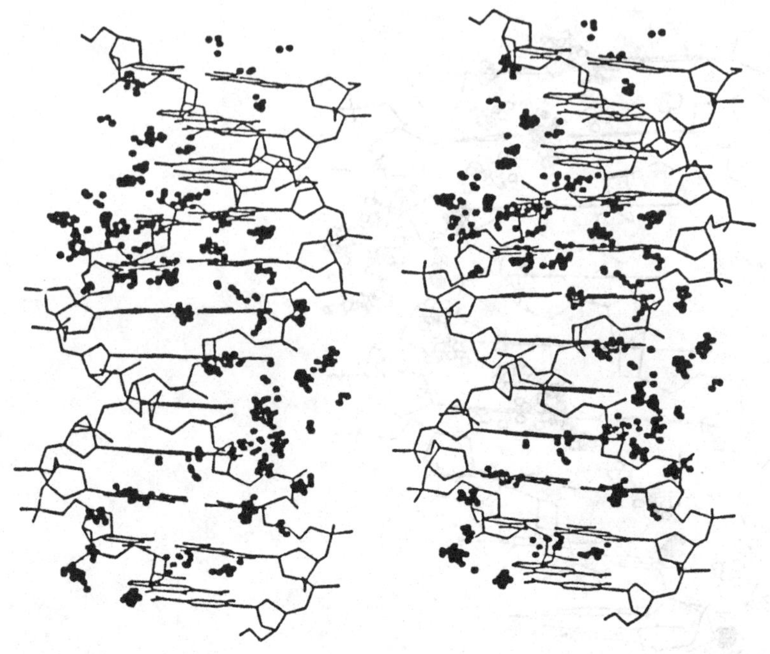

Figure 59. First shell hydration density for the major groove of d(CGCGAATTCGCG) in the canonical B-form from MC simulation by Subramanian and Beveridge. [207]

An intriguing result from the minor groove proximity analysis of the simulation is the extent to which the C–G as well as the A–T region can support an ordered water structure. The C–G region, with the floor of the minor groove lined with N2 donor groups and cytosine O2 acceptor sites, appears to nucleate localized hydration density as readily as the A–T region. The calculated ordered water network in the minor groove of our model dodecamer thus extends completely from one end of the structure to the other, with the interface between the C–G and A–T regions dealt with readily by the geometric versatility of water–water hydrogen bonding. The penetration of water into the DNA is of course clearly greater for A–T than for G–C tracts, and this feature distinguishes the 'spine of hydration' as defined by Dickerson and Drew **[103]** from the generic minor groove hydration pattern. In the rerefinement of the same B-dodecamer, **[40]** water bridges between guanine N2s and cytosine O2s were, however, detected. Marky and Breslauer **[208]** recently reported a positive binding entropy of comparable magnitude to that of d(CGCGAATTCGCG) for the binding of netropsin to the poly[d(GC)].poly[d(CG)] duplex. If the binding mechanism remains the same, this observation supports the existence of an ordered water network extending into the G–C region. Breslauer **[209]** has pointed out that

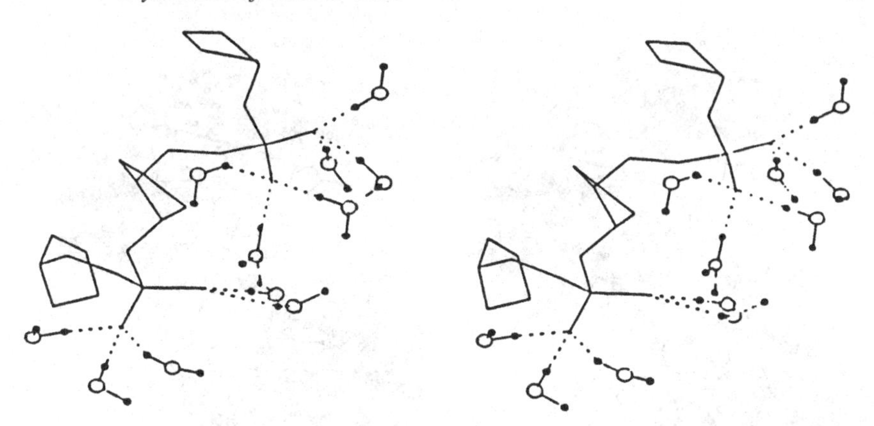

Figure 60. Calculated phosphate hydration (snapshot) in B-DNA from MC simulation by Subramanian and Beveridge, [**207**] showing the 'cones of hydration' motif.

both entropies referred to above could also be due to release of counterions as well as waters of hydration.

For the major groove (Figure 59), the calculated saturation of hydrophilic binding sites with waters of hydration in the calculation corresponds closely to the crystallographic results reported by Dickerson and coworkers. [**91, 103**] The 16 K structure and the MPD7 structure also show evidence for some various types of water bridges in the major groove. At the phosphates, the calculated results (Figure 60) show the characteristic triad of waters in the 'cone of hydration' motif at nearly every anionic oxygen. The phosphate hydration is observed to be disordered in the room temperature structure but extensive ordered solvent sites are found in the 16 K and MPD7 structures of the dodecamer, [**91, 103**] although another refinement [**40**] led to more ordered waters around phosphates for the room temperature crystal. Examining the crystallographic results with the idea of cones of hydration in mind, the observed solvation sites can nearly all be interpreted in terms of fragments of this motif in which only one or two of the three waters has turned out to be ordered. Only rarely does one see a fully ordered triad of waters of a single cone in the crystal structure (see Figure 46). The hydrophobic hydration around the thymine methyl group and the non-poplar atoms of the furanose sugar ring is calculated to be diffuse and not observed to be ordered. The water 'wedged' between the thymine methyl group and a phosphate (see Figure 33) is clearly represented in the calculation as a water bridge occurring with a frequency of 51 %.

A notable feature of the water bridge analysis of the simulation is that, with only a few exceptions, the calculated probabilities of occurrence are relatively low, i.e., $< 50\%$. The reason for this appears to be as follows: water bridges indeed belong to the energetically favored structures for an

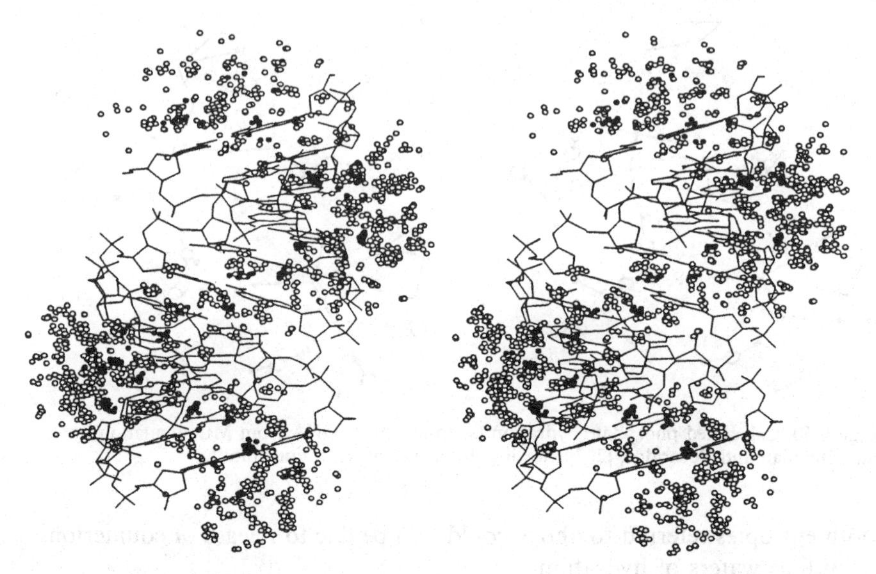

Figure 61. Hydration density in the minor groove of an A-form of d(CGCGAATTCGCG) determined from MC simulation. [**210(*a*)**]

individual water molecule, but they are not always fully ordered, since the bridging water molecules exchange with the bulk. For an assembly of water molecules at 300 K instead of absolute zero, entropy comes into the picture and destabilizes the ordered bridges with respect to forming sequential hydrogen bonds with different water molecules. This latter situation is more disordered than the water bridges and is thus entropically favorable, with little loss of energy since the number of hydrogen bonds is essentially conserved. (One could even argue that two different water molecules could give a lower energy because they can individually optimize their orientations, whereas a bridged water has to compromise a little in both bonds. But, hydrophilic atoms of the nucleic acid in their favored aqueous conformations might adopt geometrical positions minimizing such loss.) The quantitative relationship of any of this to the free energy of the system has not yet been rigorously studied.

The quantity in these MC calculations which is most applicable to the subject of relative affinities is the solute–water binding energy. The proposed ordering, given with respect to decreasing affinity for water, places the anionic phosphate oxygens in first place, followed by the major and minor groove regions. The ester oxygens in phosphate and sugar groups have correspondingly lower average binding energies to water in the calculation. It is important to note that 'affinity' is really a free energy quantity and not directly addressed in these calculations.

A MC simulation of poly(dA).poly(dT) in two conformations, the standard B-form and B′-form, [**266**] has been performed with 30 water

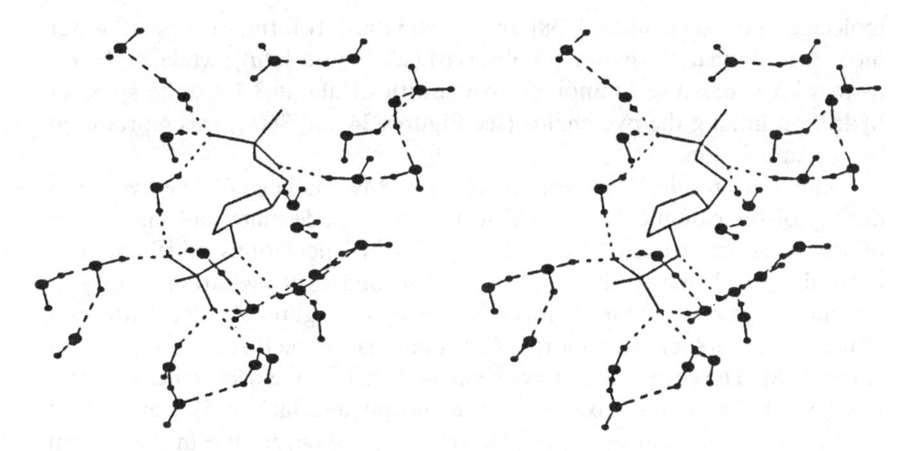

Figure 62. Calculated hydration complex for a segment of the backbone of A-form d(CGCGAATTCGCG) from MC simulation, [**210*(a)***] showing a water bridge between successive anionic oxygens. Note the bridge here involves one of the 'cone of hydration' waters.

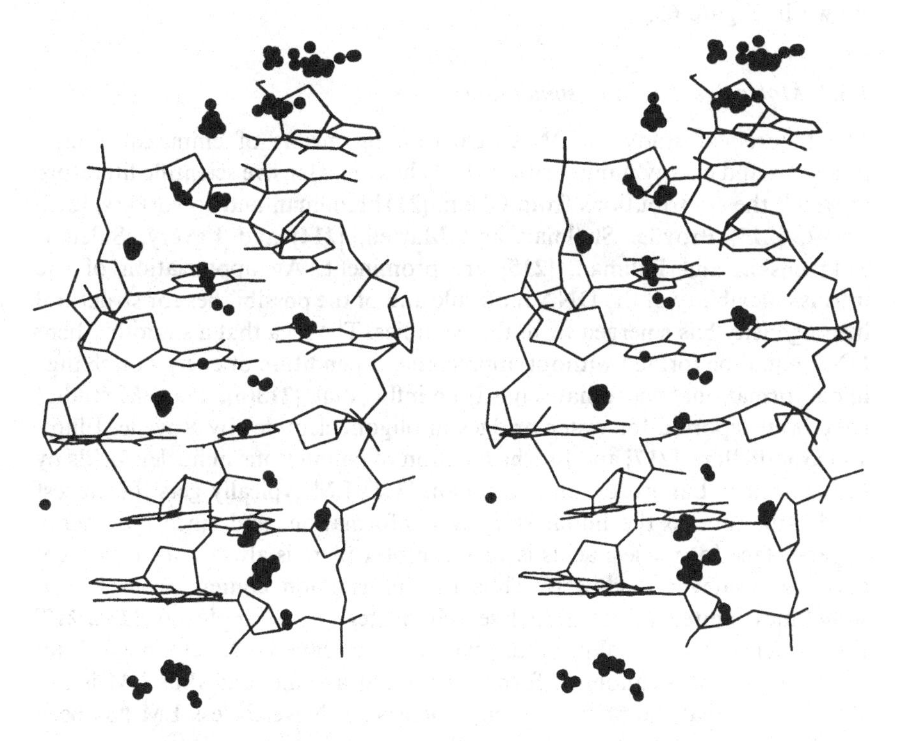

Figure 63. Hydration density in the minor groove of the Z-form of d(CGCGCG) determined from MC simulation. [**210*(b)***]

molecules per nucleotide. **[338]** In the standard B-form, strings of water molecules along each chain were observed (see Figure 36(*a*)); while, in the B′-form with a narrowed minor groove (width of about 3.7 Å), the spine of hydration linking the two chains (see Figures 34 and 36(*b*)) was reproduced by the calculations.

Preliminary results have been obtained by MC methods for the hydration density of the canonical A- and Z-forms of the dodecamer and the Z-form of the hexamer d(CGCGCG), **[210(*a*), (*b*)]** in calculations which parallel those described previously. The first and second shell hydration density in the minor groove of the A form is shown in Figure 61. The hydration structure here is somewhat more diffuse than that of the B-form in this region (Figure 58). The calculations have been analyzed for the occurrence of water bridges between anionic oxygens of the phosphates. Such bridges are indeed found (Figure 62), but the calculated frequency of occurrence in this system (aqueous hydration, i.e., 100% relative humidity) turns out to be relatively lower than expected, probably due to the entropic effects described earlier. The probability of occurrence of this bridge may increase with decreasing humidity. An account of the ordered water network in the minor groove of Z-DNA (the Z-spine) is obtained in the calculations on d(CGCGCG), as shown in Figure 63.

### 4.3. *Molecular dynamics simulations*

The theoretical study of DNA structure by means of empirical energy functions and energy minimization (EM) has an extensive scientific literature in which the contributions from Olson, **[211]** Kollman and coworkers, **[212]** Levitt, **[213]** Broyde, Stellman and Martell, **[214]** and Lavery, Sklenar, Zakrzenska, and Pullman, **[215]** are prominent. An appreciation of the intrinsic flexibility of the DNA molecule and of the possibilities for structural heterogeneity has emerged from these studies. The idea that a smoothly bent DNA could be formed without appreciable expenditure of energy or change in conformational coordinates has been influential. **[213(*a*), 216]** EM studies have been reported for water bridges in oligonucleotides by Vovelle, Elliot, and Goodfellow, **[217]** and for the location of counterions in nucleic acids by Pattabiramin, Langridge, and Kollman. **[218]** EM typically gets the closest local minimum to the initial starting conformation, and since the energy hypersurface for nucleic acids is very complex there is always one minimum close to whatever is chosen. Thus the information content from EM is sometimes limited unless global search strategies are employed. **[214, 217]** DNA structures in solution as explored to date **[219, 220]** may differ considerably from canonical forms one might assume, and thus EM is not sufficient to treat the problem comprehensively. Nevertheless EM has been an effective way to resolved steric clashes and set forth model structures, and is a customary first step in molecular dynamics (MD) calculations.

The small crystal hydrate systems relevant to nucleic acid solvation that have been studied by MD simulation to date are the cyclodextrins [**221**] and, after the MC studies cited earlier, dCpG–proflavine. [**222**] The cyclodextrins had earlier been the subject of neutron diffraction studies, [**223**] and the waters of hydration well determined with respect to both position and orientation. Koehler and coworkers carried out MD calculations using the program GROMOS [**335**] on the $\alpha$-cyclodextrin hexahydrate crystal [**221(*a*)**] and the $\beta$-cyclodextrin dodecahydrate crystal. [**221(*b*)**] They obtained generally excellent agreement with experiments on crystals, and also compared the crystal results with simulations on aqueous hydration. The calculations generally support the idea of flip-flop hydrogen-bond dynamics, [**221(*c*)**] and also provide evidence for occurrence of novel three-center hydrogen bonds [**221(*d*), (*e*)**] in the crystalline solid and aqueous solution.

The dCpG–proflavine crystal hydrate, after the MC simulation cited in the previous section, has been studied via EM and MD calculations in considerable detail by Swaminathan, Berman, and Beveridge [**222**] on a system of three crystallographic unit cells, also using GROMOS [**335**] with some modifications. The energy minimizations resulted in a hydration structure recognizable in comparison with the crystallographic results of Figure 53, but all the 'contacts' did not turn out to be hydrogen bonded. This result however only applies to 0 K. The MD calculations at 300 K can be analyzed to provide a dynamical time averaged view of the structure by means of a superposition of snapshots taken at various intervals along the time line of the simulation. The results (Figure 64) show clearly that the complete network as proposed from the crystallographic analysis is intact as 'D'-structure at 300 K. The calculated geometry for dCpG–proflavine also agrees well with experiment, reproducing even the mixed sugar pucker observed in the crystal, and generally indicate the force field to be performing quite well.

Calculations on nucleic acid helices without some kind of treatment of solvent, implicit or explicit, have not suprisingly turned out to be unstable. [**213**] A considerable number of studies have been reported in which hydration is included implicitly via a distance dependent, dielectric screening function and electrostatic effects are introduced via reduced charges on the phosphates. These implicit solvent models for nucleic acids have proved difficult to calibrate numerically, since well-defined points of reference for nucleic acids in solution are not unequivocally established yet. The following are the MD calculations reported to date on nucleic acid systems, and are seen to include both implicit treatments of solvent, and initial ventures into explicit inclusion of water and counterions together with internal degrees of freedom.

Levitt [**213**] first reported 90 ps of MD on the duplex d(CGCGAATTCGCG) and a dA.dT 24-mer with a vacuum boundary,

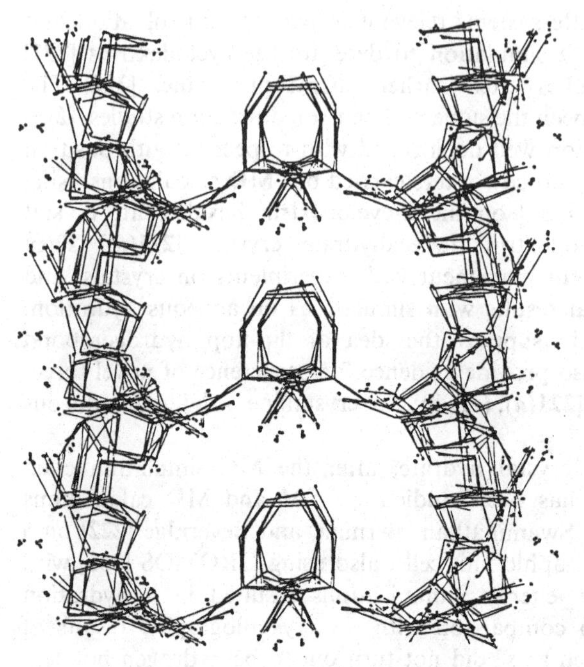

Figure 64. Calculated hydration network at 300 K for the dCpG–proflavine crystal hydrate based on MD simulations by Swaminathan *et al.* [222]

omitting electrostatic terms in the non-bonded interactions. The simulation demonstrated an overall stability with large amplitude bending and twisting in the model DNA, affecting anisotropically the major groove. Considerable informative details of the motions were analyzed and reported. Some estimates of electrostatic effects were made, but when charges were added to the model with a dielectric constant of unity the duplex unwound.

MD and normal mode analysis of the duplexes d(CGCGCG) in the B-form and normal mode analysis for d(CGCGCG) and d(TATATA) were reported by Karplus and coworkers [224] in 1983. Here the electrostatics were treated by partially reducing the negative charge on each phosphate to $-0.32$ based on a Manning-type counterion condensation model and using a distance dependent dielectric constant to mimic solvent shielding. The simulation produced a stable B-form with amplitudes of atomic fluctuation somewhat smaller than those reported from crystallography on the dodecamer, [75] and discussed in terms of possible static disorder in the crystal. Some preliminary comparisons with results on NMR order parameters and fluctuations inferred from fluorescence depolarization were given. The first [225] of a series [226] of MD simulations on *yeast t*RNA–phe was also reported in 1983 by Harvey, McCammon, and coworkers. These calculations also involved a screened charge model for the phosphates. The

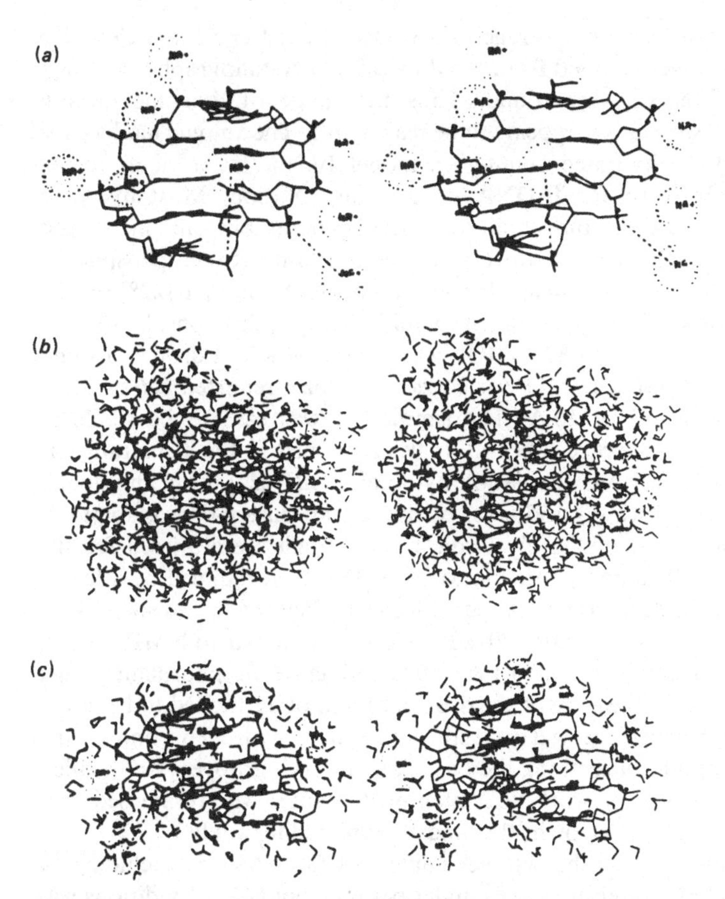


Figure 65. Stereo view of the structure of $Na^+$ d(CGCGA) at the end of the Seibel, Singh and Kollman **[228]** MD simulation: (*a*) $Na^+$ distribution, (*b*) water distribution, and (*c*) water complex within 3 Å of d(CGCGA). Reprinted with permission of the authors.

overall shape and exposed surface area remained near the X-ray value for some 32 ps of dynamics giving average atomic fluctuations reasonably close to corresponding X-ray values.

Singh, Weiner, and Kollman in 1985 **[227]** reported MD on d(CGCGA) with alternative treatments of electrostatics, considering first a fully anionic form and then explicity including 'hydrated' counterions, a +1 species with a mass and size of cation of hexahydrated sodium. The simulation was based on the AMBER force field **[336]** with a non-bonded cutoff of 12 Å and with a distance dependent dielectric screening function of $\epsilon = R$. The duration reported was 83 ps. The average structural parameters obtained for the anionic and hydrated counterion dynamics were essentially within a standard deviation of each other. A number of internal correlations between conformational parameters were pointed out. Atomic motions were found to be of the order of 1 Å. The average MD values of twist and tilt angles were close to the average obtained from the dodecamer crystallography although individual differences were found. Thus the range of dynamic motion executed by the simulation appears to be reasonable. The anionic model gave 9 bp/turn and the hydrated counterion model 10, the latter closer to the value of 10.6 bp/turn for B-DNA in aqueous solution. Most hydrated counterions remained in the vicinity of the backbone, but one, in the last 10 ps slipped into the minor groove region equidistant between phosphates.

Kollman's group followed up this simulation with another [**228**] on the sequence d(CGCGA) of larger dimensionality, with explicit consideration of eight $Na^+$ counterions and 830 TIP3P water molecules in a droplet around the DNA (Figure 65). The MD was carried out for a reported 106 ps using a non-bonded cut-off of 10 Å. The simulation was said to be stable at 10 bp/turn and the average dihedral angles remained in the same range as that found in the previous simulation by Singh *et al.* [**227**] using an implicit model for water. More tilt and twist of the central base pair was found in water, and the values of the glycosidic torsion angle were altered for the central C and G. The phosphate motion was, however, damped by a factor of 2 by the explicit water as compared with the implicit model. Sugar puckers were mixed in a statistical ratio (70–80% C2′-*endo*) similar to NMR results. All but two $Na^+$, initially at 3.1 Å, remained close to phosphate. One, however, diffused to the edge of the droplet and another looked to be in the minor groove region. The waters were observed to hydrogen bond to the expected hydrophilic sites along the base pairs in both grooves. No detailed analysis of the water network was attempted. (Note: we are advised that there may be time step error in this and the preceeding study.)

An MD simulation on the octamer duplex d(CGCAACGC) including 14 $Na^+$ and 1231 SPC water molecules under periodic boundary conditions was reported by van Gunsteren, Berendsen, Guersten, and Zwinderman in 1986 [**229**] based on the GROMOS force field. [**335**] The $Na^+$ initial positions were chosen as favorable with respect to electrostatic potential. The duration of the simulation was 80 ps, the dielectric constant equal to 1, and a twin range cut-off in which the interactions beyond 8 Å were updated less frequently was used. The simulated structure turned out to be 2.2 Å rms different from canonical B and 3.5 Å rms different from the A-form, i.e., intermediate between and not terribly greater than the rms of the B-form Dickerson dodecamer to canonical B. Some 80% of a set of two-dimensional NMR NOESY distances were satisfied by the computed average structure, considered thus to be a good candidate for solution structure of the octamer. No $Na^+$ ions were found to end up within 3.0 Å of any phosphate anion, leaving the sodium ions and phosphate groups completely hydrated, and DNA solvated by water and hydrated $Na^+$.

Nilsson and Karplus, [**230**] in the course of studies on force field development for nucleic acids, report a series of MD simulations, 5–20 ps in

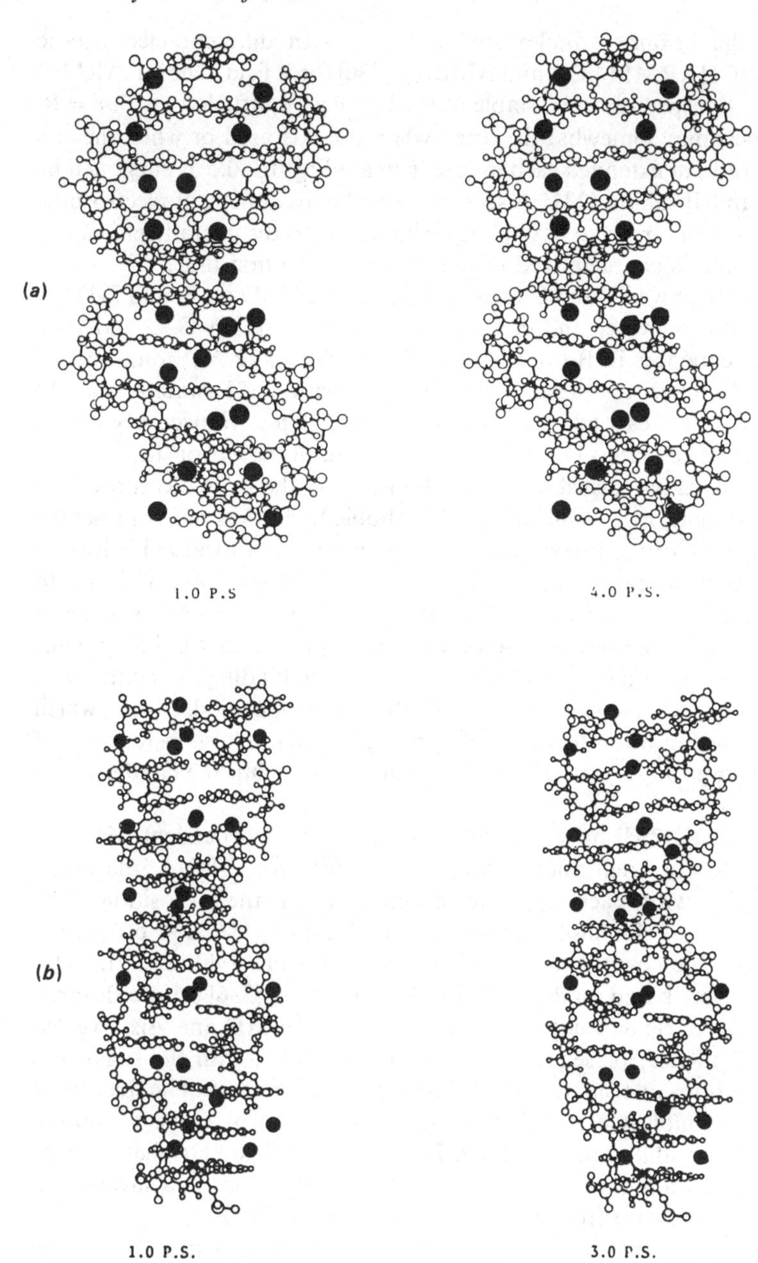


Figure 66. Stereo snapshots of the counterion positions in the MD simulations by Swami and Clementi: **[232]** (*a*) a B-form structure, and (*b*) a Z-form structure. Reprinted with permission from ref 232. Copyright © Wiley & Sons, Inc., New York.

length, of the hexamer duplex d(CGTACG) with different electrostatic options in a CHARMM [**337**] and AMBER [**336**] force field. The CHARMM [**337**] calculation proves to be stable at 9.1 bp/turn, a little low, with $\epsilon = R$. The helix becomes somewhat distorted when $\epsilon = 2$ is used or when the van der Waals' radii of extended carbons are increased by 0.3–0.5 as suggested by Jorgensen and Ibrahim. [**231**] Also the particular treatment of electrostatic truncation was observed to have a major influence on the overall behavior of the system, and is characterized in this paper for the first time.

Most recently, Swamy and Clementi [**232**] reported MD on B- and Z-DNA in the presence of water and counterions. Calculations on G–C and A–T dodecamer sequences in B-form with 1500 waters and 22 $Na^+$ ions, and on a G–C dodecamer in the Z conformation with 1851 waters and 12 counterions were carried out for 4.0 and 3.5 ps, respectively, after equilibration. The water–DNA interactions were discussed on the basis of the calculated radial distribution functions, with the gross features of a composite of ionic, hydrophilic and hydrophobic hydration effects presented to the analysis. Counterion structure defined in terms of localized solvation sites relative to the phosphate groups is presented (Figure 66) and said to confirm previous results from MC simulations. However, in 3–4 ps none of the ions has any significant chance to move, and so this point remains essentially unsubstantiated. The conclusions on site binding are contrary to that found by van Gunsteren *et al.*, [**229**] and to the model from which counterion condensation theory [**239**] springs, i.e., that a large percentage of counterions are fully hydrated and freely diffusing within the condensation domain.

Westhof *et al.* [**233**] have reported on temperature dependent MD and restrained X-xay refinement simulations for a Z-DNA hexamer, d(CGCGCG), with each cytosine brominated in the 5-position. The crystallography of this molecule was reported earlier, [**51**] and provides a basis for suitable comparison of calculations with observations. The MD calculations were based on the AMBER 3.0 force field, [**336**] using a distance dependent dielectric and neglecting counterions. The MD analysis revealed interesting evidence for sequence dependent flexibility, with the sugar and phosphate groups internal to the GpC base pair step more mobile than those for CpG. The molecule at 300 K exhibited a dynamical range of motion spanning the crystallographic ZI and ZII structures. The phosphate groups clearly exhibited a higher degree of mobility than the bases and sugars, and a larger rms difference from the crystal structure positions.

The use of restrained MD based on the CHARMM force field [**337**] along with two-dimensional NMR NOESY distance information was described in 1986 by Nilsson, Clore and Gronenborn [**234**] in a structure refinement of the oligonucleotide d(CGTACG). First a model set of canonical A and B conformational constraint distances ($< 5$ Å to mimic nuclear Overhauser effect (NOE) distances) was formed, and applied to MD simulations of the hexamer beginning with the B and A conformations, respectively. The authors demonstrated convincingly that the MD calculation beginning in A with B restraints applied transits expeditiously to B and *vice versa.* Application then of the observed NOE distances as restraints to separate MD calculations beginning in A and in B showed reasonable convergence to a proposed solution structure generally in the B-family, but showing considerable deviation from a canonical B-type structure. The strategy involved here is to allow the NOE distance constraints to pull the structure into the correct region of configuration space, and then let the force field add further details to the structure prediction. Since nucleic acid conformation is so heavily dependent on solvent, it seems likely that water and counterion effects will have to be explicitly included for this approach to be reliable in general.

In order to assess the influence of the various assumptions about force field parameters, solvation, and electrostatic effects on the behavior of MD simulations, a detailed and systematic analysis of a prodigious number of results must be undertaken. Recently Ravishanker *et al.* [235] introduced a new procedure for the analysis of the structures and MD of duplex DNA, in which comprehensive visualization of results and pattern recognition is greatly facilitated. The method involves determining the conformational and helicoidal parameters of structures entering the analysis using the 'Curves' algorithm, [236] followed by computer graphic display of the conformational wheels ('dials') for each torsion and 'windows' on the range of values assumed by a complete set of helicoidal parameters. The complete time evolution of an MD calculation can be presented using 'dials and windows' in a set of six composite figures. The method has been used to analyze an example MD calculation of 30 ps duration on d(CGCGAATTCGCG). [235] Another approach developed to aid the visualization of the relevant information from molecular computer simulations was described by Weiner, Gallion, Westhof, and Levy. [237]

### 4.4. *Ion atmosphere and solvation*

The nature of the ion atmosphere around DNA has been the subject of considerable research attention, both experimental and theoretical, in recent years. The central organizing principle is the phenomenon of 'counterion condensation': no matter how dilute the solution, a number of the counterions remain in close proximity to the DNA, compensating a large percentage of the phosphate charges and said to be 'condensed'. This follows simply from thermodynamic arguments: small ion pairing is well known to decrease with dilution due to the increasing potential for a large entropy of mixing, which favors dissociation. For polyions, the superposition of Coulomb fields on any given counterion makes enthalpic effects dominant in the equilibrium, and as a consequence a significant fraction of counterions remains associated with the DNA regardless of concentration. Release of condensed counterions is considered to be an important thermodynamic component of ligand and protein binding to DNA. [**238**]

In a series of papers, Manning [**239**] analyzed the problem and came to the vexingly simple conclusion that the percentage condensation of monovalent mobile cations on DNA was simply 76%, (88% for divalents) and independent of salt concentration. This formed the basis for 'counterion condensation theory', in which DNA was treated as a line of charges, screened from $-1$ to $-0.24$ by the fraction of condensed monovalent counterions. The remaining mobile ions form a less concentrated distribution around the screened charges at a concentration which can be suitably treated by a Debye–Hückel approach. An entropy of mixing term completes the picture. Manning provided an account of diverse observed properties of DNA via counterion condensation theory. [**239(*f*), (*g*)**]

A more theoretical, as opposed to phenomenological, approach to the problem is based on solutions to the Poisson–Boltzmann (PB) equation for simplified models of DNA. Zimm and LeBret [**240**] showed elegantly, using the PB equation, how a rod-like polyanion like DNA on increasing dilution will naturally condense counterions at a level intermediate between that of a charged sheet (100% condensation, the Gouy Chapman double layer) and that of a charged sphere (about 0%). Elaboration of the PB treatment of ion atmosphere has been advanced to a major extent by Record, Anderson and coworkers [**241, 242**] who have successfully investigated numerous aspects of the problems *vis-à-vis* thermodynamic measurements and NMR results, with a focus on the extent of salt dependence upon the fraction of condensed counterions. Limitations of the PB theory due to neglect of finite size of the mobile ions and spatial correlation have been a matter of some concern, and have been characterized by comparisons with MC calculations in several groups [**243–5**] and HNC ('hypernetted chain') theory. [**246**] General accord has been established, although the PB approach seems to underestimate the concentration by 12–18% close to the DNA. [**245(*a*)**] Both PB and MC calculations have been used to investigate Manning's hypothesis about the insensitivity of charge compensation to concentration, and found, in fact, slight but potentially significant variation. [**245(*b*)**]

Recently the theoretical studies of ion atmosphere have been extended to treat more realistic, all-atom models of DNA. Klein, Pack, and coworkers [**247**] obtained electrostatic potentials from an iterative PB solution to a combination of Coulombic potentials from the fixed macromolecular charges and the distribution of mobile charges obtained from the Boltzmann equation. However this treatment assumes $\epsilon = 80$ everywhere including inside the DNA. The results also predict significant concentration of mobile cations in the minor groove as well as along the sugar–phosphate backbone, a consequence of the superposition of transgroove anionic phosphate potentials. In contrast to the A-form, where high concentrations of ions

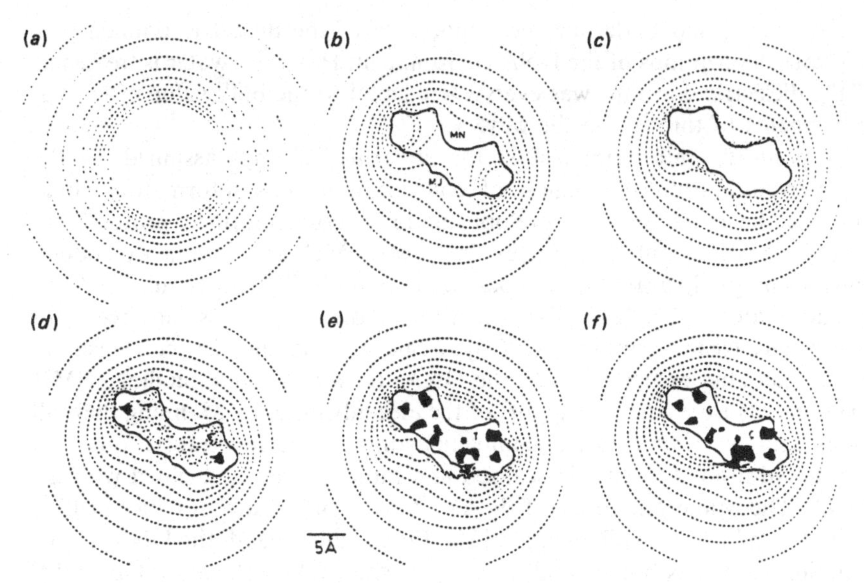


Figure 67. Two-dimensional isopotential contours in a plane perpendicular to the helix axis for various DNA models at physiological ionic strength. The dotted lines are contours at 0.5 $kT/e$ intervals from $-0.5$ to 3.0 $kT/e$. The solid shadings indicate the region of positive potential. (*a*) Isopotential contours of a linear lattice of negative charges placed 1.7 Å apart on the axis of a 9.5 Å radius cylinder with an uniform dielectric constant of 80. (*b*)–(*f*) The cross-section of the solvent-accessible surface of the DNA is shown as a thick line. The section plane lies between two successive base pairs of a standard B-DNA helix. (*b*) Only the phosphates with a 2 Å ion exclusion zone and a dielectric constant of 80 everywhere. (*c*) As for (*b*) but with a dielectric constant of 2 for the DNA. (*d*) Same as (*b*) with the partial charges of an A–T pair included. (*e*) As for (*d*) with a dielectric constant of 2 for the DNA. (*f*) As for (*e*) with the partial charges corresponding to a G–C pair. Reprinted with permission from ref. 249. Copyright © 1989 John Wiley & Sons, Inc.

accumulated into restricted locations of the major groove, in the B-form, the distributions were more diffuse in both grooves. **[247(*c*)]**

The potential influence of the dielectric boundary between the DNA ($\epsilon =$ 2–4?) and solvent at $\epsilon = 80$ was recently explored by Troll, Roitman, Conrad, and Zimm **[248]** by means of a macroscopic, non-computer simulation of duplex DNA represented by a clay model in an electrolyte tank. They found interactions of ions on the same side of the DNA to be enhanced as a consequence of a focusing of field lines by the DNA, increasing both attractive phosphate–cation and repulsive cation–cation interactions, and also that shielding is increased for charges on opposite sides of DNA by the low dielectric medium. The net effect is a diminished tendency for the residency of cations in the grooves. Jayaram, Sharp, and Honig **[249]** have most recently incorporated dielectric boundary, solvent screening, and ionic strength into solutions to the PB equations for an all-atom model of DNA (Figure 67), using a new and quite rigorous finite difference algorithm for the numerical calculation. The numerical results support the electrolyte tank observations, and underline the importance of the dielectric boundary in rigorous descriptions of the DNA environment. However, water is treated as if the dielectric constant was everywhere equal to the bulk value of 80, but refinements of this are on the way.

A point of future interest is the structure (if any) assumed by the counterions within the region of condensation; this information could provide the basis for the development of a more informed protocol for beginning MD simulation studies. In an early MD simulation, two sodium ions forming a hydrated bound pair were found bridging a guanine N7 and a phosphate group. [**250**] The situation ot the moment is, however, not completely clear. Manning [**239(*f*)**] is careful to distinguish condensation (i.e., remaining associated with the DNA) from actual site binding. NMR experiments were cited in support of the idea advanced by Manning that 'all small cations are in a state of complete hydration and free translational and rotational mobility', i.e., delocalized and rather loosely associated with the DNA. The Manning radius (of the counterion condensate) is typically 12 Å [**239(*f*)**] and quite possibly larger. [**244**] On the other hand, the DNA duplex rotation angle has been found to vary systematically with cation type, [**251**] which may require an explanation involving some degree of site binding. Multivalent cations are more disposed to site binding than monovalents. [**252**]

The NMR literature is quite extensive on cation resonance in DNA systems, [**251–4**] but the data have proved to be difficult to interpret unequivocally in terms of structure. Experiments involving $^{23}$Na NMR support in large measure the condensation hypothesis. [**253**] The nature of the quadrupole relaxation mechanism of the sodium nucleus via electric field gradients is, however, currently being argued. Record, Anderson, and coworkers [**254**] speak of a relaxation mechanism involving radial diffusion, sensitive to DNA conformation but not composition. A recent theoretical analysis by Reddy, Rossky, and Murthy [**255**] points to relaxation via ion motion in the vicinity of the polyion electrostatic potential, with negligible contribution from radial diffusion. Whatever, the theoretical situation with regard to the nature of ion distributions in nucleic acids is not yet equilibrated.

## 5. Macroscopic phenomena related to hydration

*Fluctuat nec mergitur ...*

Motto of the city of Paris

Several macroscopic phenomena have been related to the hydration properties of nucleic acids. Edelhoch and Osborne [**340**] have stressed the fact that the interactions of the nucleic acid bases with water are fundamentally different from those of the hydrocarbons or the non-polar side chains of hydrophobic amino acids. For the solution of nucleic acid bases, the enthalpy and entropy changes are positive, thus opposite in sign from those of the non-polar molecules. In fact, the solution process is enthalpy-driven. Also, nucleic acids always have positive enthalpy and entropy of denaturation, while those of proteins can be negative. Consequently, nucleic acids do not show low-temperature dissociation as found in protein aggregates like the tobacco mosaic virus. Further, while there is a large increase in heat capacity upon denaturation of proteins, the denaturation of nucleic acids leads only to a small increase in heat capacity. [**340**] Urea and guanidium chloride have profound effects on protein stability but only minor ones on nucleic acid transition temperatures, indicating that the interactions of the bases with water are less affected by these 'denaturants' than those of nonpolar molecules. [**340**] DNA denaturation occurs in a number of organic solvents because of dehydration. [**341**]

B-DNA double helices exert a mutual repulsion which is exponential with a 2.5–3.5 Å decay distance when the separation between DNA surfaces is 5–15 Å. [**342**] Such 'solvation' or 'hydration' forces arising from the structured water molecules enclosed between the facing molecular surfaces are always repulsive when the surfaces are hydrophilic, [**343**] since they reflect the work necessary for removing polarized water molecules. The decay length is insensitive to the cation type present, to the ionic strength, and to the base-pair composition. [**342**] It can thus be said it is a property of water itself. Interestingly, it was concluded that the exponential decay of the polarization excludes the possibility of detecting ordered water molecules beyond the first layer around phosphate groups by X-ray diffraction. [**342**]

### 5.1. *Hydration and static DNA curvature*

In the past few years, a wealth of observations based on various techniques has indicated that DNA molecules do not behave as straight molecules but rather as molecules with an apparent local stable curvature (for a review, see ref. 256). Although a precise atomic description of the origin of bending is still a matter of debate, sequences with runs of adenines have clearly been implicated as the main determinant of sequence-directed DNA bending. [**257, 258**] It was further shown that while the sequence 5′-AAAATTTT-3′ appears bent the sequence 5′-TTTTAAAA-3′ does not. [**259, 260**] In a cocrystal of netropsin with the B-dodecamer, Kopka and collaborators [**261**] observed that the drug had displaced the spine of hydration from the A–A–T–T center, widening the minor groove and bending back the helix axis. It had already been noticed that binding of distamycin, a close relative of netropsin, normalizes the gel mobility of K-DNA fragment, i.e., removes DNA bending. [**257**] In two recent B-DNA structures, [**95, 96**] the absence of a hydration spine is related to a larger width of the minor groove and a lack of bending of the helical axis. Computer simulations, referred to in Section 3.1, suggest that 5′-TpA-3′ steps destabilize the spine of hydration because of

Table 5. *Some average local helical parameters in AT-type sequences and the allowed contacts in the two grooves of B-DNA (those in parenthesis are either not observed or unfavorable). From data in Yoon* et al. **[265]** *The cross-strand hydrogen bonds between N6 and O4 in the major groove were first noticed in the structures of d(CGCAAAAAAGCG) and of d(CGCAAATTTGCG). After the rerefinement of the B-dodecamer d(CGCGAATTCGCG), the distances are of* 3.54 *and* 3.48 *Å for the two N6 ··· O4 contacts, quite far from accepted values for hydrogen bonding. For comparison, the same values are given for a small RNA oligomer in which there is also a pronounced propeller twist without cross-strand hydrogen bonds in the major groove.* **[267]**

| Sequence | Propeller-twist angle (°) | Twist angle (°) | Roll angle (°) | Major groove | Minor groove |
|---|---|---|---|---|---|
| B-form | | | | | |
| 5′–AA-3′ | −20 | 36 | 0 | N6···O4 | N3···W···O2 |
| 5′–AT-3′ | −15 | 32 | −5 | (N6···N6) | O2···W···O2 |
| 5′–TA-3′ | −15 | 40 | 10 | (O4···O4) | (N3···W···N3) |
| RNA oligomer | | | | | |
| 5′–AU-3′ | −17.6 | 32.3 | 7.4 | slide = 1 Å | |
| 5′–UA-3′ | −19.6 | 33.7 | 13.8 | R–R cross-strand overlap; slide = 1.5 Å | |

In the crystal structure **[268]** of d(pApTpApT), the twist of the ApT step is 31° with a propeller twist between bases of the base pairs of 13°, while there is a break in the structure at the TpA step. In the ApT mini-helix there is a water bridging the two O2 atoms of thymines.

purine–purine steric clashes and the resultant widening of the minor groove. **[106]** These observations link together DNA bending, the width of the minor groove, and the presence of the spine of hydration. More recent crystal structures of deoxyoligomers with tracts of As have shown that such sequences are unbent, but display high propeller-twist together with cross-strand hydrogen bonds between adenine N6s and thymine O4s in the major groove. **[263–5]** A general consensus holds that DNA bends at junctions between A-tract sequences and mixed-sequence DNA because of the differences in local helical parameters between the adjacent regions (see Table 5).

In short, the following scheme appears plausible: water binding in the minor groove of runs of continuous A residues leads to the formation of the spine of hydration, helping to stabilize DNA conformations with narrow minor grooves. Junctions between narrow groove sequences and mixed sequences, especially at TpA steps, would lead to bending of the DNA to prevent steric clashes and awkward stereochemistry. Drug binding in the minor groove, after displacing the water molecules of the spine of hydration, provokes an increase in the minor groove size leading to a straightening of the DNA. The spine of hydration of Z-DNA cannot induce bending because of its central position in the double helix. In B-DNA, the spine of hydration is excentric and highly sequence-dependent. Fiber diffraction gave indications that the poly(dA).poly(dT) helix is in the B-form with a distinctively narrow minor groove. [**266**]

## 5.2. *Hydration and DNA transitions*

Before discussing the various proposals made toward a better understanding of the DNA transitions, it is worth recalling the main conclusions reached by Franklin and Gosling. [**2**] First, it was clear that structure A, obtained at 75% rh, is the most highly ordered form having a high degree of crystallinity with structure B having a lower degree of order, and dehydrated DNA being disordered. Form A contains 40% by weight of water, which corresponds to about eight water molecules per nucleotide. Secondly, the lower degree of order of structure B, which is obtained at 92% rh, was explained by the existence of water layers (and sodium ions) between DNA molecules weakening the directional property of the phosphate–phosphate link, thereby shielding the helical molecules from the deforming influence of neighbouring molecules. Similarly, in crystals of oligomers, form A structures diffract generally to higher resolution than form B ones and, consequently, more water molecules per nucleotide are detected in form A crystals (4–6) than in form B crystals (less than 4). [**92**] Thus, the more highly hydrated form, the B-form, contains the fewer crystallographically determined water molecules. However, the opposition is only apparent since the more hydrated form is also the less ordered form and, consequently, less amenable to X-ray analysis. The disorder of the B-form can be static or dynamic. However, even with well-ordered B structures, the large groove volume available to water molecules would make it extremely difficult to localize the water molecules with precision without extensive ordering of the solvent structure.

Theoretical work on the hydration of DNA and the underlying causes of the B–A transition came from the solvent accessibility studies of Alden and Kim [**269**] on various conformational forms of deoxyribose and ribose duplexes. The enclosure of a canonical B-DNA model in a cylinder led to estimates of the groove volumes and established the total water enclosed by the van der Waals surface and grooves to be 19.3 waters/nucleotide. This indicated that the 20 waters/nucleotide to maintain the B-form of DNA corresponds to an intact first hydration shell plus completely filled grooves. They also found the canonical A- and B-forms of DNA to be similar in solvent accessibility to a water sized probe. Further, they noticed that, as the probe radius increases, the accessible surface for B-DNA phosphate oxygen atoms increases more than that for A-DNA phosphate oxygen atoms. Thus, phosphate oxygen atoms would be more amenable to formation of extended solvent complexes about the phosphate groups in B-DNA. This was also

Table 6. *Some accessibility values for nucleic acids. From the work of Alden and Kim* **[269]** *as well as Ponnuswamy and Anukanth.* **[270(*c*)]**

| | |
|---|---|
| Accessible surface (unfolded chain) to a probe of 1.4 Å | 490 $Å^2$/nucleotide |
| Accessible surface (buried in folded chain) to a probe of 1.4 Å of which 35% due to aliphatic carbons and 44.5% due to phosphate oxygen atoms | 317 $Å^2$/nucleotide |
| Average exposure to a probe of 1.4 Å | 173 $Å^2$/nucleotide |
| Average area covered by a water molecule[a] | 10 $Å^2$ |
| Average number of waters/nucleotide | 17 |
| Burial of water-accessible area/unit mass | 0.95 $Å^2$/dalton |
| Groove volume accessible to water | 315 $Å^3$ in A-DNA<br>579 $Å^3$ in B-DNA |
| Number of waters/nucleotide | 10.5 in A-DNA<br>19.3 in B-DNA |

| | Percentages of contact areas ($Å^2$) for a probe radius of 3 Å | | Polar exposure areas ($Å^2$) for a probe radius of 5 Å | | |
|---|---|---|---|---|---|
| | Major groove | Minor groove | Backbone | Base | Total |
| A-DNA | 24.3 | 75.7 | 10.5 | 5.2 | 15.7 |
| B-DNA | 57.2 | 42.8 | 11.7 | 6.2 | 17.9 |
| Z-DNA | 72.0 | 28.0 | 7.6 | 12.5 | 20.1 |

[a] From ref. 271.

remarked by Ponnuswamy and Anukanth. **[270(*c*)]** Table 6 contains some accessibility values useful for the present discussion.

Alden and Kim **[269]** concluded that 'under high water activity the major groove will fill and the phosphates will form more extensive hydration complexes; under low water activity, the major groove will collapse and the phosphate oxygens will become more buried'. The work of Drew and Dickerson **[103]** clearly showed that water molecules are ordered in the minor groove and positionally disordered in the major groove. It has been emphasized **[179]** that the ordered water in the minor groove, although energetically favorable, is entropically unfavorable and that more disordered water in the major groove might contribute entropically as well as energetically to the stability of B-DNA. Thus, one could view the B–A transition as a phenomenon of enthalpy–entropy compensation as a function of water activity controlled by the accessible surfaces of the two grooves. The great contact area and the large volume of the major groove of B-DNA allow for disordered water molecules contributing entropic stabilization, while the reduced area and volume of the major groove of A-DNA permit the

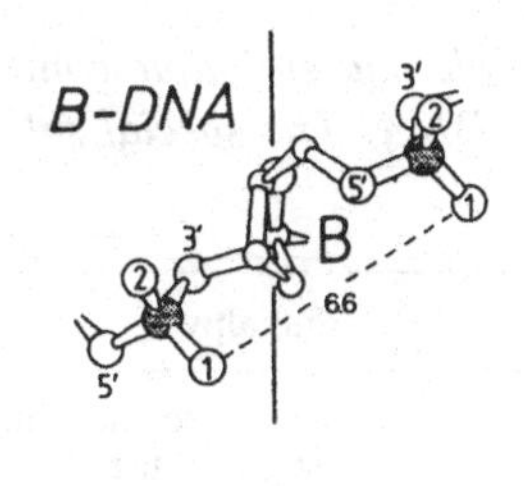


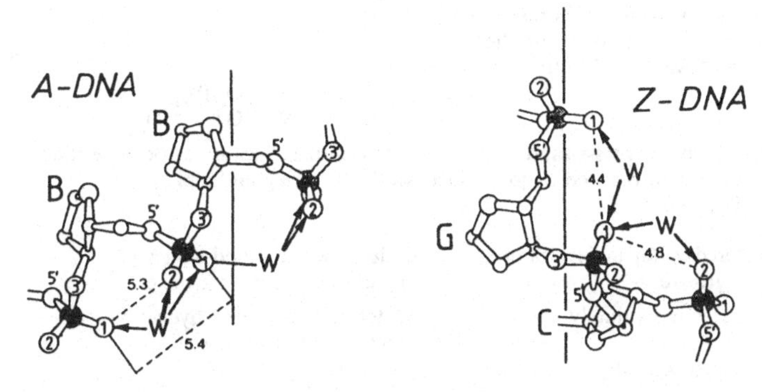


Figure 68. Drawing showing the observed water bridges between anionic phosphate oxygens in oligomers of the A- and Z-forms. The distance in B-DNA is too large for a similar bridge. This observation forms the basis of the 'economy of hydration' concept. Reprinted with permission from ref. 100. Copyright © 1986 Macmillan Magazines Ltd.

localization and organization of water molecules in hydration networks contributing enthalpic stabilization. On the other hand, the narrow minor groove of B-DNA organizes the water networks (enthalpic stabilization), while the flat concave and hydrophobic surface of A-DNA favors the release of bound water molecules and intermolecular associations between DNA molecules (entropic stabilization).

Following this scheme, the concept of 'hydration economy' suggested by Saenger and coworkers [**100**] (Figure 68) would be one aspect of the different modes of water organization around the DNA conformations. According to this concept, since water molecules can form hydrogen-bonded bridges between successive phosphates in the A- and Z-forms, but not in the B-form where the phosphate groups are individually hydrated, the hydration around A- and Z-DNA is more 'economical' and should be favored at low water activity. We suggest that the DNA molecule can modulate the order and organization of the hydrating water molecules by varying the size and accessibility of its grooves, thereby varying the enthalpic and entropic contributions to the free energy so that the free energy remains roughly constant at various water activities (Table 7). Thus the problem is one of ecology, not economy.

It has been shown that, in solution as in fibers, the water activity is the main factor controlling the B–A transition and that the equilibrium is not

Table 7. *Speculations about the entropic and enthalpic contributions to the free energy of the A- and B-forms of DNA. The special role of the YpR steps is emphasized.* [**275**]

| Form | Entropy | Enthalpy |
|---|---|---|
| A-DNA | Intermolecular associations through contacts of flat, concave, hydrophobic surfaces of the minor groove following the release of water molecules | Structured water in major groove (e.g., pentagons) |
| | | O ($P_i$)···W···O ($P_{i+1}$)<br>N3/O2 (*i*)···W···O4′ (*i*+1) |
| | Y–R steps with cross-strand overlap between purines because of positive slide (1.5 Å) lead to reduced major groove widths, favoring enthalpic contributions. | |
| B-DNA | Disordered water molecules in major groove | Ordered waters build spines of hydration in minor groove |
| | Phosphates individually hydrated | Water bridges in first hydration shell of major groove |
| | Delocalized screening by hydrated ions | |
| | Y–R steps with purine–purine clashes in the major groove because of negative slide (−0.5 Å) and propeller twist lead to increased minor groove widths and the absence of minor groove water bridges, favoring entropic contributions. | |

influenced by temperature between 0 and 30 °C (the DNA starts to melt at 30 °C. [**272, 273**] Differential scanning calorimetry studies on the dehydration of polyadenylic acid and calf thymus DNA showed that the enthalpy of the desorption process increases from a value essentially that of the vaporization of liquid water (9.7 kcal $mole^{-1}$) to approximately 16.0 kcal $mole^{-1}$ as the quantity of adsorbed water decreases from 10 to 2 waters/nucleotide. [**274**] However, within experimental error, the entropy of the adsorbed water was that for liquid water at 25 °C, i.e. 16.7 cal $mole^{-1}$ $K^{-1}$. [**274**]

The structural basis allowing this modulation of the size and accessibility of the grooves of the DNA molecules resides in the natural flexibility of the sugar conformations and in the versatility of stacking interactions. First, deoxyribose sugars can exchange almost freely between the two pucker types, C3′-*endo* and C2′-*endo* and, secondly, stacking interactions between bases are not demanding geometrically. This led Calladine and Drew [**275**] to propose a base-centered explanation of the B–A transition in DNA. It is based on the observation that Y–R dinucleotide steps are bistable (i.e., possess two equally stable arrangements) for their stacking interactions. Consequently, they are able to direct the backbone in either the A- or B-form because of the possibility of pyrimidine–purine same-strand overlap in the B-form and purine–purine cross-strand overlap in the A-form (Figure 69). The

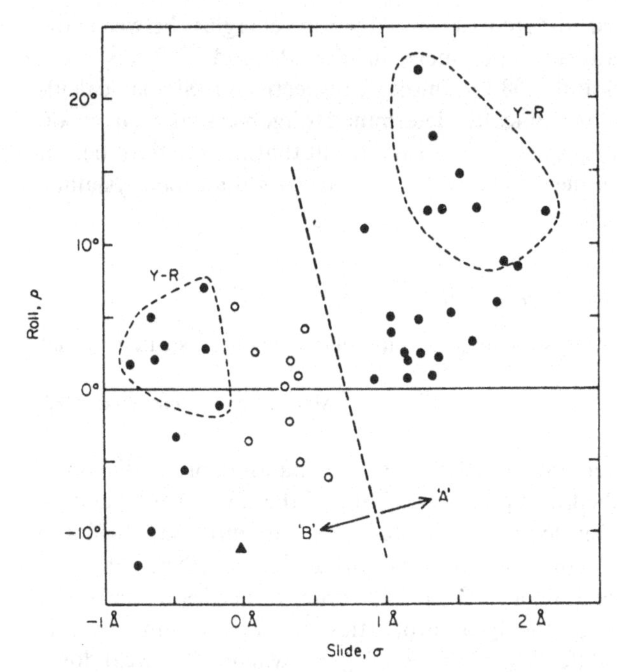


Figure 69. Plot of roll versus slide for 45 base pairs in 6 different X-ray structures of DNA oligomers. Open circles correspond to water-bridged steps. Reprinted with permission from ref. 275. Copyright © 1984 Academic Press Inc. (London) Ltd.

stacking bistability is transmitted to neighbouring steps by the passive backbone via the slide component between base pairs (relative translation along their long axis). Neutral steps, like R–Y, R–R, or Y–Y steps, can be stabilized by water bridges at low slide (A–T, A–A, T–T) in the B-form or by weak cross-strand overlap (G–C, G–G, A–G, C–C) in the A-form. It should be reminded that water bridges have not been observed in Y–R steps (see Section 3.1).

The discovery of left-handed Z-DNA [**33**] led to a wealth of theoretical work on the underlying mechanisms responsible for the transition (for reviews, see refs 276 and 277). Although the strong salt dependence of the B–Z transition induced an emphasis of the electrostatic contributions, it was also suspected [**51, 278**] and stressed [**277**] that specific structural water networks [**100**] (with or without specific ion binding) or differences in solvation energy [**279**] form far from negligible contributions to the transition. Ho, Quigley, Tilton, and Rich [**280**] using the semiempirical approach of Eisenberg and McLachlan [**281**] whereby the water-accessible areas are weighted by atomic solvation parameters, tried to clarify the fact that C5-methylated alternating d(G–C) forms Z-DNA at physiological ionic strength, [**282**] while the unmethylated polymer requires 2.5 M salt. [**283**]

However, the calculated differences in solvation energies between the unmethylated and methylated polymers in the B- and Z-forms were depressingly small (0.18 and 0.08 kcal $mole^{-1}$, respectively). Similar calculations made on crystallographically determined oligomers as well as on standard helical oligomers gave the surprising result that the standard helical structures had 1–2 kcal $mole^{-1}$ lower solvation energies than the crystalline oligomers. [**284**]

### 5.3. *The dynamics of the DNA hydration shells*

There has hardly been a novelty or a vogue among solid-state physicists that has not been applied to DNA.

Maxim Frank-Kamenetskii. [**285**]

Since the review by Texter [**286** ] in 1978, our understanding of the dynamics of the DNA hydration shells has progressed both on the experimental and on the theoretical side. Developments in Raman, far infrared, Brillouin scattering, and NMR techniques allowed for a clearer albeit far from complete view of the dynamic and thermodynamic properties of the DNA hydration shells. Although the local properties of water around specific groups in DNA may differ from those of pure water, the breakdown in primary and secondary hydration shells of the dynamic and thermodynamic properties is difficult and averages over the hydration shells are only slightly different from those of bulk water. Measurements of ultrasound velocity in dilute aqueous solutions of nucleic acid bases led to the conclusion that water molecules 'in contact' with the nucleic acid bases had properties approaching those of normal liquid. [**287**]

Some characteristics of the primary hydration shell around DNA molecules obtained with various techniques are gathered in Table 8. It appears that, up to 10–14 waters/nucleotide, the properties of the water molecules around the DNA are quite different from pure water and that, above 20–23 waters/nucleotide, the properties of the water molecules are identical to those of pure water. It should be kept in mind that Na-DNA (with 1% excess salt) undergoes the B–A transition at a relative humidity above 85%, i.e. 10–12 waters/nucleotide, and is completely in the B-form above 20 waters/nucleotide. Interestingly (or coincidentally?), it was estimated that 10.5 water molecules would fit in the grooves of A-DNA and 19.3 water molecules in the grooves of B-DNA (Table 7). While the activation energy for the relaxation in the primary hydration shell is typical of liquid water despite much longer relaxation times, in the secondary hydration shell the activation energy is 7.0 kcal $mole^{-1}$ with a relaxation time similar to that of liquid water (2 ps). [**295**] It was concluded that at 0 °C the behavior of water in the secondary hydration shell is similar to that of supercooled water. [**295**]

Table 8. *Some characteristics of the primary hydration shell*

| Number of water molecules per nucleotide | Properties, method, reference |
|---|---|
| 1 | Rotationally hindered: no contribution to dielectric loss **[288]** |
| 2 | At 10% rh, gravimetric analysis **[88]** |
| 3.5 | Do not diffuse appreciably on the time scale of neutron measurements (300 ps) **[289]** |
| 9 | Unable to take up an ice-like structure, IR **[89]** |
| 9 | Strongly bound, dielectric measurements **[290]** |
| 9.5 | Water molecules diffuse isotropically in a sphere of 9 Å diameter with as diffusion coefficient 25% that of bulk water **[289]** |
| 10 | No endothermic fusion, calorimetry **[291]** |
| 11 | Unfrozen at 238 K, NMR **[292]** |
| 14 | Unable to assume ice-like properties in terms of their dielectric behavior down to 200 K **[288]** |
| 23 | Primary hydration shell completely formed at 80% rh, relaxation time (4–80 GHz) = 40 ps activation energy = 5 kcal/mole$^{-1}$, primary hydration shell most strongly bound to the DNA itself and poorly connected to external waters; Brillouin scattering **[293]** |
| 35 | Interact with DNA (weakly or strongly): dielectric measurements **[294]** |

Raman scattering also gave with precision a relaxation time for the secondary hydration shell of 4 ps at 0 °C. **[296]** Coupled dynamics between the collective vibrational mode of the DNA and the relaxational mode of the secondary hydration shell was inferred from an analysis of the low-frequency collective modes of DNA by Raman spectorscopy. **[296]** It was suggested **[297]** that the A–B transition is driven by the lowest frequency mode of DNA which dominates the low-frequency Raman spectrum. However, later it was shown **[298, 299]** that mass-loading of the molecular subunits of DNA by bound water molecules together with the change in lattice constant accompanying the A–B transition are the main causes of the softening of the modes upon hydration. Another case where the 'hijackers failed' **[285, 300]** is that of the claim about sharp resonances in the microwave absorption spectrum of aqueous solutions of DNA. **[301]** Two groups concluded later **[302, 303]** that no resonance absorption or any form of enhanced absorption can be observed. However, the initial observations led to the development of theoretical work which emphasized the nature and the degree of interaction between solute and solvent molecules. **[304]**

Recent NMR work studied the DNA motions as a function of hydration. Earlier work **[305–7]** had shown that internal motion has limited amplitude in dehydrated A-DNA and that the amplitudes of motion increases with the degree of hydration. Using $^{31}$P-NMR **[308]** and $^{2}$H-NMR **[309]** (on DNA

Table 9. *Results of* $^{31}P$ *relaxation measurements (after ref.* 308) *and* $^{2}H$ *relaxation measurements of purine deuteron (after ref.* 309). $^{31}P$*-NMR at* 40.5 *MHz and* $^{2}H$*-NMR at* 76.8 *MHz*

| Number of water molecules per nucleotide | Properties |
|---|---|
| Below 5 | Low-amplitude with high-frequency motions of phosphates ($T_1$ = 130 s); maximum base excursion less than 10° ($T_1$ = 7.3–2.0 s) |
| 5–10 | Sharp drop in $T_1$ ($^{31}$P:4.9–1.3 s; ($^{2}$H: 0.6–0.2 s); amplitude of motion increases; new motional regime of bases |
| Up to 20 | Angular fluctuations of phosphates ±30° ($T_1$ = 0.25 s), either local or local and collective ns time scale processes; base motion coupled to backbone motion ($T_1$ ($^{31}$P)/$T_1$ ($^{2}$H) ratio constant; $T_1$ ($^{2}$H) = 0.068 s |
| Above 20 | Above 25 waters/nucleotide, $T_1$ ($^{31}$P) begins to increase again; bases and backbone undergo large-scale motion dominated by low-frequency fluctuations |

deuterated at position C8), two groups obtained converging views on the effects of hydration on DNA motions. The results are collected in Table 9.

## 5.4. *The desolvation step*

L'eau, telle une peau
Que nul ne peut blesser.
Paul Éluard

*Les animaux et leurs hommes.*
*Les hommes et leurs animaux.*
Mouillé

In many processes involving binding between biopolymers, large changes in the apparent heat capacity of the interacting systems are observed. Although several factors can contribute to the heat capacity changes, Sturtevant **[310]** showed that intramolecular vibrations are of importance. A conversion from soft internal modes to stiffer modes (resulting in a tightening of the macromolecule) would lead to a significant decrease in heat capacity. For example, stabilization of flexible loops would be expected to contribute to a decrease in heat capacity. Thus, the binding of nucleic acid oligomers to bovine pancreatic ribonuclease leads to a 25% decrease in crystallographic thermal parameters near the active site and in the binding regions. **[311]** In many instances, following the binding of a substrate or of an inhibitor there is a striking decrease in thermal parameters or mobility of a polypeptide segment involved in ligand binding. **[312–15]** Recently, the role of flexibility in DNA–protein interactions has become more and more apparent both at the level of the DNA and at the level of the protein. **[316–18]**

An important and crucial contribution to the heat capacity changes observed comes from the release of bound solvent molecules upon ligand binding. [**241**] The role of segmental flexibility in that desolvation step is unclear. A dynamic state for the solvent around flexible regions, as observed in crystallographic work and in protein simulations, could play a role. The flexibility of some of those regions might be due to unfavorable and destabilizing interactions with other regions of the macromolecule or with the solvent. In any case, several examples exist in the literature which indicate that, when bound to a macromolecule, small ligands occupy the locations previously filled by well-localized water molecules often bound to clefts or cavities with low thermal parameters. [**241, 319, 320**] A striking example can be found in the work of Kopka and collaborators [**261**] on the binding of netropsin in the minor groove of the dodecamer sequence d(CGCGAATTCGCG) after displacement of the water molecules making up the spine of hydration. The work of Breslauer and coworkers also showed that netropsin binding to polydA·polydT is entropy driven, while that to polyd(A–T)·polyd(A–T) is enthalpy driven. [**262**] This has been related to the presence of the spine of hydration in the former polymer and to its absence in the latter polymer, although differences in overall hydration and DNA conformations are also implicated. [**262(*b*)**] During MD of ribonuclease, [**321**] water molecules were found bound to the enzyme as if 'mimicking' the substrate. Rather than helping to structure the active site, such water molecules probably play more of a space-filling role. Recent calculations have shown that the presence of a cleft in the macromolecule boundary is associated with a region of enhanced electrostatic potential in the solvent medium, [**322, 323**] thus promoting binding of charged ligands or solvent molecules. It has been stressed [**324**] that ligand binding does not merely displace solvent molecules but that it may modify the electrostatic field and change the binding of solvent molecules at distant sites.

In short, it is tempting to envisage that the thermal fluctuations, by facilitating the interplay of the various physico-chemical forces, contribute to the desolvation of the interacting macromolecules and help in establishing a minimum in the free energy of complexes between macromolecules or between macromolecules and ligand(s).

## 6. Concluding remarks

The plausibility that hydration plays a role in the stability of nucleic acid helices was first suggested by Geiduschek and Gray [**344**] soon after the discovery of the double helix. Later, base stacking forces together with hydrogen bonding between complementary bases were held responsible for double helical structures in solution. In 1967, Lewin [**132**] propounded the concept that water bridges contribute greatly to the stability of the DNA double helix in solution on the basis of model building and theoretical considerations. However, the evidence was indirect and the lack of direct crystallographic experimental confirmation held the paper in respectable obscurity. Following the impetus given by the pioneering work of Drew and Dickerson in 1981, [**103**] a wealth of information has been gathered from crystal structures and the results compared with theoretical calculations. This work attempts to merge the resulting views about nucleic acid hydration.

Our present empirical knowledge of the molecular details of nucleic acid hydration comes from the crystallographic determination of ordered water positions in nucleic acid crystal hydrates. [**8, 92, 112, 223**] Nucleic acid crystals are typically 50–75 % water, a heavily hydrated state. Water, at or near liquid state density, is literally everywhere in the system not occupied by nucleic acid, the organic constituents of the crystal, and ions. Judging from what we know about the structure of liquid water and aqueous solutions, this means that all the water molecules proximal to the other constituents are (unless sterically excluded) participating in hydration complexes with anionic oxygens or with mobile ions, quasiclathrate structures surrounding hydrophobic groups, or hydrogen bonds to the hydrophilic atoms of the solutes. Furthermore, all the hydration structures can be expected to interface smoothly to bulk water in the crystal via (bent) water–water hydrogen bonds. The ordered water observed crystallographically is thus a glimpse into what is undoubtedly an extensive, robust, elaborate hydration network in the crystal. However, since many of the waters are diffusing about, the overall structure of the hydration is in general dynamical rather than static. The time averaged structure, termed by Eisenberg and Kauzmann [**325**] the 'D' (diffusionally averaged) structure, may be diffuse or disordered although considerable 'I' or 'V' structure exists on an instantaneous (snapshot) or vibrationally averaged basis, especially in an associated liquid like water.

Waters that turn out to be ordered crystallographically have time averaged positions which are localized in space. Waters participating in bridges, pinned down from two sides by the solute atoms, are probably more likely to fulfill this criterion, which accounts for the ubiquitous appearance of water bridges in crystal structures. The main intranucleotide water bridges are tabulated in Table 10 and schematically represented in Figure 70 with the thickness of the lines representing roughly the frequency of occurrence. Strong ones appear mainly in the minor groove of helical nucleic acids and should be expected to contribute significantly to the stability of the helical arrangements of nucleic acid in aqueous solution. The critical role of the sugar–water–base and base–water–base bridges in the minor groove contrasts with the versatility and mobility of the base–water–base bridges and arrangements in the major groove of helical nucleic acids. The hydration around phosphate groups of helical stuctures is characterized by cones of hydration centered on each anionic phosphate oxygen, by water bridges between anionic phosphate oxygen atoms of successive residues on the same strand, and by 5′-phosphate–water–base bridges. The 3′-phosphate–

Table 10. *Structural water in nucleic acids*

| |
|---|
| *Hydration around phosphate groups*: |
| Cones of hydration (B- and Z-forms) |
| O1P ··· W ··· O2P bridges (A- and Z-forms) |
| *5′-Phosphate–water–sugar bridges*: |
| Never seen |
| *5′-Phosphate–water–base bridges*: |
| B-DNA: methyl ··· W ··· phosphate |
| A-DNA and RNA: N7(purines) ··· W ··· phosphate |
| *3′-Phosphate–water–sugar bridges*: |
| Helical conformations: never seen |
| Present in non-helical conformations (turns, ...) |
| *3′-Phosphate–water–base bridges*: |
| Right-handed helical structures: never seen |
| Purine in *syn*: Z-DNA, G*anti*–A*syn* in B-DNA |
| *Other phosphate–water–base bridges*: |
| Unusual pairs: A–A, A–G (*t*RNAs, RNA–drug complexes) |
| Z-DNA: N7(G) ··· (hydrated ion) ··· 5′-phosphate of C |
| *Sugar–water–base bridges*: |
| In minor groove of all helical forms (A, B, and Z) |
| Periodicity → Spines of hydration (B and Z) |
| *Base–water–base bridges*: |
| Purines: N6/O6 ··· W ··· N7 and N2 ··· W ··· N3 |
| Interresidue (inter- and intrastrand) versatile |

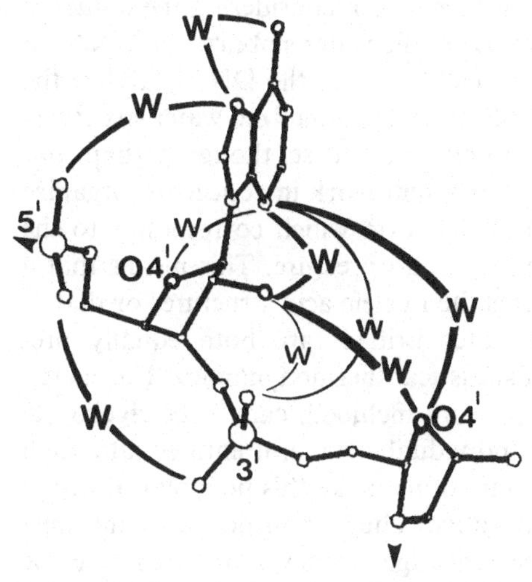


Figure 70. Schematic view of the observed internucleotide water bridges with the thickness of the lines representing roughly the frequency of occurrence. From ref. 345.

water–base and 3′-phosphate–water–sugar water bridges around unusual nucleotide conformations (e.g., *syn* conformation of the base or the *gauche minus–gauche plus* conformation of the sugar–phosphate backbone) stress the role of water bridges in non-standard conformations. Around non-canonical base pairs (e.g., A–G, A–A, U–G pairs) as well, new phosphate–water–base bridges, appear systematically.

However, it is important not to think of crystallographically ordered water as 'the' hydration, neglecting, first, the fact that these water molecules exchange with bulk water [7] and, secondly, the existence of the 'non-crystallographic' water. This 'disordered' water, in close proximity to the nucleic acid, is capable of interacting energetically with the solute and contributing to the thermodynamics of the system as much as the ordered water. The structure in the non-crystallographic water is just dynamic rather than static.

These considerations result in a *caveat* to those interested in making connections between hydration, and particularly ordered water structure, and the thermodynamic stability of the system: just because a hydration pattern can be located crystallographically does not imply it is responsible for the thermodynamic stability of the system. It may indicate an energetically favorable local region of structure, but stability is a matter of the free energy of the system and is only dominated by energy at 0 K, and global energy at that. A confusion between factors unequivocally responsible for thermodynamic stability, and apparent structural correlations thereof, pervades structural biology literature and particularly that of nucleic acid hydration at the moment.

There is also a 'chicken and egg' problem when considering the influence of hydration on nucleic acid structure. Does the water stabilize the DNA via bridges, 'spines', filaments and networks? Or, does the DNA stabilize the water, providing a trellis on which these elaborate organized water structures nucleate? The answer is fortunately simple in this case, though perhaps not too informative. All components of the system work in concert to organize the collective molecular assembly into the form which corresponds to the lowest free energy at a given temperature and pressure. This means that a statement like 'a water bridge stabilizes the nucleic acid structure' or that 'a nucleic acid structure nucleates a water bridge' are both equally presumptuous from the point of view of statistical thermodynamics. The energy of the system, when cooperative effects are included, cannot be rigorously partitioned into contributions from individual structural entities (although the approximations introduced in calculations make this possible). Entropy is unequivocally a property of the collective. Thus a holistic viewpoint must be adopted. It is important that the language we develop to describe the structural chemistry and biology of nucleic acid hydration is thermodynamically as correct as possible. Right now this requires that we keep the total organization of the system in mind as well as the structure of individual components.

The evidence is that water is an integral part of helical as well as non-helical structures of nucleic acids [**8, 345**] and, thus, that the development of an adequate description of the solvation and electrostatic properties of the highly charged nucleic acids is essential for correct modeling and simulation of nucleic acid structures.

**Acknowledgments**

We are glad to dedicate this work to Professor M. Sundaralingam for his sixtieth birthday.

A lot of the ideas and of the drive to pursue this work through a collaboration between crystallographers and theoreticians originate in a series of CECAM workshops organized in Orsay (summer of 1983 and 1986, fall of 1987) following the encouragement and the support of C. Moser. We enjoyed the stimulating discussions and we thank all the members of the '*Mort Subite*' group. Discussions with Professor Helen Berman have been particularly helpful.

We wish to thank the authors who gave us permission to reproduce figures as well as the editors of Academic Press, American Chemical Society, Adenine Press, Butterworth, IRL Press, Macmillan Magazines, New York Academy of Sciences, and Springer-Verlag for giving us permission to reproduce published material. E.W. is thankful to Y. Timsit and L. Jaeger for figures and to V. Fritsch for help with the bibliography.

This research was supported by grants to DLB from the National Institute of Health (GM-37909) and the National Science Foundation (CHE-8696117), and the Office of Naval Research (N-00014-87-K-0312). Support from Merck, Sharpe and Dohme Research Laboratories, and the Bristol Myers Corporation via a Connecticut Cooperative High Technology Research Development Grant is also gratefully acknowledged.

**Appendix: Nucleic acid conformations**

In this appendix, some general aspects of the nomenclature and of the stereochemistry of nucleic acids, helpful for an understanding of the review, are gathered. Nucleic acids are made of three chemical entities: heterocyclic bases, five-membered sugar rings, and phosphate groups. The sugar rings, or furanose rings, together with the phosphate groups form the sugar–phosphate backbone with the aromatic bases constituting the side chains of the polynucleotide. The heterocyclic bases are either pyrmidines (Y: cytosine (C), uracil (U), thymine (T), ...) or purines (R: guanine (G), adenine (A), ...). The uracil base does not normally occur in DNA, where it is replaced by thymine (5-methyl-uracil), but is common in RNA in which thymine can also occur. The bases associate via horizontal hydrogen bonding and via vertical stacking interactions. The classical Watson–Crick base pairs are shown on Figures A1 and A2, together with the presently accepted nomenclature. Unusual or non-canonical base pairs (between G and U, for example) are frequent, and possibly structurally as well as functionally important in

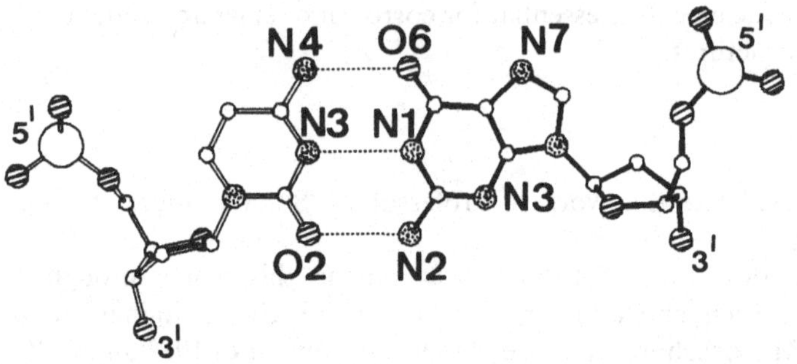

Figure A1. Watson–Crick base pair between G and C.

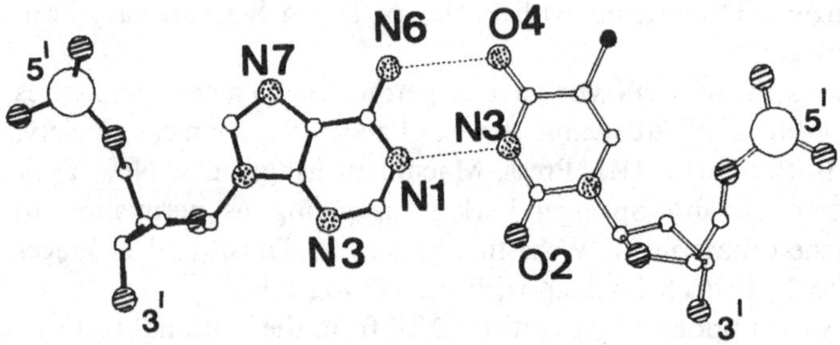

Figure A2. Watson–Crick base pair between A and T.

ribonucleic acids (RNA); while they lead to mismatches and mutations in deoxyribonucleic acids (DNA). Hoogsteen base pairs [334] between A and U (or G and N3-protonated C) are linked via hydrogen bonds involving the N7 and N6 (or O6) sites of purine bases and either O4 (or N4) and N3 (*cis* pairing) or O2 and N3 (*trans* pairing) of pyrimidine bases (see Figure 40).

In RNA, the five-membered furanose ring is a $\beta$-D-ribose, while it is a $\beta$-D-2′-deoxyribose in DNA. The sugar ring is not planar [328] but puckered with one atom out of the plane defined by the other atoms (envelope puckering) or with two atoms in opposite directions out of the plane (twist puckering). The displacement can be towards the C5′-exocyclic group (*endo* puckering) or away from it (*exo* puckering). The displacement is about 0.5 Å, which corresponds to a maximum torsion angle in the ring around 39°. Through pseudorotation, [329] i.e., via concerted changes of the torsion angles around the sugar ring, the sugar varies the atom(s) displaced and, consequently, its pucker. Figure A3 displays the two main types of pucker, C2′-*endo* (top) and C3′-*endo* (below). Stereochemically, [30] these two main pucker types give rise to the main nucleic acid helices for DNA: A-DNA (C3′-*endo*) and B-DNA (C2′-*endo*). RNA helices have always the C3′-*endo* pucker. However, single-stranded regions (forming loops and turns) in RNAs do display the C2′-*endo* pucker.

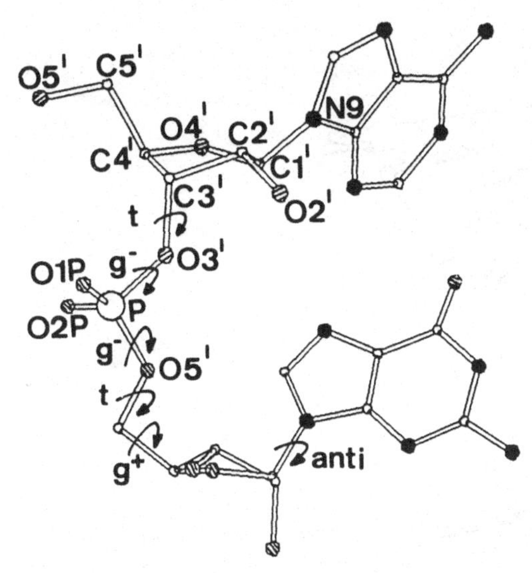


Figure A3. Nomenclature of the sugar–phosphate backbone with the values for the various torsion angles in helices indicated (*trans*: torsion angle is 180°; *gauche-plus*: +60°; *gauche-minus*: −60°). Both bases are in the *anti* orientation, but the ribose sugar at the top is C2′-*endo* and the one below is C3′-*endo*. The view shown does not correspond to any observed structure, but was chosen for maximum clarity.

Table A1. *Summary of helical parameters of the main nucleic acid helices. First line, from fiber work*; **[326]** *second line averages and/or variations observed in monocrystals. For Z-DNA, numbers refer to C and G residues.* **[33, 122]**

| Parameter | RNA | A-DNA | B-DNA | Z-DNA |
|---|---|---|---|---|
| Residues/turn | 11 | 11 | 10 | 12 |
| Axial rise/turn | 2.8 Å | 2.6 Å | 3.4 Å | 3.5/4.1 Å |
| | ±0.2 Å | ±0.4 Å | ±0.4 Å | ±0.2 Å |
| Pitch | 30.9 Å | 28.5 Å | 34.0 Å | 45.0 Å |
| Rotation/residue | 32.7° | 32.7° | 36.0° | −52°/−8° |
| | ±3.0° | ±5.9° | ±4.3° | ±1.3° |
| Propeller twist | −13.8° | −13.8° | −1.2° | −4.4° |
| | −(18.6±3.5°) | −(15.4±6.2°) | −(11.7±4.8°) | ±2.8° |
| Inclination to overall axis | 16.7° | 22° | 2.4° | 9° |
| | 17.1±3.9° | 13.0±1.9° | −2.0±4.6° | ±0.7° |
| Displacement to overall axis | −4.4 Å | −4.4 Å | +0.6 Å | −0.5/+1.5 Å |
| Pucker type | C3′-*endo* | C3′-*endo* | C2′-*endo* | both |
| Distance P···P | 5.8 Å | 5.8 Å | 6.9 Å | 5.9 Å |
| Minor groove width | 11.3 Å | 10.7 Å | 5.7 Å | 5.3/1.9 Å |
| Major groove width | 4.1 Å | 2.8 Å | 11.4 Å | 9.6/11.2 Å |

Under conditions of low humidity and high salt molarity, DNA can adopt the right-handed C-conformation **[326(*c*)]** with a rotation/residue of 38.6°, a pitch of 31.0 Å, and 9.33 residues/turn. The homopolymer polyd(A–T) adopts preferentially the right-handed D-conformation **[332, 333]** with a rotation/residue of 45°, a pitch of 24.3 Å, and 8 residues/turn.

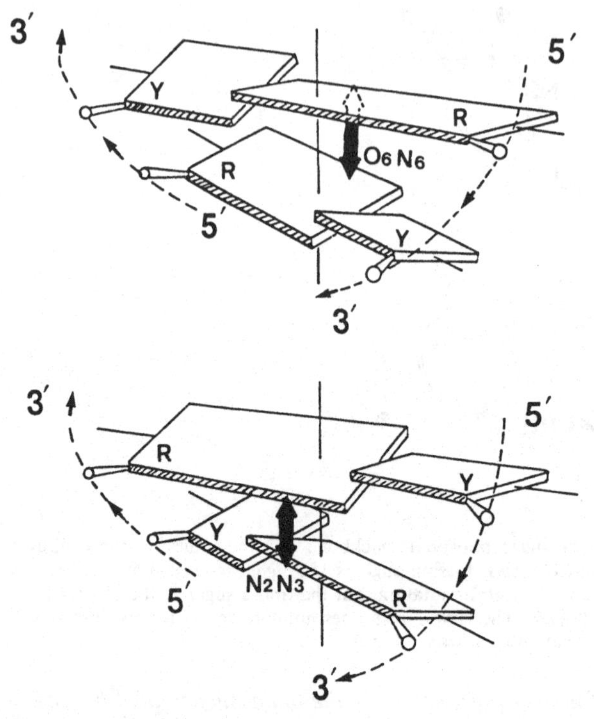


Figure A4. Schematic view of the propeller twisting between bases of the same base pair and the consequences for contacts in the major groove in 5′-RpY-3′ sequences and in the minor groove in 5′-YpR-3′ sequences. In homopolymer sequences, especially 5′-ApA-3′ sequences, propeller-twisting leads to cross-strand contacts in the major groove, which have been assigned to bifurcated hydrogen bonds, between N6(A) and O4(T) [263].

Table A1 gives the helical parameters of the main types of nucleic acid helices. The helical domains for the various torsion angles along the sugar–phosphate backbone are given on Figure A3, together with the atomic nomenclature. Most commonly, the base is oriented in the *anti* orientation with respect to the base (torsion angle O4′–C1′–N9–C4 around 180° in purines and torsion angle O4′–C1′–N1–C2 around 180° in pyrimidines) with a tendency for the *syn* orientation in deoxyguanosine derivatives. In the *anti* orientation, the C8 of purine bases or the C6 of pyrimidine bases are over the sugar ring such that there is the possibility for C8–H···O5′ or C6–H···O5′ interactions. [30]

The Z-DNA helix [33] is more complex because the repeating unit is a dinucleotide and not only a nucleotide as in the other types of helices, with alternating sugar pucker and orientation of the base with respect to the sugar: C2′-*endo*-*anti*-pyrimidine and C3′-*endo*-*syn*-purine.

X-ray analysis of nucleic acid oligomers has unveiled a wealth of information on the fine details of nucleic acid structures. One important

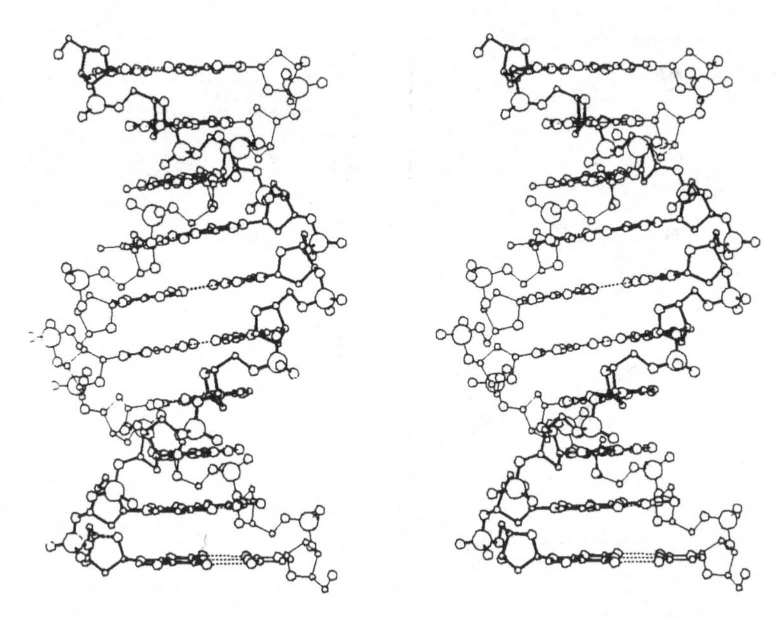

Figure A5. Stereo view of a B-type DNA helix. Note how the base pairs are roughly perpendicular to the helical axis. The view is down the minor groove in which case the 5′-end of the right strand is at the top and the 3′-end below.

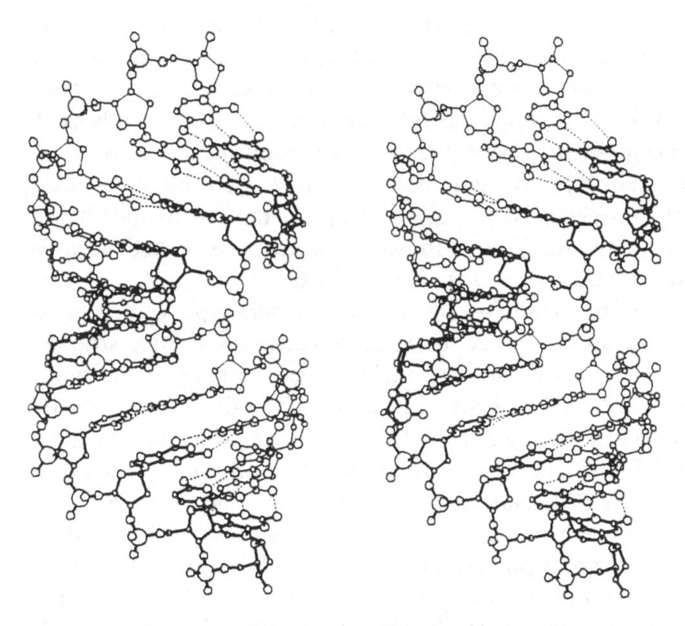

Figure A6. Stereo view of an A-type DNA helix. The base pairs are displaced from the helical axis and form a cylindrical hollow around the axis.

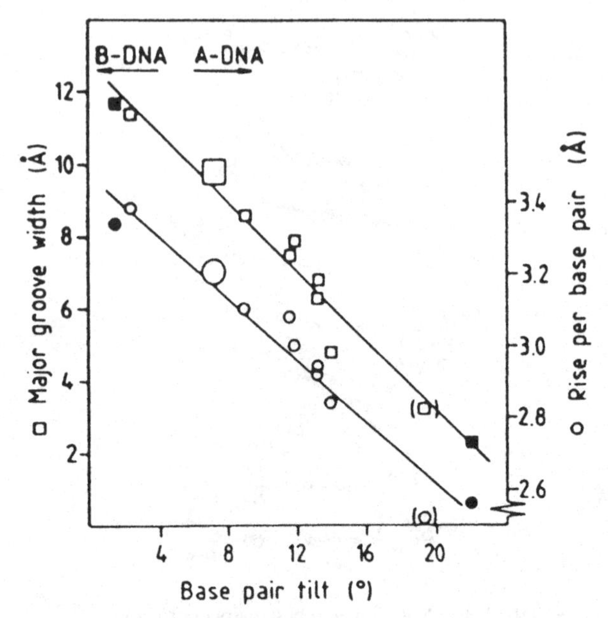


Figure A7. Plot of major groove width versus base pair tilt in various X-ray structures of A- and B-form oligmers. Filled symbols are for fiber diffraction models. [**326**] Reprinted with permission from ref. 101. Copyright © 1987 IRL Press Ltd.

parameter is the propeller-twisting between bases of the same base pair (Figure A4). The origins of propeller-twist are unclear. It was thought that propeller-twisting leads to more efficient stacking between neighboring bases on the same strand. But, non-coplanarity of bases in base pairs already exists in crystals of nucleosides where the average propeller-twist is 14.7°, independently of the type of intermolecular hydrogen bonding. [**330**] However, while the propeller-twist is always negative in right-handed helical structures, it is roughly equally distributed between positive and negative values in structures of nucleic acid fragments. [**330**]

Figures A5 and A6 shows stereo views of the two main types of helices: B-DNA and A-DNA, respectively. A subtle chiral feature of double-stranded nucleic acid helices should be stressed. [**331**] When looking in the minor groove of B-DNA (or A-DNA, RNA), the 5′ to 3′ chain direction on the left goes from bottom to top (Figure 36). On the other hand, when looking in the minor groove of Z-DNA, the 5′ to 3′ chain direction on the left goes from top to bottom (Figure 35). Naturally, in both cases, the chain direction should be reversed when looking in the major groove. It should be emphasized that these canonical A- and B-forms of DNA are models of two extreme conformations derived from X-ray diffraction of fibres [**326**] and that intermediate forms are frequently observed in crystal structures. The plot shown in Figure A7 illustrates the continuum of possible structures between the two canonical A- and B-forms: as the rise per base pair increases from 2.6 Å to 3.4 Å, the width of the major groove increases and the tilt (or inclination to overall helical axis) of the base pairs decreases from 22° to 2°. [**101**] In B-DNA, bases on the same strand stack on top of each other with no interstrand overlap. However, in A-DNA (and RNA), bases in sequences 5′–Y–R–3′ display a pronounced interstrand overlap of the purine bases, in contrast to 5′–R–Y–3′ sequences which have only intrastrand stacking. This sequence dependence stems from the inclination of the bases with respect to the helical axis and the concomitant lateral slide of the base pairs (see Figure 69) with respect to each other.

## Note added in proof

Among the new crystallographic structures which have appeared recently, some are noticeable for the aspects described in the present review. In the crystal structure of d(CGATCG) complexed with daunomycin [**346**] the hydration environment is observed to be slightly different from that of the complex with d(CGTACG), since the sodium ion linking the drug and the nucleic acid is replaced by two water molecules. In the crystal structure of the A-DNA octamer d(GTGTACAC), the binding of a spermine molecule in the major groove of the helix could be described at 2.0 Å resolution. [**347**] All the interactions are with the polar atoms of the bases, either directly or via water molecules. The interactions of polyamines with Z-DNA in the crystals have been described [**358**] and the following conclusions reached. The polyamine, not always fully extended, binds in the minor groove region, with the terminal amino nitrogen often penetrating it and the central imino nitrogen atom making strong hydrogen bonds to successive anionic phosphate oxygen atoms. Also concerning Z-DNA hydration, a Z-DNA structure containing U–A sequences showed that such sequences do not disrupt the spine of hydration in the minor groove like T–A sequences. [**359**] Concerning the hydration of B-DNA, the crystal structure of d(CCAACGTTGG) [**360**] indicated a narrow minor groove with evidence of a spine of hydration in a mixed A/T and C/G region. The full description of the hydration around a RNA helix, r(UUAUAUAUAUAUAA), has been described at 2.25 Å (Fig. 71). MC computations on the hydration around poly(dG)·poly(dC) and poly(G–C) have appeared. [**348**] MC estimates of the distribution of ions around DNA with electrostatic potentials modified so as to take into account the large discontinuity of the dielectric constant at the surface of the DNA molecule indicate a marked tendency for the counterions to leave the major groove and to cluster around the phosphates. [**349**] On the theoretical side, a systematic approach to the problem of solvation thermodynamics of biomolecules has been presented [**350**] and the salt-induced conformational B- to Z-forms transition has been analyzed using the site–site Ornstein–Zernike equation. [**351**] Further Raman studies confirmed that the lowest

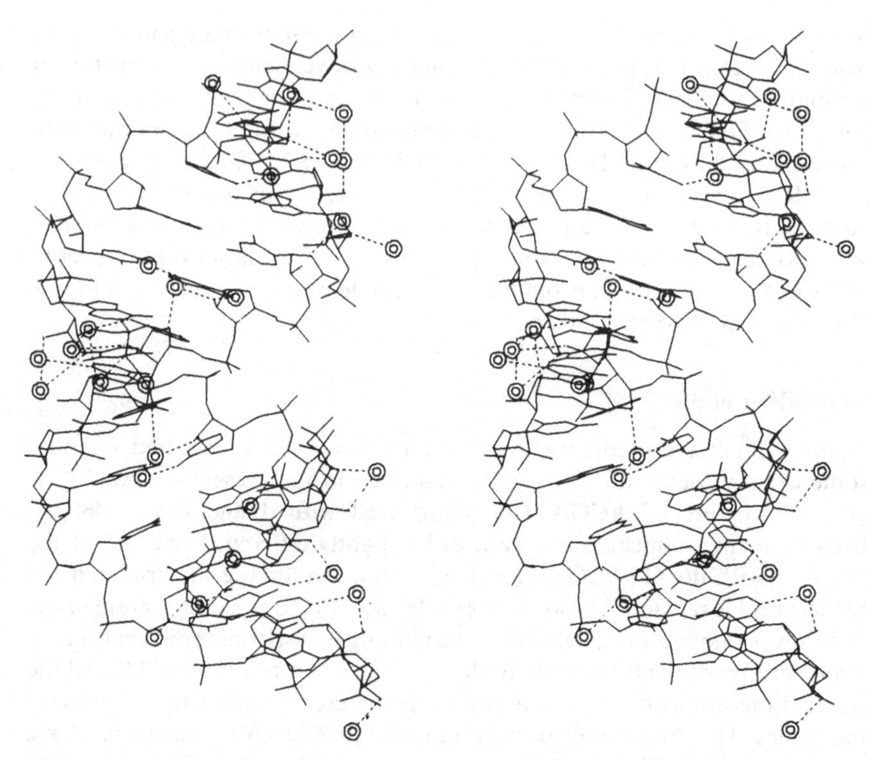

Figure 71. Stereo view of the hydration around the minor groove of the RNA helix r(UUAUAUAUAUAUAA). Adapted from ref. 361 with permission from the authors.

frequency band (below 30 $cm^{-1}$) reflects the state of hydration of the helix rather than intrahelical conformation. **[352]** Softening of this helix mode is, apparently, linearly related to the degree of hydration of the DNA helix. Another remarkable Raman scattering study led to the following conclusions: **[353]** (i) Dehydration beyond 2–3 waters/nucleotide disrupts the DNA structure substantially. (ii) The tetrahedral hydrogen-bond network of the water molecules is completely destroyed at 80% rh. (iii) Thus, the primary hydration shell, defined as those water molecules forming a different structure from that of bulk water, does not have an ice-like structure and contains about 15 waters/nucleotide (this number is consistent with the values given in Table 8 if the 3 waters/nucleotide at 0% rh are included). A dielectric study, using thermally stimulated depolarization currents, concluded that, up to a water content of 25% w/w, all the water molecules are irrotationally bound to the DNA structure. **[354]** A new technique, which measures the relative changes in ultrasonic velocity at frequencies of about 7 MHz, indicated an anomalously high level of hydration for the poly(dA)·poly(dT) polymer. **[355]** At the level of the oligomer, it could be shown **[355]** that the sequence d(AAAATTTT) had a decreased hydration compared to that of the sequence d(AAAAAAAA). Such results tend to confirm the determining role of hydration in DNA bending (see Section 5.1). The first solid-state $^2$H NMR of the DNA oligomer d(CGCGAATTCGCG) concluded that, at all levels of hydration, there is a rapid small-amplitude libration of the bases ($\tau_c < 1$ ns, 6–10° amplitude) with, at 80% rh, a lower motion ($\tau_c \approx 10$–$100\ \mu$s) of small amplitude (5°). **[356]** Further, this study found no evidence for large-amplitude motion on a nanosecond time scale, as previously inferred. **[357]**

## References


**1.** A. Sayre. In *Rosalind Franklin & DNA*, W. W. Norton, 1978,pp. 125–6.
**2.** (*a*) R. E. Franklin and R. G. Gosling. *Acta Cryst.* **6**, 673–7(1953); (*b*) R. E. Franklin and R. G. Gosling. *Acta Cryst.* **6**, 678–85(1953).
**3.** W. Saenger. *Principles of Nucleic Acid Structure*. Springer-Verlag: New-York, pp. 368–384, 1984.
**4.** J. Maddox. *Nature* **333**, 11(1988).
**5.** G. A. Jeffrey. *Acct. Chem. Res.* **2**, 344–52(1969).
**6.** H. F. Savage. *Water Sci. Rev.* **2**, 67–148(1986).
**7.** H. F. Savage and A. Wlodawer. *Methods in Enzymology* **127**, 162–83(1986).
**8.** E. Westhof. *Ann. Rev. Biophys. Biophys. Chem.* **17**, 125–44(1988).
**9.** A. C. Dock, B. Lorber, D. Moras, G. Pixa, J.-C. Thierry and R. Giege. *Biochimie* **66**, 179–201(1984).
**10.** W. Cochran. *Acta Cryst.* **4**, 81–92(1951).
**11.** J. N. Broomhead. *Acta Cryst.* **4**, 92–100(1951).
**12.** M. Sundaralingam. *Acta Cryst.* **21**, 495–505(1966).
**13.** G. H.-Y. Lin, M. Sundaralingam and S. K. Arora. *J. Amer. Chem. Soc.* **93**, 1235–41(1971).
**14.** G. H.-Y. Lin and M. Sundaralingam. *Acta Cryst.* **B27**, 961–9(1971).
**15.** T. P. Haromy, J. Raleigh and M. Sundaralingam. *Biochemistry* **19**, 1718–22(1980).
**16.** J. R. Clark. *Rev. Pure Appl. Chem.* **13**, 50–90(1963).
**17.** G. A. Jeffrey. In *Landolt-Börnstein, Numerical Data and Functional Relationships in Science and Technology*, New Series, Group VII, Vol. 1, Nucleic Acids, Subvolume b, Crystallographic and Structural Data II, Springer-Verlag, Berlin, 1989, pp. 277–342.
**18.** S. T. Rao and M. Sundaralingam. *J. Amer. Chem. Soc.* **92**, 4963–70(1970).
**19.** D. Rohrer and M. Sundaralingam. *J. Amer. Chem. Soc.* **92**, 4956–62(1970).
**20.** D. G. Watson, D. J. Sutor and P. Tollin. *Acta Crystallogr.* **19**, 111–16(1965).
**21.** E. Shefter, M. Barlow, R. A. Sparks and K. N. Trueblood. *Acta Crystallogr.* **B25**, 895–900(1965).
**22.** R. Parthasarathy, T. Srikrishnan, and S. L. Ginell. In *Biomolecular Stereodynamics I* (ed. R. H. Sarma). Adenine Press: New York, 1981, pp. 261–7.
**23.** M. Sundaralingam, T. P. Haromy and P. Prusiner. *Acta Cryst.* **B38**, 1536–40(1982).
**24.** J. M. Rosenberg, N. C. Seeman, R. O. Day and A. Rich. *J. Molec. Biol.* **104**, 109–144(1976).

**25.** N. C. Seeman, J. M. Rosenberg, F. L. Suddath, J. J. P. Kim and A. Rich. *J. Molec. Biol.* **104**, 109–44(1976).
**26.** H. M. Berman, W. Stallings, H. L. Carrell, J. P. Glusker, S. Neidle, G. Taylor and A. Achari. *Biopolymers* **18**, 2405–29(1979).
**27.** E. Westhof, S. T. Rao and M. Sundaralingam. *J. Mol. Biol.* **142**, 331–61(1980).
**28.** E. Westhof, P. Dumas and D. Moras. *Biochimie* **70**, 145–65(1988).
**29.** P. H. Bolton and D. R. Kearns. *Biochim. Biophys. Acta* **517**, 329–36(1978).
**30.** M. Sundaralingam. In *Conformations of Biological Molecules and Polymers. The Jerusalem Symposia on Quantum Chemistry and Biochemistry* (eds. V. E. D. Bergmann and B. Pullman), Vol. 5, 1973, pp. 417–56.
**31.** J. V. Silverton, W. Limm and H. T. Miles. *J. Amer. Chem. Soc.* **104**, 1081–7(1982).
**32.** C. E. Bugg, J. M. Thomas, M. Sundaralingam and S. T. Rao. *Biopolymers* **10**, 175–219(1971).
**33.** A. H.-J. Wang, G. J. Quigley, F. J. Kolpak, J. L. Crawford, J. H. van Boom, G. A. van der Marel and A. Rich. *Nature* **282**, 680–6(1979).
**34.** J. Emerson and M. Sundaralingam. *Acta Cryst.* **B36**, 1510–13(1980).
**35.** M. Sundaralingam and J. Abola. *J. Amer. Chem. Soc.* **94**, 5070–6(1972).
**36.** S. T. Rao and M. Sundaralingam. *J. Amer. Chem. Soc.* **91**, 1210–17(1969).
**37.** H. M. Berman, A. Sowri, S. L. Ginell and D. L. Beveridge. *J. Biomol. Struct. Dyn.* **5**, 1101–10(1988).
**38.** R. Gerdil. *Acta Cryst.* **14**, 333–44(1961).
**39.** U. Thewalt, C. E. Bugg and R. E. Marsh. *Acta Cryst.* **B27**, 2358–63(1971).
**40.** E. Westhof. *J. Biomol. Struct. Dyn.* **5**, 581–600(1987).
**41.** G. A. Jeffrey, H. Maluszynska and J. Mitra. *Int. J. Biol. Macromol.* **7**, 336–48(1985).
**42.** C. E. Bugg and R. E. Marsh. *J. Mol. Biol.* **25**, 67–74(1967).
**43.** T. Srikrishnan, S. M. Fridey, and R. Parthasarathy. *J. Amer. Chem. Soc.* **101**, 3739–44(1979).
**44.** Y. Yamagata, Y. Suzuki, S. Fujii, T. Fujiwara and K. Tomita. *Acta Crystallogr.* **B35**, 1136–42(1979).
**45.** K. N. Trueblood, P. Horn and V. Luzzati. *Acta Cryst.* **14**, 965–82(1961).
**46.** M. Coll, X. Solans, M. Font-Altaba and J. A. Subirana. *J. Biomol. Struct. Dyn.* **4**, 797–811(1987).
**47.** W. B. T. Cruse, E. Egert, O. Kennard, G. B. Sala, S. A. Salisbury and M. A. Viswamitra. *Biochemistry* **22**, 1833–39(1983).
**48.** B. Ramakrishnan and M. A. Viswamitra. *J. Biomol. Struct. Dyn.* **6**, 511–23(1988).
**49.** G. J. Quigley, A. H. J. Wang and A. Rich. *American Crystallographic Association*, Abstract PB58, June 1986.
**50.** A. H.-J. Wang, T. Hakoshima, G. A. van der Marel, J. H. van Boom and A. Rich. *Cell* **37**, 321–31(1984).
**51.** B. Chevrier, A. C. Dock, B. Hartmann, M. Leng, D. Moras, M. T. Thuong and E. Westhof. *J. Molec. Biol.* **188**, 707–19(1986).
**52.** B. Ramakrishnan and M. A. Viswamitra. In press.
**53.** F. Jacob. In *La Statue Intérieure* (ed. O. Jacob), Seuil, 1987, pp. 353.
**54.** C. M. Frey and J. Stuehr. In *Metal Ions in Biological Systems* (*ed. H. Sigel*), *Vol.* 1. *Marcel Dekker*: *New York*, 1974, *pp.* 51–116.

**55.** D. Porschke. *Nucleic Acids. Res.* **6**, 883–98(1979).
**56.** V. Swaminathan and M. Sundaralingam. *CRC Critical Reviews in Biochemistry* **6**, 245–336(1979).
**57.** L. G. Marzilli, T. J. Kistenmacher and G. L. Eichhorn. In *Nucleic Acid-Metal Ion Interactions* (ed. T. G. Spiro). John Wiley & Sons: New York, 1980, pp. 180–250.
**58.** J. Hogle, M. Sundaralingam and G. H.-Y. Lin. *Acta Cryst.* **B36**, 564–70(1980).
**59.** H. Sternglanz, E. Subramanian, J. C. Jr. Lacey and C. E. Bugg. *Biochemistry* **15**, 4797–802(1976).
**60.** P. Swaminathan and M. Sundaralingam. *Acta Crystallogr.* **B36**, 2590–7(1980).
**61.** N. Camerman and J. K. Fawcett. *J. Molec. Biol.* **107**, 601–4(1976).
**62.** H. Einspahr, W. J. Cook and C. E. Bugg. *Biochemistry* **20**, 5788–94(1980).
**63.** S. C. Jain and H. M. Sobell. *J. Molec. Biol.* **68**, 1–20(1972).
**64.** H. M. Berman and P. R. Young. *Ann. Rev. Biophys. Bioeng.* **10**, 87–114(1981).
**65.** S. Neidle and H. M. Berman. *Prog. Biophys. Mol. Biol.* **41**, 43–66(1983).
**65.** S. Neidle and H. M. Berman. *Prog. Biophys. Mol. Biol.* **41**, 43–66(1983).
**66.** S. Neidle, L. H. Pearl and J. V. Skelly. *Biochem. J.* **243**, 1–13(1987).
**67.** D. Suck, P. C. Manor and W. Saenger. *Acta Cryst.* **B32**, 1727–37(1976).
**68.** A. H.-J. Wang, J. Nathans, G. A. van der Marel, J. H. van Boom and A. Rich. *Nature* **276**, 471–4(1978).
**69.** S. Neidle, H. M. Berman and H. S. Shieh. *Nature* **288**, 129–33(1980).
**70.** G. J. Quigley, A. H.-J. Wang, G. Ughetto, G. A. van der Marel, J. H. van Boom and A. Rich. *Proc. Natl. Acad. Sci. USA* **77**, 7204–8(1980).
**71.** A. H.-J. Wang, G. Ughetto, G. J. Quigley and A. Rich. *Biochemistry* **26**, 1152–63(1987).
**72.** S. Vijay-Kumar, T. D. Sakore, and H. M. Sobell. *J. Biomol. Struct. Dyn.* **2**, 333–44(1984).
**73.** F. Takusagawa, M. Dabrow, S. Neidle and H. M. Berman. *Nature* **296**, 466–9(1982).
**74.** A. H.-J. Wang, G. Ughetto, G. J. Quigley, T. Hakoshima, G. A. van der Marel, J. H. van Boom and A. Rich. *Science* **225**, 1115–21(1984).
**75.** A. V. Fratini, M. L. Kopka, H. R. Drew and R. E. Dickerson. *J. Biol. Chem.* **257**, 14686–707(1982).
**76.** J. L. Finney. In *Water: A Comprehensive Treatise* (ed. F. Franks), Vol. 6. Plenum Press: New York, 1979, pp. 47–122.
**77.** H.-S. Shieh, H. M. Berman, S. Neidle, G. Taylor and M. Sanderson. *Acta Cryst.* **B38**, 523–31(1982).
**78.** P. Swaminathan, E. Westhof and M. Sundaralingam. *Acta Cryst.* **B38**, 515–22(1982).
**79.** T. Hakoshima, S. Toda, S. Sugio, K. Tomita, S. Nishikawa, H. Morioka, K. Fuchimura, T. Kimura, S. Uesugi, E. Ohtsuka and M. Ikehara. *Protein Engin.* **2**, 55–61(1988).
**80.** I. T. Weber & T. A. Steitz. *J. Molec. Biol.* **198**, 311–26(1987).
**81.** D. J. Filman, J. T. Bolin, D. A. Matthews and J. Kraut. *J. Biol. Chem.* **257**, 13663–72(1982).
**82.** D. Suck, A. Lahm and C. Oefner. *Nature* **332**, 465–8(1988).
**83.** Z. Otwinowski, R. W. Schevitz, R.-G. Zhang, C. L. Lawson, A. Joachimiak, R. Q. Marmorstein, B. F. Luisi and P. B. Sigler. *Nature* **335**, 321–9(1988).

**84.** F. A. Cotton, E. E. Hazen and M. J. Legg. *Proc. Natl. Acad. Sci. USA* **76**, 2551–5(1979).
**85.** D. L. Ollis, P. Brick, R. Hamlin, N. G. Xuong and T. A. Steitz. *Nature* **313**, 762–6(1985).
**86.** K. Namba, R. Pattanayek and G. Stubbs. Private communication.
**87.** G. T. Grant, E. R. Morris, D. A. Rees, P. J. C. Smith, and D. Thom. *FEBS Letters* **32**, 195–8(1973).
**88.** (*a*) M. Falk, K. A. Hartmann Jr. and R. C. Lord. *J. Amer. Chem. Soc.* **84**, 3843–6(1962); (*b*) *ibid.* **85**, 387–91(1963); (*c*) *ibid.* **85**, 391–4(1963).
**89.** M. Falk, A. G. Poole and C. G. Goymour. *Can. J. Chem.* **48**, 1536–42(1970).
**90.** B. Wolf and S. Hanlon. *Biochemistry* **14**, 1661–70(1975).
**91.** M. L. Kopka, A. V. Fratini, H. R. Drew and R. E. Dickerson. *J. Mol. Biol.* **163**, 129–46(1983).
**92.** E. Westhof. *Int. J. Biol. Macro.* **9**, 186–92(1987).
**93.** S. Sprang, R. Scheller, D. Rohrer and M. Sundaralingam. *J. Amer. Chem. Soc.* **100**, 2867–71(1978).
**94.** O. Kennard, W. B. T. Cruse, J. Nachman, T. Prange, Z. Shakked and D. Rabinovich. *J. Biomol. Struct. Dyn.* **3**, 623–47(1986).
**95.** W. B. T. Cruse, S. A. Salisbury, T. Brown, R. Cosstick, F. Eckstein and O. Kennard. *J. Mol. Biol.* **192**, 891–905(1986).
**96.** G. G. Privé, U. Heinemann, S. Chandrasegaran, L. S. Kan, M. L. Kopka and R. E. Dickerson. *Science* **203**, 498–504(1987).
**97.** E. N. Baker and R. E. Hubbard. *Prog. Biophys. Mol. Biol.* **44**, 97–179(1984).
**98.** E. Chargaff. *Heraclitean Fire*, Warner Books Ed, 1978, p. 13.
**99.** W. N. Hunter, T. Brown and O. Kennard. *J. Biomol. Struct. Dyn.* **4**, 173–91(1986).
**100.** W. Saenger, W. N. Hunter, and O. Kennard. *Nature* **324**, 385–8(1986).
**101.** U. Heinemann, H. Lauble, R. Frank and H. Blöcker. *Nucleic Acids Res.* **15**, 9531–50(1987).
**102.** G. J. Quigley. *Transactions ACA* **22**, 121–30(1986).
**103.** H. R. Drew and R. E. Dickerson. *J. Mol. Biol.* **151**, 535–56(1981).
**104.** M. Coll, A. H.-J. Wang, G. A. van der Marel, J. H. van Boom and A. Rich. *J. Biomol. Struct. Dyn.* **4**, 157–72(1986).
**105.** C. R. Calladine. *J. Mol. Biol.* **161**, 343–52(1982).
**106.** V. P. Chuprina. *Nucleic Acid Res.* **15**, 293–311(1987).
**107.** B. N. Conner, C. Yoon, J. L. Dickerson and R. E. Dickerson. *J. Mol. Biol.* **174**, 663–95(1984).
**108.** M. M. Teeter. *Proc. Natl. Acad. Sci. USA* **81**, 6014–18(1984).
**109.** T. Brown, G. Kneale, W. N. Hunter and O. Kennard. *Nucl. Acids. Res.* **14**, 1801–9(1986).
**110.** P. S. Ho, C. A. Frederick, G. J. Quigley, G. A. van der Marel, J. H. van Boom, A. H.-J. Wang and A. Rich. *EMBO J.* **4**, 3617–23(1985).
**111.** W. N. Hunter, G. Kneale, T. Brown, D. Rabinovich and O. Kennard. *J. Mol. Biol.* **190**, 605–18(1986).
**112.** (*a*) O. Kennard. *J. Biomol. Struct. Dyn.* **3**, 205–26(1986); (*b*) O. Kennard. *Transactions ACA* **22**, 131–49(1987).
**113.** G. Kneale, T. Brown, O. Kennard and D. Rabinovich. *J. Mol. Biol.* **186**, 805–14(1985).

**114.** A. H.-J. Wang, R. V. Gessner, G. A. van der Marel, J. H. van Boom and A. Rich. *Proc. Natl. Acad. Sci. USA* **82**, 3611–15(1985).

**115.** E. Westhof and M. Sundaralingam. *Biochemistry* **25**, 4868–78(1986).

**116.** J. Petruska, L. C. Sowers, and M. F. Goodman. *Proc. Natl. Acad. Sci. USA* **83**, 1559–62(1986).

**117.** K. Kim and M. S. Jhon. *Biochim. Biophys. Acta* **565**, 131–47(1979).

**118.** M. M. Teeter, G. J. Quigley and A. Rich. In *Nucleic Acid-Metal Ion Interactions* (ed. T. G. Spiro). John Wiley & Sons: New York, 1980, pp. 145–78.

**119.** R. V. Gessner, G. J. Quigley, A. H.-J. Wang, G. A. van der Marel, J. H. van Boom and A. Rich. *Biochemistry* **24**, 237–40(1985).

**120.** P. S. H. Ho, C. A. Frederick, D. Saal, A. H.-J. Wang and A. Rich. *J. Biomol. Struct. Dyn.* **4**, 521–34(1987).

**121.** R. G. Brennan, E. Westhof, and M. Sundaralingam. *J. Biomol. Struct. Dyn.* **3**, 649–65(1986).

**122.** A. H.-J. Wang, G. J. Quigley, F. J. Kolpak, G. A. van der Marel, J. H. van Boom and A. Rich. *Science* **211**, 171–6(1981).

**123.** P. Dumas. Ph.D. Thesis, Université Louis Pasteur, Strasbourg (France), 1986.

**124.** R. Lavery, A. Pullman, B. Pullman and M. De Oliviera. *Nucleic Acids Res.* **8**, 5095–111(1980).

**125.** V. N. Bartenev, Eu. I. Golovamov, K. A. Kapitanova, M. A. Mokulskii, I. I. Volkova and I. Ya. Skuratovskii. *J. Mol. Biol.* **169**, 217–34(1983).

**126.** J. W. Pflugrath and F. A. Quiocho. *J. Mol. Biol.* **200**, 163–80(1988).

**127.** O. Sinanoglu and S. Abdulnar. *Fed. Amer. Soc. Exp. Biol.* **24**, 5–7(1965).

**128.** D. M. Crothers and D. I. Ratner. *Biochemistry* **7**, 1823–7(1968).

**129.** (*a*) C. Bartzsch, H.-J. Hofman and C. Weiss. *Studia Biophysica* **93**, 197–200 (1983); (*b*) C. Bartzsch, C. Weiss and H.-J. Hofmann. *J. Pract. Chem. (Leipzig)* **326**, 407–14(1984).

**130.** J. T. Egan, S. Nir, R. Rein and R. MacElroy. *Int. J. Quantum Chem. QBS* **5**, 443–6(1978).

**131.** J. Emsley, D. J. Jones and R. E. Overill. *J.C.S. Chem. Comm.* 476–8(1982).

**132.** S. Lewin. *J. Theoret. Biol.* **17**, 181–212(1967).

**133.** H. Berthod and A. Pullman. *Theoret. Chim. Acta (Berlin)* **47**, 59–66(1978).

**134.** A. Pullman, H. Berthod and N. Gresh. *Chem. Phys. Lett.* **33**, 11–14(1975).

**135.** B. Pullman, A. Pullman, H. Berthod and N. Gresh. *Theoret. Chim. Acta (Berlin)* **40**, 93–111(1975).

**136.** H. Berthod and A. Pullman. *Chem. Phys. Lett.* **46**, 249–52(1977).

**137.** A. Pullman, B. Pullman and H. Berthod. *Theoret. Chim. Acta (Berlin)* **47**, 175–92(1978).

**138.** G. N. J. Port and A. Pullman. *FEBS Lett.* **31**, 70–4(1973).

**139.** A. Pullman and B. Pullman. *Quart. Rev. Biophys.* **7**, 505–66(1975).

**140.** A. Pullman. In *Structure and Conformation of Nucleic Acids and Protein-Nucleic Acid Interactions*, (eds M. Sundaralingan and S. T. Rao). University Park Press: Baltimore, 1975, pp. 457–84.

**141.** A. Pullman. In *Environmental Effects on Molecular Structures and Properties* (ed. B. Pullman). Reidel Publishing Co.: Dordrecht, Holland, 1976, pp. 1–15.

**142.** C. Giessner-Prettre, F. Ribas Prado, and B. Pullman. *Nucleic Acid. Res.* **4**, 3229–38(1977).

**143.** A. Goldblum, D. Perahia and A. Pullman. *FEBS Lett.* **91**, 213–15(1978).

**144.** A. Pullman and D. Perahia. *Theoret. Chim. Acta* **48**, 29–36(1978).

**145.** B. Pullman, S. Miertus and D. Perahia. *Theoret. Chim. Acta (Berlin)* **50**, 317–25(1979).

**146.** J. Langlet, C. Giessner-Prettre, B. Pullman, P. Claverie and D. Piazzola. *Int. J. Quantum Chem.* **18**, 421–37(1980).

**147.** B. Pullman. In *Water and Metal Cations in Biological Systems* (eds B. Pullman and K. Yagi). Japan Scientific Societies Press: Tokyo, 1980, pp. 1–12.

**148.** D. Perahia, M. S. Jhon and B. Pullman. *Biochim. Biophys. Acta* **474**, 349–62(1977).

**149.** R. Lavery and B. Pullman. *Studia Biophysica* **92**, 99–110(1982).

**150.** R. Lavery, A. Pullman and B. Pullman. *Theoret. Chim. Acta (Berlin)* **62**, 93–106(1982).

**151.** R. Lavery, A. Pullman and B. Pullman. *Biophysical Chem.* **17**, 75–86(1983).

**152.** (*a*) A. Pullman, B. Pullman and R. Lavery. In *Nucleic Acids: The Vectors of Life* (eds. B. Pullman and J. Jortner). D. Reidel Publishing Co.: Dordrecht Holland, 1983, pp. 75–88; (*b*) A. Pullman, B. Pullman and R. Lavery. *J. Mol. Struct.* **93**, 85–91(1983).

**153.** (*a*) M. de Oliviera Neto. *J. Comput. Chemistry* **7**, 617–28 (1986); (*b*) M. de Oliviera Neto. *J. Comput. Chemistry* **7**, 629–39(1986).

**154.** R. Scordomaglia, F. Cavallone, and E. Clementi. *J. Am. Chem. Soc.* **99**, 5545–50(1977).

**155.** G. Corongiu and E. Clementi. *Gazz. Chim. It.* **108**, 273–306(1978).

**156.** G. Corongiu and E. Clementi. *Gazz. Chim. It.* **108**, 687–91(1978).

**157.** E. Clementi, G. Corongiu and F. Lejl. *J. Chem. Phys.* **70**, 3726–9(1979).

**158.** O. Matsuoka, E. Clementi and M. Yoshimine. *J. Chem. Phys.* **64**, 1351–61(1976).

**159.** H. Popkie, H. Kistenmacher and E. Clementi. *J. Chem. Phys.* **59**, 1325–36(1973).

**160.** H. Kistenmacher, G. C. Lie, H. Popkie and E. Clementi. *J. Chem. Phys.* **61**, 546–61(1974).

**161.** G. Corongiu and E. Clementi. To be submitted (ref. 72 of 162 below).

**162.** E. Clementi and G. Corongiu. *J. Chem. Phys.* **72**, 3979–92(1980).

**163.** E. Clementi. In *Structure and Dynamics, Nucleic Acids and Proteins* (eds E. Clementi & R. H. Sarma). Adenine Press: New York, 1983, pp. 321–64.

**164.** (*a*) J. E. Del Bene. *J. Mol. Struct.* **108**, 179–97(1984); (*b*) J. E. Del Bene. *J. Comput. Chem.* **2**, 188–99(1981); (*c*) J. E. Del Bene. *ibid.* **2**, 200–6(1981); (*d*) J. E. Del Bene. *ibid.* **2**, 416–21(1981); (*e*) J. E. Del Bene. *ibid.* **2**, 422–32(1981); (*f*) J. E. Del Bene. *J. Phys. Chem.* **86**, 1341–7(1982); (*g*) J. E. Del Bene. *J. Chem. Phys.* **76**, 1058–63(1982); (*h*) J. E. Del Bene. *J. Comput. Chem.* **4**, 226–33(1983).

**165.** J. S. Kwiatkowski and B. Pullman. *Theoret. Chim. Acta (Berlin)* **42**, 83–6(1976).

**166.** J. S. Kwiatkowski, D. Perahia and B. Pullman. *Int. J. Quantum. Chem. QCS* **12**, 249–56(1978).

**167.** (*a*) J. S. Kwiatkowski and B. Szczodrowska. *Chem. Phys.* **27**, 389–98(1978). (*b*) J. S. Kwiatkowski. In *Biomolecular Structure Conformation, Function and Evolution* (ed. R. Srinivasan). Pergamon Press: Oxford, 1980, pp. 311–19.

**168.** J. S. Kwiatkowski and A. Tempczyk. *Chem. Phys.* **85**, 397–402(1984).

**169.** J. S. Kwiatkowski and M. Berndt. In *Theoretical Chemistry of Biological Systems* (ed. G. Nary-Szabo). Elsevier: Amsterdam, 1985.

**170.** T. Yamaguchi and C. Nagata. *J. Theoret. Biol.* **69**, 693–707(1977).

**171.** V. I. Danilov, M. R. Sharafudinof, G. G. Malenkov and V. I. Poltev. *Doklady Akad. Nauk. USSR* **268**, 485–8(1982).

**172.** V. I. Danilov and I. S. Tokokh. *FEBS Lett.* **173**, 347–50(1984).

**173.** V. I. Danilov, I. S. Tolokh, and V. I. Poltev. *FEBS Lett.* **171**, 325–8(1984).

**174.** V. I. Danilov, I. S. Tolokh, V. I. Poltev and G. G. Malenkov. *FEBS Lett.* **167**, 245–8(1984).

**175.** A. Pohorille, S. K. Burt, and R. D. MacElroy. *J. Am. Chem. Soc.* **106**, 402–9(1984).

**176.** A. Pohorille, L. R. Pratt, S. K. Burt and R. D. MacElroy. *J. Biomol. Str. Dyn.* **1**, 1257–80(1984).

**177.** P. K. Mehrotra and D. L. Beveridge. *J. Am. Chem. Soc.* **102**, 4287–94(1980).

**178.** M. Mezei and D. L. Beveridge. *Methods in Enzymology* **127**, 21–47(1986).

**179.** D. L. Beveridge, P. V. Maye, B. Jayaram, G. Ravishanker and M. Mezei. *J. Biomol. Struct. Dyn.* **2**, 261–70(1984).

**180.** P. V. Maye. Ph.D. Thesis, City University of New York (1986).

**181.** B. Jayaram. Ph.D. Thesis, City University of New York (1986).

**182.** (*a*) B. Jayaram, M. Mezei and D. L. Beveridge. *J. Comput. Chem.* **8**, 917–21(1987); (*b*) B. Jayaram, M. Mezei and D. L. Beveridge. *J. Am. Chem. Soc.* **110**, 1691–4(1988).

**183.** B. Jayaram, G. Ravishanker and D. L. Beveridge. *J. Phys. Chem.* **92**, 1032(1988).

**184.** P. L. Howell. Ph.D. Thesis, Birkbeck College, University of London (1986).

**185.** H. S. Shieh, H. M. Berman and S. Neidle. *Nucleic Acids Res.* **8**, 85–97(1980).

**186.** M. Mezei, D. L. Beveridge, H. M. Berman, J. M. Goodfellow, J. L. Finney and S. Neidle. *J. Biomol. Str. Dyn.* **1**, 287–97(1983).

**187.** S. Neidle. Private communication.

**188.** M. Mezei and D. L. Beveridge. *J. Comput. Chem.* **5**, 523–7(1984).

**189.** K. S. Kim, G. Corongiu and E. Clementi. *J. Biomol. Struct. Dyn.* **1**, 263–85(1983).

**190.** K. S. Kim and E. Clementi. *J. Am. Chem. Soc.* **107**, 227–34(1985).

**191.** K. S. Kim and E. Clementi. *J. Phys. Chem.* **89**, 3655–63(1985).

**192.** J. M. Goodfellow. *J. Theoret. Biol.* **107**, 261–74(1984).

**193.** J. M. Goodfellow, P. L. Howell and F. Vovelle. *Ann. NY Acad. Sci.* **482**, 179–94(1986).

**194.** (*a*) E. Clementi and G. Corongiu. *Int. J. Quantum Chem.* **16**, 897–915(1979); (*b*) E. Clementi and G. Corongiu. *Biopolymers* **18**, 2431–50(1979); (*c*) E. Clementi and G. Corongiu. *Gazz. Chim. Ital.* **109**, 201–5(1979); (*d*) E. Clementi and G. Corongiu. *Chem. Phys. Lett.* **60**, 175–8(1979).

**195.** (*a*) G. Corongiu and E. Clementi. *Biopolymers* **20**, 551–71(1981); (*b*) G. Corongiu and E. Clementi. *Biopolymers* **20**, 2427–83(1981).

**196.** E. Clementi. *IBM J. Res. Devel.* **25**, 315(1981).

**197.** (*a*) E. Clementi and G. Corongiu. In *Biomolecular Stereodynamics* (ed. R. H. Sarma), Vol. 1. Adenine Press; New York, 1981, pp. 209ff; (*b*) E. Clementi and G. Corongiu. *Ann. NY Acad. Sci.* **367**, 83(1981).

**198.** E. Clementi, G. Corongiu, M. Gratarola, P. Habitz, C. Lupo, P. Otto and D. Vercauteren. *Int. J. Quantum Chem. QCS* **16**, 409(1982).

**199.** (*a*) E. Clementi and G. Corongiu. *Biopolymers* **21**, 763–77(1982); (*b*) E. Clementi and G. Corongiu. In *Current Aspects of Quantum Chemistry 1981* (ed. R. Carbo). Elsevier Scientific Publishing Co.: Amsterdam, 1982, pp. 331ff; (*c*) E. Clementi and G. Corongiu. *Int. J. Quantum Chem.* **22**, 595(1982).

**200.** P. Otto, E. Clementi and G. Corongiu. *J. Chem. Phys.*, submitted.

**201.** E. Clementi and G. Corongiu. *J. Biol. Phys.* **11**, 33–42(1983).

**202.** H. M. Berman, A. Sowri, S. L. Ginell and D. L. Beveridge. *J. Biomol. Str. Dyn.* **5**, 1101–10(1988).

**203.** P. S. Subramanian, S. Pitchumani, D. L. Beveridge and H. M. Berman. *Biopolymers*, submitted (1988).

**204.** B. Hingerty, E. Subramanian, S. D. Stellman, T. Sato, S. Broyde and R. Langridge. *Acta Cryst.* **B32**, 2998–3013(1976).

**205.** A. Aggarwal, S. A. Islam, R. Kuroda, M. R. Sanderson, S. Neidle, and H. M. Berman. *Acta Cryst.* **B39**, 98–104(1983).

**206.** P. S. Subramanian, G. Ravishanker and D. L. Beveridge. *Proc. Natl. Acad. Sci.* (*USA*) **85**, 1836–40(1988).

**207.** P. S. Subramanian and D. L. Beveridge. *J. Biomol. Struct. Dyn.* **6**, 1093–122(1989).

**208.** L. A. Marky and K. J. Breslauer. *Proc. Natl. Acad. Sci. USA* **84**, 4359–63(1987).

**209.** K. J. Breslauer. Private communication.

**210.** (*a*) P. S. Subramanian and D. L. Beveridge. MS in preparation. (*b*) P. S. Subramanian and D. L. Beveridge, MS in preparation.

**211.** W. K. Olson. In *Topics in Nucleic Structure*, Part II, S. Neidle, ed., Macmillan Publishers Ltd., London (1982) pp. 1–79.

**212.** P. A. Kollman, P. K. Weiner and A. Dearing. *Biopolymers* **20**, 2583–621(1981).

**213.** (*a*) M. Levitt. *Proc. Natl. Acad. Sci. USA* **75**, 640–4(1977). (*b*) M. Levitt. *Cold Spr. Harbor Symp. Quant. Biol.* **47**, 251–62(1983).

**214.** S. Broyde, S. Stellman and R. Wartell. *Biopolymers* **17**, 533–53(1975).

**215.** (*a*) R. Lavery, H. Sklenar, K. Zakrenska and B. Pullman. *J. Biomol. Struct. Dyn.* **3**, 989–1014(1986); (*b*) *ibid.* **3**, 1015–31(1986); (*c*) *ibid.* **3**, 1155–70(1986).

**216.** W. K. Olson, A. R. Srinivasan, M. A. Cueto, R. Torres, R. C. Maroun, J. Cicariello and J. L. Nuss. *Biomolecular Stereodynamics IV* (ed. R. H. Sarma). Adenine Press: New York, 1985.

**217.** F. Vovelle, R. J. Elliot, and J. M. Goodfellow. *Int. J. Biol. Macromol.* **11**, 39–42(1989).

**218.** N. Pattabiramin, R. Langridge, and P. A. Kollman. *J. Biomol. Str. Dyn.* **1**, 1525–33(1984).

**219.** B. Ried. *Quart. Rev. Biophys.* **20**, 1–34(1987).

**220.** D. J. Patel, L. Shapiro and D. J. Hare. *Quart. Rev. Biophys.* **20**, 35–112(1987).

**221.** (*a*) J. E. H. Koehler, W. Saenger and W. F. van Gunsteren. *Eur. Biophys.* **15**, 197–210(1987); (*b*) *ibid.* **15**, 211–24(1987); (*c*) in press (1988); (*d*) J. E. H. Koehler, W. Saenger and W. F. van Gunsteren. *J. Mol. Biol.* in press (1989); (*e*) J. E. H. Koehler, W. Saenger and W. F. van Gunsteren. *J. Biomol. Str. Dyn.* **6**, 181–98(1989).

**222.** S. Swaminathan, H. M. Berman and D. L. Beveridge. MS in preparation.

**223.** W. Saenger. *Ann. Rev. Biophys. Biophys. Chem.* **16**, 93–114(1987).

**224.** B. Tidor, K. K. Irikura, B. R. Brooks and M. Karplus. *J. Biomol. Struct. Dyn.* **1**, 231–52(1983).

**225.** M. Prabhakaran, S. C. Harvey, B. Mao and J. A. McCammon. *J. Biomol. Struct. Dyn.* **1**, 357–69(1983).

**226.** S. C. Harvey, M. Prabhakaran, B. Mao and J. A. McCammon. *Science* **223**, 1189–91(1984).

**227.** U. C. Singh, S. J. Weiner and P. A. Kollman. *Proc. Natl. Acad. Sci.* (*USA*) **82**, 755–9(1985).

**228.** G. L. Seibel, U. C. Singh and P. A. Kollman. *Proc. Natl. Acad. Sci. USA* **82**, 6537–40(1985).

**229.** W. F. van Gunsteren, H. J. C. Berendsen, R. G. Guersten, H. R. Zwinderman. *Ann. NY Acad. Sci.* **482**, 287–303(1986).

**230.** L. Nilsson and M. Karplus. *J. Comput. Chem.* **7**, 591–7(1986).

**231.** W. L. Jorgensen and M. Ibrahim. *J. Am. Chem. Soc.* **103**, 3976–85(1981).

**232.** K. Swamy and E. Clementi. *Biopolymers* **26**, 1901–27(1987).

**233.** E. Westhof, B. Chevrier, S. L. Gallion, P. K. Weiner and R. M. Levy. *J. Mol. Biol.* **190**, 699–712(1986).

**234.** L. Nilsson, G. M. Clore, A. M. Gronenborn, A. J. Brunger and M. Karplus. *J. Mol. Biol.* **188**, 455–75(1986).

**235.** G. Ravishanker, S. Swaminathan, D. L. Beveridge, R. Lavery and H. Sklenar. *J. Biomol. Struct. Dyn.* **6**, 669–99(1989).

**236.** (*a*) R. Lavery and H. Sklenar. *J. Biomol. Struct. Dyn.* **6**, 63–91(1988); (*b*) *ibid.* **6**, 655–67(1989).

**237.** P. K. Weiner, S. L. Gallion, E. Westhof and R. M. Levy. *J. Mol. Graphics* **4**, 203–7(1985).

**238.** M. T. Record, Jr., C. F. Anderson, P. Mills, M. Mossing and J. H. Roe. *Adv. Biophys.* **20**, 109–35(1985).

**239.** (*a*) G. S. Manning. *J. Chem. Phys.* **51**, 924–33(1969); (*b*) G. S. Manning. *ibid.* **51**, 934–8(1969); (*c*) G. S. Manning. *ibid.* **51**, 3249–53(1969); (*d*) G. S. Manning. *Biophys. Chem.* **7**, 95–102(1977); (*e*) G. S. Manning. *ibid.* **9**, 65–70(1978); (*f*) G. S. Manning. *Quart. Rev. Biophys.* **11**, 179–246(1978); (*g*) G. S. Manning. *Acct. Chem. Res.* **12**, 443–49(1979).

**240.** B. H. Zimm and M. LeBret. *J. Biomol. Str. Dyn.* **1**, 461–71(1983).

**241.** M. T. Record, C. F. Anderson and T. Lohman. *Quart. Rev. Biophys.* **11**, 103–78(1978).

**242.** (*a*) C. F. Anderson and M. T. Record, Jr. *Ann. Rev. Phys. Chem.* **33**, 191–222(1982); (*b*) C. F. Anderson and M. T. Record, Jr. In *Structure and Dynamics: Nucleic Acids and Proteins* (eds E. Clementi and R. H. Sarma). Adenine Press: New York, 1983 pp. 301–19.

**243.** M. LeBret and B. H. Zimm, *Biopolymers* **23**, 271–85(1984).

**244.** C. S. Murthy, R. J. Bacquet and P. J. Rossky. *J. Phys. Chem.* **89**, 701–10(1985).

**245.** (*a*) P. A. Mills, C. F. Anderson and M. T. Record, Jr. *J. Phys. Chem.* **89**, 3984–94(1985); (*b*) P. A. Mills, M. D. Paulsen, C. F. Anderson and M. T. Record, Jr. *Chem. Phys. Lett.* **129**, 155–8(1986).

**246.** R. Bacquet and P. J. Rossky. *J. Phys. Chem.* **88**, 2660–9(1984).

**247.** (*a*) B. J. Klein and G. R. Pack. *Biopolymers* **22**, 2331–52(1982); (*b*) G. R. Pack

and B. J. Klein. *ibid.* **23**, 2801–23(1983); (*c*) G. R. Pack, L. Wong and C. V. Prasad. *Nucleic Acids Res.* **14**, 1479–93(1986).
**248.** M. T. Troll, D. Roitman, J. Conrad and B. H. Zimm. *Macromolecules* **19**, 1186(1986).
**249.** B. Jayaram, K. Sharp and B. Honig. *Biopolymers*, **28**, 975–93(1989).
**250.** W. K. Lee, Y. Gao and E. W. Prohfsky. *Biopolymers* **23**, 257–70(1984).
**251.** L. Nordenskiold, D. K. Chang, C. F. Anderson and M. T. Record, Jr. *Biochemistry* **23**, 4309–17(1984).
**252.** S. Forsén, T. Drakenberg and H. Wennerstrom. *Quart. Rev. Biophys.* **19**, 83–114(1987).
**253.** C. F. Anderson, M. T. Record, Jr. and D. A. Hart. *Biophysical Chemistry* **7**, 301–6(1978).
**254.** (*a*) M. L. Bleam, C. F. Anderson and M. T. Record, Jr. *Proc. Nat. Acad. Sci.* (*USA*) **77**, 3085–9(1980). (*b*) M. L. Bleam, C. F. Anderson and M. T. Record. *Biochemistry* **22**, 5418–25(1983).
**255.** M. R. Reddy, P. J. Rossky and C. S. Murthy. *J. Phys. Chem.*, **91**, 4923–33(1987).
**256.** W. K. Olson, M. H. Sarma, R. H. Sarma and M. Sundaralingam (eds). *Structure and Expression*, Vol. 3. Adenine Press: New York.
**257.** H.-M. Wu and D. M. Crothers. *Nature* **308**, 509–13(1984).
**258.** H.-S. Koo, H.-M. Wu and D. M. Crothers. *Nature* **320**, 501–6(1986).
**259.** P. J. Hagerman. *Nature* **321**, 449–50(1986).
**260.** A. M. Burhoff and T. D. Tullius. *Nature* **331**, 455–6(1988).
**261.** (*a*) M. L. Kopka, C. Yoon, D. Goodsell, P. Pjura and R. E. Dickerson. *Proc. Natl. Acad. Sci. USA* **82**, 1376–80(1985); (*b*) M. L. Kopka, C. Yoon, D. Goodsell, P. Pjura and R. E. Dickerson. *J. Mol. Biol.* **183**, 553–63(1985).
**262.** (*a*) L. A. Marky, J. Curry and K. J. Breslauer. In *Molecular Basis of Cancer, Part B* (ed. R. Rein), Vol. 172. Alan R. Liss: New York, 1985, pp. 155–73; (*b*) K. J. Breslauer, D. P. Rementa, W.-Y. Chou, R. Ferrante, J. Curry, D. Zaunczkowski, J. G. Snyder and L. A. Marky. *Proc. Natl. Acad. Sci. USA* **84**, 8922–6(1987).
**263.** H. C. M. Nelson, J. T. Finch, B. F. Luisi and A. Klug. *Nature* **330**, 221–6(1987).
**264.** M. Coll, C. A. Frederick, A. H.-J. Wang and A. Rich. *Proc. Natl. Acad. Sci. USA* **84**, 8385–9(1987).
**265.** C. Yoon, G. G. Privé, D. S. Goodsell and R. E. Dickerson. *Proc. Natl. Acad. Sci. USA* **85**, 6332–6(1988).
**266.** D. G. Alexeev, A. A. Lipanov and I. Ya. Skuratovskii. *Nature* **325**, 821–3(1987).
**267.** A. C. Dock-Bregeon, B. Chevrier, A. Podjarny, D. Moras, J. S. deBear, G. R. Gough, P. T. Gilham and J. E. Johnson. *Nature* **335**, 375–8(1988).
**268.** M. A. Viswamitra, Z. Shakked, P. G. Jones, G. M. Sheldrick, S. A. Salisbury and O. Kennard. *Biopolymers* **21**, 513–33(1982).
**269.** C. J. Alden and S. H. Kim. *J. Mol. Biol.* **132**, 411–34(1979).
**270.** (*a*) P. K. Ponnuswamy and P. Thiyagarajan. *Biopolymers* **17**, 2503–18(1978); (*b*) P. Thiyagarajan and P. K. Ponnuswamy. *Biopolymers* **18**, 2233–47(1979); (*c*) P. K. Ponnuswamy and A. Anukanth. *Current Trends in Life Sciences* **XII**, 31–7(1984).

**271.** J. A. Rupley, E. Gratton, and G. Careri. *TIBS* 18–22(1982).
**272.** G. Malenkov, L. Minchenkova, E. Minyat, A. Schyolkina and V. I. Ivanov. *FEBS Letters* **51**, 38–42(1975).
**273.** V. I. Ivanov, L. E. Minchenkova, E. E. Minyat, M. D. Frank-Kamenetskii and A. K. Schyolkina. *J. Mol. Biol.* **87**, 817–33(1974).
**274.** H. W. Hoyer, M. Chow and V. Gary. *J. Colloid and Interface Sc.* **80**, 132–5(1981).
**275.** C. R. Calladine and H. R. Drew. *J. Mol. Biol.* **178**, 773–82(1984).
**276.** A. Rich, A. Nordheim and A. H.-J. Wang. *Annu. Rev. Biochem.* **53**, 791–846(1984).
**277.** (*a*) T. M. Jovin, D. M. Soumpasis and L. P. McIntosh. *Annu. Rev. Phys. Chem.* **38**, 521–60(1987); (*b*) D. M. Soumpasis and T. M. Jovin. In *Nucleic Acids and Molecular Biology* (eds F. Eckstein and D. M. J. Lilley). Springer-Verlag: Berlin, 1987, pp. 85–111.
**278.** M. J. Behe, S. C. Szu, E. Charney and G. Felsenfeld. *Biopolymers* **24**, 289–300(1985).
**279.** J. B. Matthew and F. M. Richards. *Biopolymers* **23**, 2743–59(1984).
**280.** P. S. Ho, G. J. Quigley, R. F. Tilton Jr. and A. Rich. *J. Phys. Chem.* **92**, 939–45(1988).
**281.** D. Eisenberg and A. D. McLachlan. *Nature* **319**, 199–203(1986).
**282.** M. J. Behe and G. Felsenfeld. *Proc. Natl. Acad. Sci. USA* **78**, 1619–32(1981).
**283.** F. M. Pohl and T. M. Jovin. *J. Mol. Biol.* **57**, 375–95(1972).
**284.** E. Westhof. Unpublished results.
**285.** M. Frank-Kamenetskii. *Nature* **328**, 108(1987).
**286.** J. Texter. *Prog. Biophys. Molec. Biol.* **33**, 83–97(1978).
**287.** (*a*) V. A. Buckin. *Mol. Biol. (USSR)* **21**, 512–25(1987); (*b*) V. A. Buckin. *Biophys. Chem.* **29**, 283–92(1988).
**288.** T. E. Cross and R. Pethig. *Int. J. Quant. Chem: QBS* **10**, 143–52(1983).
**289.** L. J. Schreiner, M. M. Pintar, A. J. Dianoux, F. Volino and A. Rupprecht. *Biophys. J.* **53**, 119–22(1988).
**290.** K. R. Foster, M. A. Stuchly, A. Kraszewski and S. S. Stuchly. *Biopolymers* **23**, 593–9(1984).
**291.** G. M. Mrevlishvili. *Biofizika* **12**, 180–91(1977).
**292.** I. D. Kuntz, T. S. Brassfield, G. D. Law and G. V. Purcell. *Science* **163**, 1329–31(1969).
**293.** N. J. Tao, S. M. Lindsay and A. Rupprecht. *Biopolymers* **26**, 171–88(1987).
**294.** A. Bonicontro, A. Di Biasio and F. Pedone. *Biopolymers* **25**, 241–7(1986).
**295.** N. J. Tao, S. M. Lindsay and A. Rupprecht. *Biopolymers* **27**, 1655–71(1988).
**296.** Y. Tominaga, M. Shida, K. Kubota, H. Urabe, Y. Nishimura and M. Tsuboi. *J. Chem. Phys.* **83**, 5972–5(1985).
**297.** (*a*) J. M. Eyster and E. W. Prohofsky. *Biopolymers* **16**, 965–82(1977); (*b*) K. V. Devi Prasad and E. W. Prohofsky. *J. Biomol. Struct. Dyn.* **3**, 551–8(1985).
**298.** A. Wittlin, L. Genzel, F. Kremer, S. Häseler, A. Poglitsch and A. Rupprecht. *Phys. Rev. A* **34**, 493–500(1986).
**299.** S. M. Lindsay, S. A. Lee, J. W. Powell, T. Weidlich, C. Demarco, G. D. Lewen, N. J. Tao and A. Rupprecht. *Biopolymers* **27**, 1015–43(1988).
**300.** J. Maddox. *Nature* **324**, 11(1986).

**301.** G. C. Edwards, C. C. Davis, J. D. Saffer and M. L. Swicord. *Phys. Rev. Lett.* **53**, 1284–7(1984).
**302.** C. Gabriel, E. H. Grant, R. Tata, P. R. Brown, B. Gestblom and E. Noreland. *Nature* 328, 145–6(1987).
**303.** K. R. Foster, B. R. Epstein and M. A. Gealt. *Biophys. J.* **52**, 421–5(1987).
**304.** (*a*) L. L. van Zandt. *Phys. Rev. Lett.* **57**, 2085–7(1986); (*b*) M. E. Davis and L. L. van Zandt. *Phys. Rev. A* **37**, 888–901(1988).
**305.** H. Shindo, J. B. Wooten, B. H. Pheiffer and S. B. Zimmerman. *Biochemistry* **19**, 518–26(1980).
**306.** S. J. Opella, W. B. Wise and J. A. DiVerdi. *Biochemistry* **20**, 284–90(1981).
**307.** B. T. Hall, W. P. Rotwell, J. S. Waugh and A. Rupprecht. *Biochemistry* **20**, 1881–7(1981).
**308.** M. T. Mai, D. E. Wemmer and O. Jardetzky. *J. Am. Chem. Soc.* **105**, 7149–52(1983).
**309.** R. Brandes, R. R. Vogt, R. L. Vold and D. R. Kearns. *Biochemistry* **25**, 7744–51(1986).
**310.** J. M. Sturtevant. *Proc. Natl. Acad. Sci. USA* **74**, 2236–40(1977).
**311.** A. McPherson, G. D. Brayer, and R. D. Morrison. *J. Mol. Biol.* **189**, 305–27(1986).
**312.** D. Ringe and G. A. Petsko. *Prog. Biophys. Molec. Biol.* **45**, 197–235(1985).
**313.** A. Wlodawer, M. Miller and L. Sjolin. *Proc. Natl. Acad. Sci. USA* **80**, 3628–31(1983).
**314.** R. J. Read and M. N. G. James. In *Proteinase Inhibitors* (eds A. J. Barrett & J. Salvesen). Elsevier: Amsterdam, 1986, pp. 301–36.
**315.** K. Suguna, E. A. Padlan, C. W. Smith, W. D. Carlson and D. R. Davies. *Proc. Natl. Acad. Sci. USA* **84**, 7009–13(1987).
**316.** (*a*) M. Ptashne. *Nature* **322**, 607–700(1986); (*b*) J. E. Anderson, M. Ptashne and S. C. Harrison. *Nature* **326**, 846–50(1987); (*c*) G. B. Koudelka, S. C. Harrison and M. Ptashne. *Nature* **326**, 886–9(1987).
**317.** M. Robertson. *Nature* **327**, 464–6(1987).
**318.** K. Struhl. *Cell* **49**, 295–7(1987).
**319.** C. C. F. Blake, W. C. A. Pulford and P. J. Artymiuk. *J. Mol. Biol.* **167**, 693–723(1983).
**320.** M. N. G. James, A. R. Sielecki, G. D. Brayer, L. T. J. Delbaere and C. A. Bauer. *J. Mol. Biol.* **144**, 43–88(1980).
**321.** A. T. Brunger, C. L. Brooks and M. Karplus. *Proc. Natl. Acad. Sci. USA* **82**, 8458–62(1986).
**322.** J. Warwicker and H. C. Watson. *J. Mol. Biol.* **157**, 671–9(1982).
**323.** R. J. Zauhar and R. S. Morgan. *J. Mol. Biol.* **186**, 815–20(1985).
**324.** K. Zakrzewska, R. Lavery and B. Pullman. *Nucl. Acids Res.* **12**, 6559–74(1984).
**325.** D. Eisenberg and W. Kauzmann. *Structure and Properties of Water*, Oxford University Press, 1959.
**326.** (*a*) S. Arnott and D. W. L. Hukins. *Biochem. Biophys. Res. Commun.* **47**, 1504–10(1972); (*b*) S. Arnott, D. W. L. Hukins and S. D. Dover. *ibid.* **48**, 1392–9(1972); (*c*) S. Arnott and E. Selsing. *J. Mol. Biol.* **98**, 265–9(1975).
**327.** T. J. Richmond. *J. Mol. Biol.* **178**, 63–89(1984).
**328.** M. Sundaralingam. *Biopolymers* **7**, 821–60(1969).

**329.** C. Altona and M. Sundaralingam. *J. Amer. Chem. Soc.* **94**, 8205–12(1972).
**330.** C. C. Wilson. *Nucleic Acid. Res.* **15**, 8577–91(1987).
**331.** R. C. Hopkins. *Science* **211**, 289–91(1981).
**332.** D. R. Davies and R. L. Baldwin. *J. Mol. Biol.* **6**, 251–5(1963).
**333.** A. Mahendrasingam, V. T. Forsyth, R. Hussain, R. J. Greenall, W. J. Pigram and W. Fuller. *Science* **233**, 195–7(1986).
**334.** K. Hoogsteen. *Acta Crystallogr.* **16**, 907–16(1963).
**335.** W. H. van Gunsteren, H. J. C. Berendsen, J. Hermans, W. G. J. Hol and J. P. M. Postma. *Proc. Natl. Acad. Sci. USA* **80**, 4315–19(1983).
**336.** (*a*) P. K. Weiner and P. A. Kollman. *J. Comput. Chem.* **2**, 287–97(1981); (*b*) S. J. Weiner, P. A. Kollman, D. T. Nguyen and D. A. Case. *J. Comput. Chem.* **7**, 230–8(1986).
**337.** B. R. Brooks, R. E. Bruccoleri, B. D. Olafson, D. J. States, S. Swaminathan and M. Karplus. *J. Comput. Chem.* **4**, 187–95(1983).
**338.** V. I. Poltev, A. V. Teplukin and V. P. Chuprina. *J. Biomol. Struct. Dyns.* **6**, 575–86(1988).
**339.** V. T. Forsyth, A. Mahendrasingam, W. J. Pigram, R. J. Greenall, K. Bellamy, S. A. Mason and W. Fuller. Submitted for publication.
**340.** H. Edelhoch and J. C. Osborne, Jr. *Adv. Protein Chem.* **30**, 183–250(1976).
**341.** W. C. Johnson, Jr. and J. D. Girod. *Biochim. Biophys. Acta* **353**, 193–9(1974).
**342.** D. C. Rau, B. Lee and V. A. Parsegian. *Proc. Natl. Acad. Sci. USA* **81**, 2621–5(1984).
**343.** J. N. Israelachvili and P. M. McGuiggan. *Science* **241**, 795–800(1988).
**344.** E. P. Geiduschek and I. Gray. *J. Amer. Chem. Soc.* **78**, 535–56(1956).
**345.** E. Westhof. In *Water and Ions in Biomolecular Systems*. Birkhäuser Verlag: Basel, 1990, pp. 11–18.
**346.** M. H. Moore, W. N. Hunter, B. Langlois d'Estaintot and O. Kennard. *J. Mol. Biol.* **206**, 693–705(1989).
**347.** S. Jain, G. Zon and M. Sundaralingam. *Biochemistry* **28**, 2360–4(1989).
**348.** F. Eisenhaber, V. G. Tumanyan, F. Eisenmenger and W. Gunia. *Biopolymers* **28**, 741–61(1989).
**349.** J. Conrad, M. Troll and B. H. Zimm. *Biopolymers* **27**, 1711–32(1989).
**350.** (*a*) A. Ben-Naim, K. L. Ting and R. L. Jernigan. *Biopolymers* **28**, 1309–25(1989); (*b*) *ibid.* **28**, 1327–37(1989).
**351.** F. Hirata and R. M. Levy. *J. Phys. Chem.* **93**, 479–84(1989).
**352.** O. P. Lamba, A. H. J. Wang and G. J. Thomas Jr. *Biopolymers* **28**, 667–78(1989).
**353.** N. Tao, S. M. Lindsay and A. Rupprecht. *Biopolymers* **28**, 1019–30(1989).
**354.** A. Anagnostopoulou-Konsta, D. Daouki-Diamanti, P. Pissis and E. Sideris. In *Proceedings of the 6th International Symposium on Electrets* (eds. D. K. Das-Gupta and A. W. Pattullo). Oxford University Press: Oxford, 1988, pp. 271–75.
**355.** (*a*) V. A. Buckin, B. I. Kankiya, N. V. Bulichov, A. V. Lebedev, I. Ya. Gukovsky, V. P. Chuprina, A. P. Sarvazyan and A. R. Williams. *Nature* **340**, 321–2(1989); (*b*) V. A. Buckin, B. I. Kankiya, A. P. Sarvazyan and H. Uedaira. *Nucleic Acids Res.* **17**, 4189–203(1989).
**356.** A. Kintanar, W. C. Huang, D. C. Schindele, D. E. Wemmer and G. Drobny. *Biochemistry* **28**, 282–93(1989).

**357.** D. E. Kearns. *Methods Stereochem. Anal.* **9**, 301–47(1987).

**358.** K. Tomita, T. Hakoshima, K. Inubushi, S. Kunisawa, H. Ohishi, G. A. van der Marel, J. H. van Boom, A. H.-J. Wang and A. Rich. *J. Mol. Graphics* **7**, 71–5(1989).

**359.** G. Zhou and P. S. Ho. ACA Abstracts PA38, Seattle July 23–9, (1989).

**360.** G. G. Privé and R. E. Dickerson. ACA Abstracts HB1, Seattle July 23–9, (1989).

**361.** A. C. Dock-Bregeon, B. Chevrier, A. Podjarny, J. Johnson, J. S. de Bear, G. R. Gough, P. T. Gilham and D. Moras. *J. Molec. Biol.* **209**, 459–74 (1989).

# The solvation of amino acids and small peptides

T. H. LILLEY

*Chemistry Department, The University, Sheffield S3 7HF*

If I were called in
To construct a religion
I should make use of water. Philip Larkin, 1954

## 1. Introduction

The naturally occurring amino acids exist in living organisms in both their free forms and in condensed forms as peptides and proteins. From a structural viewpoint there is a marked contrast in the range of types found for the free amino acids, of which some hundreds have been identified, and the rather limited, although still structurally diverse group of amino acid residues found in the peptides and proteins of higher living creatures. The diversity of structural types found in nature is further widened if lower organisms such as fungi and bacteria are considered and in such organisms peptides are found which have amino acid residues containing as examples, $\alpha,\beta$ unsaturated residues, cyclic structures of various types and unusual chiral centres. [1] Almost all of the work of a physico-chemical nature which has been reported in the literature has, however, been directed towards studies on the 20 primary, protein amino acids (actually there are only 19 amino and 1 imino acid – proline). In some ways this is unfortunate because of the structural varieties which exist but given the propensity of most physical chemists to avoid synthesis or natural product isolation, is not too surprising. Consequently, however, most of the information presented and discussed here is directed towards the amino acids which are present in the proteins of higher organisms. For the convenience of the reader the names, structural formulae and abbreviations for these amino acids (and their residues in peptides and proteins) are given in Table 1.

The point referred to above regarding the diversity of molecular structure is clear from this table. Superimposed upon this is that, with the exception of glycine, all of these amino acids contain a chiral centre at the $\alpha$ carbon atom and two of them, isoleucine and threonine have $\beta$ carbons which are also asymmetric centres. All of the acids apart from cysteine are of the (S)-configuration but almost universally they are still referred to as the L-$\alpha$-amino acids. An interesting point which has not attracted much attention is

Table 1. *Names, structural formulae and abbreviations of the primary protein amino acids*

| Name | $R{-}CH(NH_2)CO_2H$<br>R | IUPAC—IUB abbreviation<br>Three letter | <br>One letter |
|---|---|---|---|
| Alanine | $CH_3$ | ala | A |
| Arginine | NH<br>$\backslash\backslash$C—NH—$(CH_2)_3$<br>/<br>$NH_2$ | arg | R |
| Asparagine | $H_2N{-}CO{-}CH_2$ | asn | N |
| Aspartic acid | $HO{-}CO{-}CH_2$ | asp | D |
| Cysteine | $HS{-}CH_2$ | cys | C |
| Glutamic acid | $HO{-}CO{-}(CH_2)_2$ | glu | E |
| Glutamine | $H_2N{-}CO{-}(CH_2)_2$ | gln | Q |
| Glycine | H | gly | G |
| Histidine | N—— $CH_2$<br>NH | his | H |
| Isoleucine | $CH_3CH_2CH(CH_3)$ | ile | I |
| Leucine | $(CH_3)_2CHCH_2$ | leu | L |
| Lysine | $H_2N{-}(CH_2)_4$ | lys | K |
| Methionine | $CH_3{-}S{-}(CH_2)_2$ | met | M |
| Phenylalanine | $C_6H_5{-}CH_2$ | phe | F |
| Proline[a] | HN—— $CO_2H$ | pro | P |
| Serine | $HO{-}CH_2$ | ser | S |
| Threonine | $CH_3CH(OH)$ | thr | T |
| Tryptophan | $CH_2$<br>N<br>H | trp | W |
| Tyrosine | $C_6H_4(OH){-}CH_2$ | tyr | Y |
| Valine | $(CH_3)_2CH$ | val | V |

[a] Imino acid – the complete structure is given.

that although proteins as they are formed at the ribosome contain only L-residues, some racemisation does occur to the D-forms. The process is sufficiently slow so that it can be used to date on a long time scale the ages of proteins. [2]

A further feature not evident from the structures written in Table 1 is that a tautomerisation equilibrium exists in any phase between the uncharged structures and zwitterionic forms resulting from intramolecular proton transfer

$$R{-}CH(NH_2)CO_2H \rightleftharpoons R{-}CH(NH_3^+)CO_2^-$$

The position of equilibrium for this transfer depends markedly on the environment [**3–5**] in which the molecule finds itself and some estimates for glycine in the gas phase and in water serve to illustrate how the medium can moderate solute properties. The equilibrium constant for the tautomerisation in the gas phase has been estimated to be approximately $10^{-21}$ [**3**] whereas in water at ambient temperature the estimate is about $4 \times 10^{6}$. [**5**] In other words in the gas phase the equilibirum is to all intents completely in favour of the neutral molecule whereas, in contrast, the zwitterionic species is greatly dominant in water. This is rather an extreme example but it does indicate that deductions about the behaviour of solutes in a solvent from calculated or indeed experimental behaviour in the gas phase need to be treated with some caution.

The predominant presence of the zwitterionic form for both amino acids and peptides in water has been known for many years [**9**] and was established by consideration of their large dipole moments. It must be stressed, however, that the dipole moments of solutes in condensed media are by no means easy to establish without ambiguity. They are deduced from measurements of the dielectric properties of solutions and in particular from the concentration variation of the relative permittivity the dielectric increment. This is then transposed using theories such as those of Onsager [**10**], Buckingham [**11**] or Böttcher [**12**] to give the dipole moment. However, each of these theoretical approaches uses a model in which the solution is assumed to be composed of a dilute dipolar solute immersed in a dielectric continuum. This is, of course, incorrect at the molecular level and one consequence is that, although there is general agreement that the dielectric increments of amino acids and peptides are large, there are considerable variations in the reported values of dipole moments and these arise mainly from their model dependence. [**13–17**]

There are strong indications from dielectric measurements [**18, 19**] on methyl substituted amino acids in different solvents that the position of the tautomerisation equilibrium depends largely if not entirely on the permittivity of the solvent – in other words is principally determined by the free energy loss to the dielectric rather than by specific effects, so that in solvents of low permittivity the uncharged form tends to be the more preferred. This shift in the tautomerisation equilibria almost certainly will have implications for the transfer properties of amino acids which are discussed later.

Investigations, both theoretical and experimental, into the properties of solutes in solution fall into two distinct but nevertheless related categories. There are those studies which are directed towards explaining, or more frequently rationalising, the behaviour of systems in which the solute is 'infinitely dilute' in the solvent medium and there are those investigations which are concerned with interactions occurring between solvated solute molecules in the chosen solvent medium. From both biochemical and chemical viewpoints this second type of investigation is the more important since the major problems which need explanation arise from interactions and reactions between solvated molecules. However, since one must be concerned

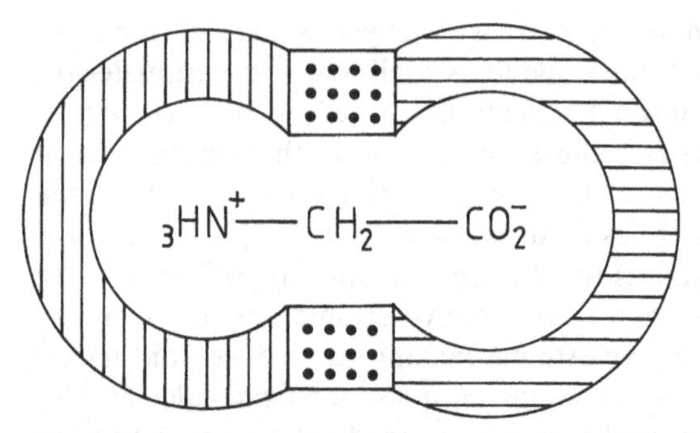


Figure 1. Pictorial representation of the solvation of glycine in water. The solvation regions of the amino, methylene and carboxylate groups are represented by vertical hatching, dots and horizontal hatching respectively.

with events which occur between solvated species, an adequate description of solvation and particularly hydration is a necessary prerequisite for the delineation of interactive propensities and reactivities.

Before beginning our discussion of the solvation properties of the amino acids it is worth stressing yet again that the molecules with which we are concerned are polyfunctional. The necessary consequence of this is that in any solvent medium their solvation regions will be heterogeneous. If we consider, for example, glycine dissolved in and surrounded by water then the molecular situation which seems to prevail can be represented diagrammatically as shown in Figure 1. Intuitively one would expect there to be three principal sub-levels of solvational organisation; one associated with the carboxylate group, another associated with the amino group and a third associated with the apolar methylene group. The possibility, or rather as will be shown later, the probability also exists that these three regions will also exhibit mutual perturbations. Clearly if one is dealing with molecules with more complex structures, such as, for example, tryptophan, then other types of solvation will need to be considered. Three other points should also be considered which are not evident from pictorial representations such as that shown in Figure 1. Firstly, irrespective of the solute or solvent, one is dealing with a dynamic rather than a static situation and there must be continuous interchange, on a relatively short time scale, between the solvent in its various solvational microenvironments and the 'bulk' solvent. [6] Secondly, the solute and the solvent peripheral to it will be undergoing various molecular motions (rotations, vibrations and librations) all of which will moderate experimentally observed properties. Thirdly, although the perturbations in the solvent on forming the solvation region are usually the focus of attention, there must necessarily also be some complementary effect of the solvent on the intramolecular events occurring within the solute. An extreme example of this sort was mentioned above for tautomerisation equilibria but is also seen, for example, in the temperature and solvent dependence of *cis–trans* equilibria of prolyl residues. [7, 8]

There are several reasons for investigating the behaviour of amino acids but it is undoubtedly true that the most common justification for such studies is to acquire information which could be assembled to give the properties of poly- and oligopeptides. The goal here is certainly worthwhile and its achievement is in many ways essential if any predictions on the properties of even the denatured (i.e., non-assembled, biologically inactive) forms of proteins are to be made.

A major problem in attempting to obtain molecular rationalisations of the properties of amino acids and small peptides arises not only from their structural diversity but also from the fact that in condensed media they are zwitterionic in nature. The fact that one has a distinct charge distribution within a given molecule in itself must contribute in a significant way to measured properties. The literature in this area abounds with articles which invoke what might be called 'solvent structural' explanations of the properties of amino acids and indeed many other groups of compounds. Now it is undoubtedly true that such effects must play a role in the properties of solutes in solution and particularly in aqueous solutions but the indications are that some authors are of the view that these are the only physico-chemical effects which ever need to be considered. Two examples using information from the recent literature, one from information on amino acids and the other from information on some related amides, will be used a little later to illustrate that considerable progress can be made without invoking solvent structural considerations.

## 2. Theoretical studies

Theoretical work on the solvation of amino acids and peptides falls into two broad categories. There are those studies which address either in an empirical or *ab initio* quantum mechanical way the energy of a relatively small group of molecules e.g., one amino acid species plus one water molecule and there are those studies of the computer simulation type in which the number of molecules of the solvent is relatively large but empirical potentials, based in part at least on quantum mechanical information, are used.

The quantum mechanical investigations although in principle important and necessary if significant progress on molecular understanding is to be made, do at the moment have difficulties associated with them. The major problem arises from the level at which the calculation is performed. The 'best' results are obtained by using very extensive basis sets but unfortunately the price to be paid is that the more extensive the basic set the longer and hence more expensive is any computation. Consequently for molecules of any complexity (and complex here means relatively few atoms) information is only available from either semi-empirical calculations or calculations using small basis sets and in which electron correlation effects are neglected. As a consequence the energies of interaction will be incorrect to an undetermined extent. This problem is compounded also by the impossibility, at present at

Table 2. *The binding energies of single water molecules to glycine at the carboxylate and amino ends of the molecule. The hydrogen-bond lengths were fixed at* 0.275 *nm*

| Binding of water to | Binding energy (kJ mol$^{-1}$) Semi-empirical[a] | *Ab initio*[b] |
|---|---|---|
| Carboxylate | −62 | −74 |
| Amino | −47 | −87 |
| Bridging | −66 | −85 |
| Water | | −49 |

[a] CNDO 2; [b] LCAO–MO–SCF (minimal Gaussian basis set)

least, of obtaining suitable experimental information (see later) which might be used to guide modification to theoretical procedures.

To illustrate the difficulties involved some results **[20]** are presented in Table 2 of the binding energies of glycine with single water molecules at different locations about the amino acid. In each instance the hydrogen-bond length was fixed at 0.275 nm and only the most stable of the various water–glycine positions are given. It is apparent from this that although the semi-empirical and *ab initio* calculations give results of the same order of magnitude, there are very large differences observed. (It is perhaps worth mentioning here that the term *ab initio* is sometimes misunderstood. It does not mean that answers derived are exact but rather that within the chosen model no quantity is parameterised. It necessitates the computation of all of the many-centre integrals and in this regard differs from semi-empirical approaches in which approximations at various levels of sophistication are used to reduce the computational time needed. **[21]**)

Almost all of the calculations which have any significant degree of quantum mechanical sophistication have been directed towards glycine and even for this, prototypical species such as the formate ion to represent the carboxylate group and the ammonium ion to simulate the amino group have been used. **[22, 23]** A fairly extensive investigation using a minimal basis set has been performed **[24]** for both glycine and serine interacting with water in various locations about the amino acids and the interaction energies obtained. The results obtained from this study, at least as far as the global minimum energy is concerned, agree quite well with those of Kokpol *et al.* **[20]** but are considerably different to those of Pullman and coworkers. **[23]**

The general conclusion one can come to is that as far as the calculations on amino acids interacting with water are concerned, there are few surprises. The water does indeed interact relatively strongly when appropriately oriented and positioned near charged hydrogen-bonding donor or acceptor centres and interacts relatively weakly (i.e., weak compared to the water–water interaction) when near apolar moieties. It would seem that there are still considerable uncertainties about the exact energies associated with any interaction. It is important that such studies should be pursued and that workers should not only use large basis sets in their calculations but also need to take into account superposition errors [**25–8**] and polarisation effects. Both of these can make very significant contributions to interaction energies. It is obvious that the problems involved are formidable and to make matters worse the information which is needed, so that ultimately links can be made with experiment, is not only the global minimum energies but rather knowledge about the amino acid–water energy hypersurface. [**24**] It is only when this is available that the links between quantum mechanics to experimental properties via statistical mechanics will be made.

Exploratory investigations along these lines have been attempted not only for amino acids but also amidic species [**29–33**] where the focus has been the peptide group–water interaction. As an illustration of what has been done but more importantly to illustrate the way things will be done in the future we will consider the results of one fairly extensive study. [**34**] It should be stressed that at the moment, the details of the calculations and hence of the conclusions drawn must be incorrect since they are based on fairly approximate potential energy functions for both solute–solvent and solvent–solvent interactions. There are also the difficulties arising from relatively short (on the macroscopic time scale) sampling periods and approximations in the statistical mechanics.

The basis of the calculation is the simulation of the properties of solutions by the use of the Monte Carlo concept. In the particular example chosen the hydration of glycine was investigated by considering 1 zwitterionic glycine molecule and 215 water molecules in an ensemble at the appropriate density and 25 °C. It was assumed that the glycine–water parameterised potential function [**24, 35**] was appropriate and that the water–water interactions could be described using an earlier [**36, 37**] potential. The various potential functions were also assumed to be pairwise additive. The simulation which was proposed used a total of $1.5 \times 10^6$ configurations of the system but only the last million were used to obtain ensemble average quantities. For the system the partial molar internal energy of transfer of glycine from the gas phase was calculated to be $-326(\pm 54)$ kJ mol$^{-1}$. It is not possible to verify this experimentally for reasons which will be mentioned below and all one can say is that it is perhaps about what would be expected.

The other information which is obtained from simulation studies is of a structural nature and is expressed in terms of radial distribution functions which give a measure of the probability of finding one species in proximity to another. The integral of these using specified lower and upper distance limits gives the coordination number of say water molecules about either groups on the solute or the entire solute. For glycine the results obtained [**34**] are given in Table 3. One of the more interesting aspects of the calculation

Table 3. *Coordination numbers and energies of interaction with water of groups on glycine [34]*

| Group | Coordination number | Mean energy of interaction ($kJ\ mol^{-1}$) |
|---|---|---|
| $NH_3^+$ | 3.2 | −58 |
| $CO_2^-$ | 5.1 | −47 |
| $CH_2$ | 6.1 | −3 |

Scheme 1

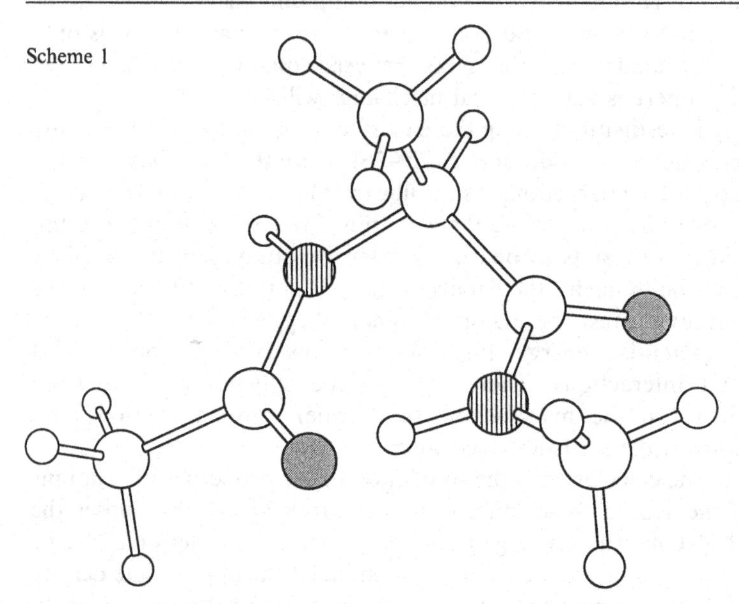

is the way it illustrates that orientations, deduced from quantum mechanics to be energetically very stable such as bridging structures between say $NH_3^+$, $H_2O$ and $CO_2^-$, occur very infrequently when the 'solvent' is present in great excess. There also are indications that the solvation of the methylene group is perturbed to a considerable extent by the presence of the peripheral ionic groups. Earlier studies **[29, 38]** using the same ideas and approximations indicated that the coordination number of each methyne fragment was about 5 which leads to a predicted value of roughly 10 for the methylene group in glycine and this contrasts with the value found **[34]** of about 6.

Investigations have been presented **[31–3]** on the energetics and dynamics of the uncharged *N*-acetyl-*N*-methylalaninamide, ALAMe,

$CH_3CONHCH(CH_3)CONHCH_3$

in simulated **[39]** water. The most comprehensive part of the overall study involved the hydration of ALAMe which was constrained to be in the $C_7$ configuration (see Scheme 1).

As well as getting information on the coordination of the molecule it was concluded that the motions of water in proximity to the various apolar methyl groups are considerably retarded both rotationally and translationally relative to the bulk solvent. In contrast the water near the polar residues behaves much more like the bulk water which is probably simply indicative of the gross similarity of the hydrogen-bonding between waters or between water and carbonyl or amino groups. The enthalpy of solution of the solute was estimated to be about $-28$ kJ mol$^{-1}$ and from the calculations this arises because of the balance between an attractive interaction of approximately $-95$ kJ mol$^{-1}$ for primary peptide–water interactions and a decrease in the binding energy of water molecules. At the time the work was performed suitable experimental information was not available to compare with the net enthalpy of solution but some recent work **[40, 42]** can now be used to make comparison. As yet there is only information available on the primary amide of *N*-acetyl-alanine but the result for the enthalpy of sublimation **[40]** of this is 115($\pm$3) kJ mol$^{-1}$ and the enthalpy of solution from the solid into water is **[41, 42]** close to 5 kJ mol$^{-1}$. Hence the molar enthalpy change for transferring the compound, $CH_3CONHCH(CH_3)CONH_2$, from the gas phase to water is about $-110$ kJ mol$^{-1}$. As will be shown below the enthalpy of transfer of a methylene group from the vapour into water is exothermic to the extent of about $-5$ kJ mol$^{-1}$ and so it is expected that the enthalpy of solvation of the alanyl 'peptide' studied by simulations will be of the order of $-115$ kJ mol$^{-1}$. This is in very marked contrast to the calculated value given above and indicates the still fairly large gulf which exists between experiment and calculation. Other examples like this on simple amides **[30]** serve to illustrate the same point.

## 3. Thermodynamic investigations

### 3.1. *Thermodynamic formalism*

A considerable amount of the information which is available on the amino acids is of a thermodynamic nature and so it is convenient to outline the formalism which is currently most commonly used. **[43, 44]**

If we consider a solution containing $n_s$ moles of a solute and $n_w$ moles of a solvent at constant temperature and pressure the Gibbs energy of the system is

$$G = n_w \mu_w + n_s \mu_s \tag{1}$$

where $\mu_i$ is the chemical potential of component $i$. The chemical potential of the solvent is customarily written in terms of the molar Gibbs energy of the pure solvent ($\mu_w^\circ$) and the solvent activity in the solution ($a_w$)

$$\mu_w = \mu_w^\circ + RT \ln a_w \tag{2}$$

but is also often expressed using the molal osmotic coefficient ($\phi$)

$$\mu_w = \mu_w^\circ - RT M_w m_s \phi \tag{3}$$

In this $m_s$ and $M_w$ are the solute molality and (kg) molecular weight of the solvent respectively. In view of the fact that when using the molal scale the quantity of solvent is fixed there is some utility in using what can be termed the partial specific Gibbs energy of the solvent ($G_w$) and this can then be written as

$$G_w = G_w^\circ - RTm_s\phi \tag{4}$$

where $G_w^\circ$ is the Gibbs energy of 1 kg of the pure solvent.

The chemical potential of the solute is usually expressed rather differently in that hypothetical standard states are almost invariable chosen and, for example, if the solution composition is represented by the molality of the solute then

$$\mu_s = \mu_s^\infty + RT\ln(a_s/m^\ominus) \tag{5}$$

In this $a_s$ is the activity of the solute and $m^\ominus$ is the standard state molality (this is invariably chosen to take the value of 1 mol kg$^{-1}$). Alternatively Equation (5) can be written in terms of the molality and and activity coefficient ($\gamma_s$)

$$\mu_s = \mu_s^\infty + RT\ln(m_s/m^\ominus) + RT\ln\gamma_s \tag{6}$$

The activity coefficient is defined in such a way that it approaches unity as the molality approaches zero.

When dealing with solutions at finite concentrations it is frequently very convenient to use apparent molar thermodynamic functions of the solute. In a formal sense these are defined as the difference between the solution property, $X$, say, and the property the solution would have if it was behaving ideally, $X^{\text{ideal}}$.

$$X_s^{\text{apparent}} = (X - X^{\text{ideal}})/m_s \tag{7}$$

Apparent molar functions are related to partial molar properties, $X_s$, by

$$X_s = X_s^{\text{apparent}} + \mathrm{d}X_w^{\text{ideal}}/\mathrm{d}m_s + \mathrm{d}X_s^{\text{apparent}}/\mathrm{d}\ln m_s \tag{8}$$

Relationships can also be made between these apparent molar properties and thermodynamic excess functions. These latter are of the general form

$$X^{\text{excess}} = X - X^{\text{ideal}} \tag{9}$$

and so we have

$$X_s^{\text{apparent}} = X_s^{\text{ideal}} + X^{\text{excess}}/m_s \tag{10}$$

and

$$X_s = X_s^{\text{ideal}} + \mathrm{d}X_s^{\text{ideal}}/\mathrm{d}\ln m_s + \mathrm{d}X_w^{\text{ideal}}/\mathrm{d}m_s + \mathrm{d}X^{\text{excess}}/\mathrm{d}m_s \tag{11}$$

In experimental practice the apparent molar free energy is seldom if ever used but the corresponding enthalpies and volumes (and derivatives of these) are

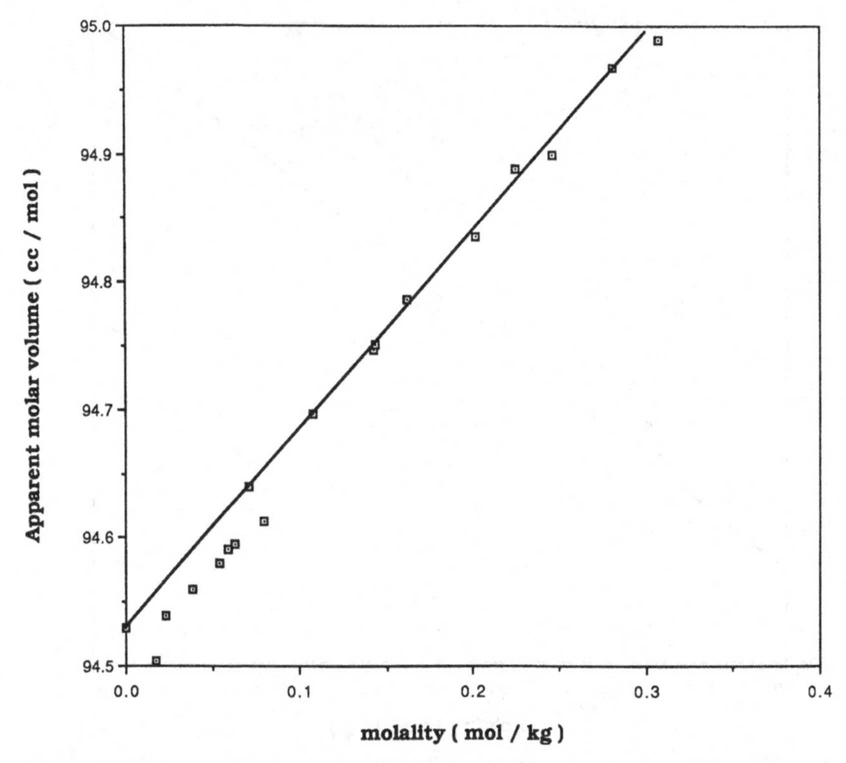


Figure 2. The apparent molar volume of alanylglycine as a function of molality in water at 25 °C. The line was drawn from a weighted least-squares analysis of the results.

used very frequently. The apparent molar volume, usually but not invariably represented by the symbol $\phi_V$, and its temperature and pressure derivatives can be determined absolutely but the corresponding enthalpy, often termed the relative apparent molar enthalpy, $\phi_L$, can only be derived relative to its standard state value although its derivatives can be obtained absolutely.

There are good theoretical reasons **[45–8]** for expressing the excess functions for solutes which have no net charge as polynomials in the solute molality. For a system containing a single solute

$$X^{\text{excess}} = x_{\text{ss}} m_{\text{s}}^2 + x_{\text{sss}} m_{\text{s}}^3 + \ldots \tag{12}$$

where the $x$ terms represent interactions between the solvated solute species. Consequently by combining Equations (10) and (12) we can write for example the following expansions for the apparent molar volume, and heat capacity

$$V_{\text{s}}^{\text{apparent}} = V_{\text{s}}^{\infty} + v_{\text{ss}} m_{\text{s}} + \text{higher order terms} \tag{13}$$

$$C_{p,s}^{\text{apparent}} = C_{p,s}^{\infty} + c_{\text{ss}} m_{\text{s}} + \text{higher order terms} \tag{14}$$

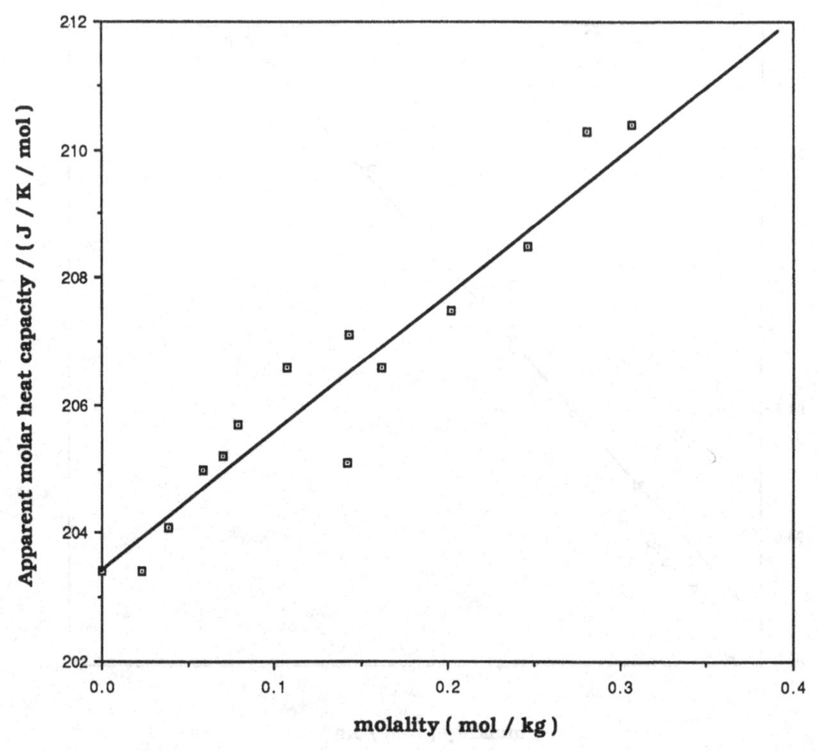


Figure 3. The apparent molar heat capacity of alanylglycine as a function of molality in water at 25 °C.

Expressions such as these are important in a solvation context since extrapolation of experimental data to infinite dilution is placed on a firm basis and it is the extrapolated information which relates to the solvation of the solute under consideration. Illustrations of the way in which expressions such as these can be used are given in figures 2 and 3 in which are plotted the apparent molar volumes and heat capacities in water of the dipeptide alanylglycine. [**49**] Incidentally among other things, some of which will be mentioned later, these figures show the sort of precision in apparent molar properties which can be achieved using modern instrumentation.

The standard state partial molar free energy cannot be obtained but the free energy difference between two phases can be deduced by, for example, solubility measurements. For equilibrium between say a pure substance and a solution of that substance in a particular solvent we have

$$\mu_A^{\text{pure}} = \mu_A^{\text{solution}} \tag{15}$$

and this can be written as

$$\mu_A^{\circ} = \mu_A^{\infty} + RT\ln(m_A^{\text{sat}}\gamma_A^{\text{sat}}/m^{\ominus}) \tag{16}$$

where the superscript stresses that saturation values are being considered. Activity coefficients for the solute can be obtained from independent measurements and a short extrapolation to the saturation molality allows the second term on the rhs of Equation (16) to be evaluated. Consequently the difference between the standard state chemical potentials for the solute in the solvent and in the pure substance can be obtained. If the same procedure is carried out using another solvent one can then obtain for example the difference

$$\mu_A^\infty \text{(in solvent 1)} - \mu_A^\infty \text{(in solvent 2)} \tag{17}$$

As a 'thought' experiment this corresponds to the chemical potential change on transferring one mole of the solute, at infinite dilution so that solute–solute interactions are absent, from solvent 2 to solvent 1. It is often colloquially termed the 'free energy of transfer' from solvent 2 to solvent 1.

Corresponding 'transfer functions' can be defined for other partial molar properties such as the enthalpy, heat capacity and volume.

### 3.2. *Partial molar properties*

Superficially the most direct method of assessing the solvational properties of the amino acids would be to determine experimentally the changes in the properties for the process

Amino acid in dilute gas ⟶ Amino acid infinitely dilute in a solvent

The free amino acids and peptides are, however, generally extremely involatile and, for example, the equilibrium vapour pressure of glycine above an aqueous 1 M solution at ambient temperatures would only be about $2 \times 10^{-8}$ Pa which corresponds to a molarity in the gas phase of less than $10^{-14}$ mol $l^{-1}$. Monitoring such pressures or concentrations with any precision is quite beyond current experimental procedures. In principle it would be possible to determine the changes for the above process by obtaining information on

Amino acid (solid) ⟶ Amino acid in dilute gas

and

Amino acid (solid) ⟶ Amino acid infinitely dilute in a solvent

The second of these can be determined but the sublimation process is experimentally inaccessible not only because of the involatility of the acids but for two further reasons. Firstly, the amino acids usually do not have sharp melting temperatures but rather decompose over a range of temperatures. This obviates the possibility of monitoring the vapour pressures of the solids at higher temperatures and extrapolating to obtain information pertinent to ambient temperatures. Secondly, as was mentioned earlier, the acids in the vapour phase and in solution exist in different tautomeric forms so even if it were possible to obtain measures of the changes arising during sublimation, a significant, but difficult to quantify contribution to the changes, would arise from the intramolecular proton transfer.

In view of these difficulties there have been two main ways in which experimental approaches to solvation have been addressed. One way has been to adopt reductionist approaches and attempt to explain the properties of the amino acids and peptides in terms of their constituent chemical groups. The other way has been to consider transfer properties of these compounds from one solvent to another. Both methods have their drawbacks and deficiencies.

### 3.3. *Chemical potentials*

It was mentioned earlier that chemical potentials cannot be determined absolutely but that differences are accessible via solubility and activity coefficient determinations. The thermodynamic formalism has been outlined above (see Equation (17)). In many of the studies which are reported in the literature the activity coefficients are not known and assumed to be unity and so the partial molar free energy of transfer is estimated using the ratio of the solubilities in the two solvents. A point which should be mentioned which is particularly important for the free energy functions and the entropy function is that the choice of the scale to represent solution composition must be considered. In what was written earlier the procedures used were demonstrated using a molality representation but in dealing with transfers from one solvent to another it is more customary for a Henry's Law standard state to be adopted. This means that solution compositions are expressed using mole fractions. However, the adoption of this scale is also arbitrary and indeed there do seem to be good reasons [**50**] generally for using neither the molality nor the mole fraction scales but rather the molarity or some other solute number density scale. However, for two given solvents the choice of standard states does not change the relative order of the free energies of transfer since any scales are related by a calculable factor, and so in a comparative sense the choice is unimportant.

The idea of studying amino acid solubilities and hence free energies of transfer, in different solvents has been used by several groups of investigators. Most of the impetus for these studies has come not so much from interest in the solvation of the acids as such but rather as a means of gaining insight into the propensity for amino acid residues to be 'buried' in folded globular proteins. To this end a variety of solvent systems has been adopted to simulate the interior environment of proteins. From a strictly hydration viewpoint the free energies of transfer from an 'inert' solvent to water would appear to be the most satisfactory. 'Inert' is, of course, a relative term but of the solvents which have been used hexane seems to be the one which is

Table 4. *Partial molar free energies of transfer of amino acids from hexane to water at 25 °C. The values are given relative to glycine*

| Amino acid | $\Delta_{\mathrm{trans}}\mu^{\infty}$ (kJ mol$^{-1}$) |
|---|---|
| ala | −1.76 |
| arg | −6.02 |
| asp | 5.19 |
| glu | 4.10 |
| gly | 0 |
| his | 2.4 |
| ile | 11.09 |
| leu | 11.30 |
| lys | 1.05 |
| phe | 10.42 |
| pro | 1.05 |
| ser | −3.22 |
| thr | −0.17 |
| val | 7.03 |

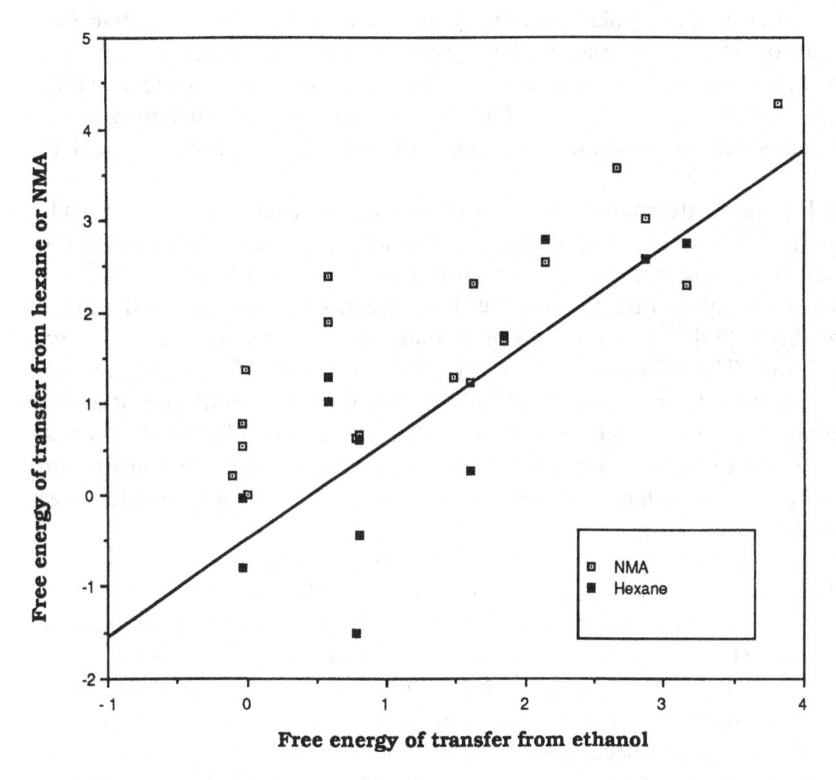


Figure 4. Partial molar free energies of transfer (in kcal mol$^{-1}$) of amino acids from three solvent systems to water.

Table 5. *Ranking orders of free energies of transfer of amino acids containing apolar side chains. The transfers are from the solvent specified to water*

| | Solvent | | |
|---|---|---|---|
| Amino acid | Hexane | Ethanol | *N*-Methylacetamide |
| ala | 1 | 2 | 2 |
| gly | 2 | 1 | 1 |
| ile | 5 | 6 | 4 |
| leu | 6 | 4 | 5 |
| phe | 4 | 5 | 6 |
| val | 3 | 3 | 3 |

most suitable and in Table 4 the free energies of transfer of amino acids from hexane to water are presented. Broadly speaking what is apparent from this is that the residues which are apolar in character exhibit positive free energies of transfer i.e., are apparently more comfortable in the apolar solvent and those which contain polar portions have a tendency to have negative free energies of transfer. Neither of these gross features is unexpected. However, there are some individual deviations from this overall behaviour which clearly indicate that specific, residue structural effects are contributing to the side chains' relative solvation properties. It is difficult to be more precise than this.

In Figure 4 information on the free energies of transfer of amino acids from three solvent systems, *N*-methylacetamide, **[52]** hexane **[51]** and ethanol, **[53]** to water are compared. There does seem to be a broad correlation between the values obtained for the three systems but one can hardly agree with those **[52]** who have claimed that 'the differences are not very significant'. The differences are by any criterion very significant and are such as to, in the author's opinion, completely negate any quantitative attempts to obtain measures of the intrinsic stabilities of the folded forms of proteins. Even if one only considers those residues which are apolar in constitution then the ranking orders of the free energies of transfer are not consistent (see Table 5).

An alternative approach to that described above has been promoted by Wolfenden, Andersson, Cullis and Southgate. **[54]** In this, attention is directed towards the side chains of the amino acid residues and for example hydrogen (H—H) was chosen to model the glycyl (H—CH) side chain and methane ($CH_3$—H) chosen to model the alanyl ($CH_3$—CH) side chain. The basis of the method is that the free energies of transfer from the gas phase to water for complex molecules are additive functions of the constituent groups. Now it is certainly true that this is approximately so **[55, 56]** but the best available data **[57]** for hydrocarbon gases indicates, even for these, the strictly

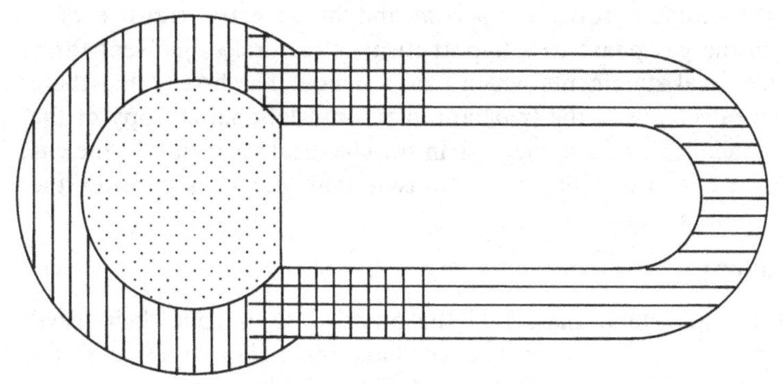

Figure 5. Diagrammatic representation of the solvation of a solute with more than one functionality. One functionality is indicated by the dotted region and the other by the open region. The solvation regions are represented by vertical and horizontal lines and the mutual perturbation of the two is indicated by cross-hatching.

non-additive nature of the transfers. This can be illustrated by considering for example n-butane. If component group additivity was applied then the free energy of transfer from gas to water of this would be given by:

(i) twice the value for ethane minus the value for hydrogen, or
(ii) the sum of the values for methane and propane minus the value for hydrogen, or
(iii) twice the value for methane plus the value for ethane minus twice the value for hydrogen.

The results obtained (using the same standard states as Wolfenden *et al.*) are: (i) 23.4, (ii) 24.7 and (iii) 22.7 kJ $mol^{-1}$ respectively. These should be compared with the experimental value [**57**] for butane of 25.6 kJ $mol^{-1}$.

There are several reasons which could account for these variations between the calculated and experimental results. One possible reason is that in the aqueous phase the solvation region peripheral to a given atom on the hydrocarbon depends to some extent at least on the solvation characteristics of adjacent atoms on the backbone. A pictorial representation of this is shown in Figure 5. Such mutual perturbations must contribute to some extent in any molecular situation and will almost certainly be particularly significant for apolar residues in water in view of the rather specific and structured nature of the hydrophobic hydration about such residues. It is envisaged that rather marked interactions between local solvation regions should also be present for amino acids and peptides, particularly for atoms and groups close to the charged centres.

Another quite different contribution to which little attention has been paid of late is fundamental and arises particularly for the free energies and entropies of transfer of solutes from one medium to another. This stems from changes, not so much in the solvation, but rather from the skeletal and, in particular, the translational motions of the molecules considered. If we neglect for the moment solvation *per se* and imagine the transfer of a molecule from the gas phase to a hypothetical 'structureless' solvent there are some categorical statements which can be made. The first is that there must be a large decrease in the (partial molar) translational entropy of the solute and a concomitant large increase in the chemical potential. In the gas phase the molar entropy arising from translational sources is given by the Sackur–Tetrode equation

$$S_{\mathrm{m,gas}}\,(\text{translation}) = R \ln [e^{5/2}(2\pi mkT/h^2)^{3/2}(kT/P)] \tag{18}$$

In this $m$ is the molecular mass and the other symbols have their usual meanings. If a 1 M standard state is adopted and the molar mass (molecular weight), $M$, is expressed in g mol$^{-1}$ then at 25 °C this becomes

$$S_{\mathrm{m,gas}}\,(\text{translation})\,(\mathrm{J\,K^{-1}\,mol^{-1}}) = 82.16 + 1.5\,R \ln M \tag{19}$$

If we consider the extreme situation where all translational motion of the solutes is lost on transfer to the gas phase then the changes in chemical potentials for the hydrocarbons considered above are

butane 39.6; propane 38.6; ethane 37.1;
methane 34.8; hydrogen 27.1,

(all in kJ mol$^{-1}$). Consequently if we look at the additivity relationships which were considered above then the following results are obtained (the numbering used is the same as that used above):

(i) 47.1, (ii) 46.3, and (iii) 52.5 kJ mol$^{-1}$.

The variations from the value calculated for butane are clearly not only not constant but also differ markedly. Now this is rather an extreme example and one would certainly expect there to be some, but incalculable at present, compensatory contributions from translational motions of the solutes within the solvent.

A similar point can be made with regard to rotational contributions in the gas phase. These will usually make a smaller contribution to the chemical potential changes on transfer than do the translations but nevertheless since the contributions from rotational sources depend upon the three principal moments of inertia of the molecule considered these also cannot be expected to be strictly atom or indeed group additive.

The conclusion that one must draw from considerations such as these is that strictly, group additivity approaches for solvation properties such as the free energy and entropy of transfer do not have a firm theoretical basis and any such approach must have an unquantifiable element of empiricism associated with it. There is nothing wrong with empirical approaches but in using them one must take care not to claim too much.

Turning now to the information which has been obtained [**54**] from the free energies of transfer from the gas phase to water, Table 6 lists the available

Table 6. *The free energies of transfer of amino acid side chains from the gas phase to water*

| Amino acid | Model compound | Free energy of transfer[a] (kJ $mol^{-1}$) |
|---|---|---|
| ala | Methane | 8.4 |
| arg | n-Propylguanidine | −83.4 |
| asn | Acetamide | −40.5 |
| asp | Acetic acid | −45.8 |
| cys | Methanethiol | −5.2 |
| glu | Propionic acid | −42.7 |
| gln | Propionamide | −39.2 |
| gly | Hydrogen | 9.8 |
| his | 4-Methylimidazole | −43.0 |
| ile | n-Butane | 8.7 |
| leu | i-Butane | 9.7 |
| lys | n-Butylamine | −39.8 |
| met | Methyl ethyl sulphide | −6.2 |
| phe | Toluene | −3.2 |
| ser | Methanol | −21.2 |
| thr | Ethanol | −20.4 |
| trp | 3-Methylindole | −24.6 |
| tyr | Cresol | −25.6 |
| val | Propane | 8.2 |

[a] This term has been called [54] the 'hydration potential'. Some of the values given in this column have been recalculated and differ a little from those given in the original paper. For those residues which have ionisable groups corrections were made for the extent of ionisation.

information for amino acid side chains. Some of the qualitative statements which were made earlier for transfers from non-aqueous solvents to water also apply to these results in that the transparently hydrophilic side chains have the most negative free energies of transfer. Again, however, there is little evidence of any quantitative correlations for transfers from different media. This is illustrated in Table 7. The overall poor correlation between the results obtained for the different systems reinforces the point made earlier about the difficulty of choosing a solvent to represent a protein interior. There are clearly specific features associated with any solvent and the very crude classification which is sometimes made and which implicitly assumes that all non-aqueous solvents have the same characteristics is obviously erroneous.

An interesting and in many ways quite remarkable relationship has been found [**54**] between the free energies of transfer of side chains from the gas phase and one aspect of protein organisation.

Some years ago the idea was introduced [**60**] of using a 'rolling-sphere' model to represent one aspect of the water–protein interface. The idea was simple in that the water molecule was represented by a sphere which was in contact with the surface of a protein. Using X-ray structure information in conjunction with van der Waals' radii the sphere was rolled across the surface

Table 7. *Ranking orders of the free energies of transfer of amino acid residues from various media to water*

| | Ranking order[a] | | | | |
|---|---|---|---|---|---|
| Amino acid | A | B | C | D | E |
| ala | 6 | 9= | 6= | 16 | 15 |
| arg | 4 | 8 | 4= | 1 | 1 |
| asn | 10 | 4 | 2 | 5 | 4 |
| asp | 15 | 6 | 6= | 2 | 5= |
| cys | 9 | 11 | 11 | 13 | 16= |
| glu | 12 | 7 | 8 | 4 | 8= |
| gln | 2 | 1 | 1 | 7 | 3 |
| gly | 1 | 5 | 3 | 19 | 12 |
| his | 5 | 9= | 4= | 3 | 7 |
| ile | 13 | 18 | 17 | 17 | 19 |
| leu | 16 | 15 | 19 | 18 | 15 |
| lys | 8 | 12 | 9 | 6 | 2 |
| met | 14 | 13 | 13 | 12 | 14 |
| phe | 17 | 17 | 18 | 14 | 16= |
| ser | 3 | 2= | 10 | 10 | 9 |
| thr | 7 | 2= | 12 | 11 | 10 |
| trp | 19 | 19 | 15 | 9 | 11 |
| tyr | 18 | 16 | 16 | 8 | 5= |
| val | 11 | 14 | 14 | 15 | 18 |

[a] The ranking order is such that the most negative free energies of transfer have the higher numbers. The columns refer to transfers to water from: A, *N*-methylacetamide; [52] B, ethanol; [53] C, surface; [58] D, gas; [54] E, protein 'residue accessibilities'. [59]

and hence the solvent accessible regions in the surface could be mapped. The concept has been used fairly often to explore different aspects of proteins and other solutes in solution. Chothia [59], for example, used the model with a sample of some 12 globular proteins to assess the numbers of amino acid residues of each type which are accessible to the rolling-sphere over 95% of their surface areas. Any such residues were considered to be 'buried' in the protein interior. By comparing the numbers found to be in this category for each type of residue with the numbers of completely exposed residues in the protein amino acid set, it was then possible to deduce an approximate (and statistical) distribution coefficient ($K_d$) for the process:

Residue in protein interior ⟶ Residue on protein surface

The corresponding standard free energy change for the transfer from interior to exterior ($\Delta\mu^\circ = -RT\ln K_d$) was then assumed to reflect the propensity for amino acid residues to be present at the surface of globular proteins. A similar approach with somewhat different 'burial' criteria has also been described. [61] The correlation between the free energies of transfer and the transfers from the gas phase is shown in Figure 6. There is obviously

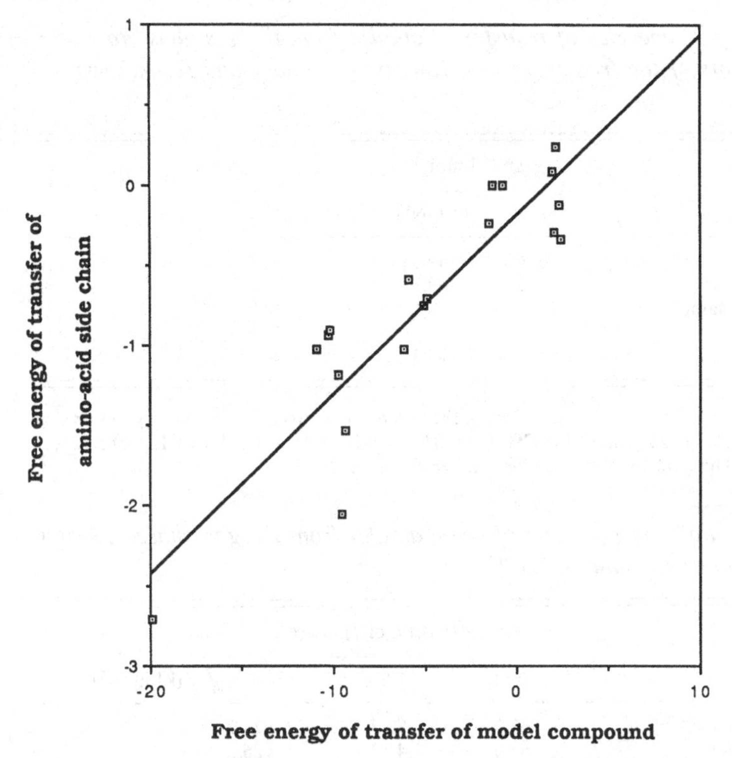


Figure 6. Correlation between the free energies (in kcal $mol^{-1}$) of transfer of amino acid residues between the interior and surface of globular proteins and of model compounds from the gas phase to water.

a considerable amount of scatter in the correlation but nevertheless the overall trend is quite obvious. This is not to say the agreement is good and certainly if one looks at the details of the correlation there are some disturbing features. This is particularly so for residues which, from intuitive criteria, would be considered to be hydrophobic. For the residues gly, ala, val and leu the fractions of protein residues considered to be 'buried' are 0.36, 0.38, 0.54 and 0.45 respectively whereas the free energies of transfer of the side chains from the gas phase are 9.8, 8.4, 8.2 and 9.7 kJ $mol^{-1}$. It is apparent both from Figure 6 and results such as these that too much credence should not be placed in quantitative deductions from the correlation, which has been observed, in spite of its qualitative utility.

Prior to the above work on amino acid side chains the same experimental technique was used [62] to obtain information on the hydration of the primary and secondary peptide groups using simple amides to model the groups. The results obtained for the distribution of amides between the gas phase and water are given in Table 8.

The transfer information can be treated in the same way as that used above

Table 8. *The free energies of transfer of amides from the gas phase to water and assessments of the free energy of transfer of primary and secondary peptide groups*

| | $\Delta_{trans}\mu^{\infty}$ (kJ mol$^{-1}$) | | |
|---|---|---|---|
| Amide | Amide | -CONH- | -CON= |
| Acetamide | −40.6 | −49.0[a] | |
| *N*-methylacetamide | −42.2 | −49.8[b] | |
| *N*,*N*-dimethylacetamide | −35.8 | 49.2[c] | −49.9[d] |
| | | | −46.2[e] |
| Mean | | −49.3 | −46.5 |

Peptide group contributions estimated from the difference between the free energy of transfer (molar scale) of the amide and of (*a*) $CH_4$ (*b*) $C_2H_6$ (*c*) $2CH_4-H_2$ (*d*) $C_2H_6+CH_4-0.5\ H_2$ (*e*) $3CH_4-1.5\ H_2$. The data for the gases were taken from ref. 57.

Table 9. *Enthalpies of solvation of some amides from the gas phase to water and to tetrachloromethane at* 25 °C

| | $-\Delta_{solv}H^{\infty}$ (kJ mol$^{-1}$) | | |
|---|---|---|---|
| Amide | Water | $CCl_4$ | $\Delta_{trans}H^{\infty}$ (kJ mol$^{-1}$) |
| *N*-Methylformamide | 63.9 | 39.7 | 24.2 |
| *N*-Ethylformamide | 67.6 | 41.3 | 26.3 |
| *N*-Methylacetamide | 73.7 | 41.4 | 32.3 |
| *N*-Ethylacetamide | 80.4 | 50.4 | 30.0 |
| *N*-Propylacetamide | 85.5 | 55.4 | 30.1 |
| *N*-Butylacetamide | 89.7 | 58.3 | 31.5 |
| *N*-Methylpropionamide | 79.8 | 49.8 | 30.0 |
| *N*,*N*-Dimethylformamide | 63.2 | 43.4 | 19.8 |
| *N*,*N*-diethylformamide | 70.0 | 49.1 | 20.8 |
| *N*,*N*-Dipropylformamide | 75.7 | 53.6 | 22.1 |
| *N*,*N*-Dimethylacetamide | 71.6 | 48.0 | 23.6 |
| *N*,*N*-Diethylacetamide | 79.1 | 54.7 | 24.3 |
| *N*,*N*-Dimethylpropionamide | 75.3 | 53.6 | 21.7 |

to determine the free energies of transfer of peptide groups and these are included in this table. Apparently the primary peptide group interacts more strongly with water than does the tertiary peptide group and even given the difficulties referred to earlier the indications are that the C=O group interacts somewhat more strongly, in a free energetic sense, than does the N—H group. This does seem to be in accordance generally with information from other sources.

The free energetic information has recently been complemented by some important work by Barone, Della Gatta and their coworkers. [**63**, **64**] The enthalpies of vaporisation of a variety of amides have been obtained using an

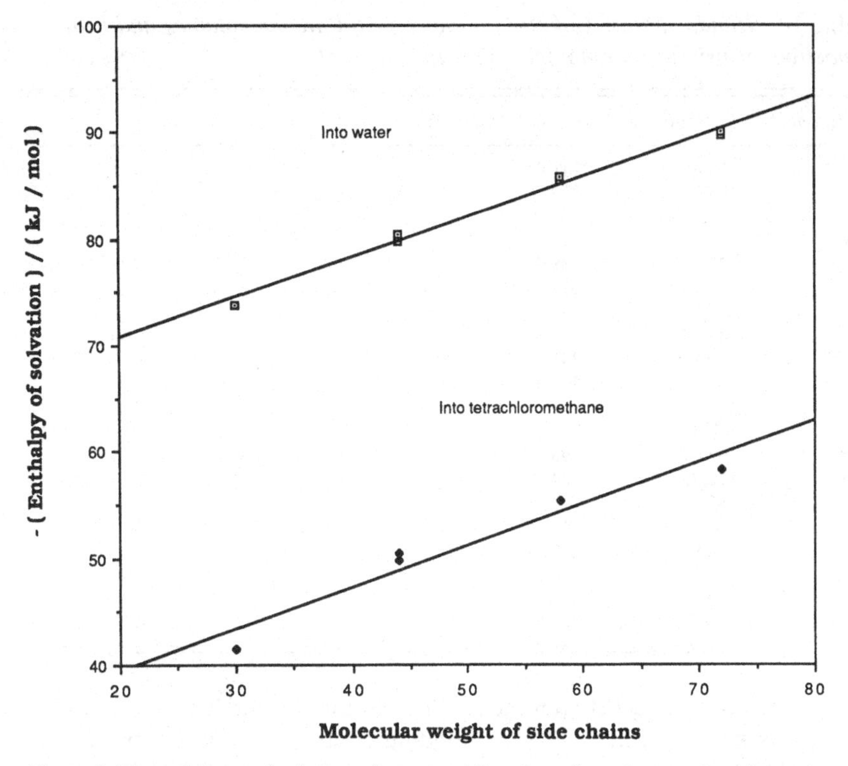


Figure 7. The enthalpies of solution of some amidic solutes from the gas phase into water and into tetrachloromethane.

effusion cell which has been incorporated into a microcalorimeter. In some independent experiments they also determined the enthalpies of solution at infinite dilution of the amides into water and into tetrachloromethane. Combination of the results from the two sets of experiments allowed the enthalpies of solvation of the amides from the gas phase into the two solvents to be obtained. The results for amides containing the secondary amide (primary peptide) and tertiary amide (secondary peptide) groups are presented in Table 9 and shown in a graphical form in Figure 7. The most obvious feature of the results obtained is that for all solutes and both solvents the solvation process gives rise to relatively large and negative enthalpy changes i.e., is thermochemically favourable. It is also quite apparent that for any given amide the transfer from the vapour is more exothermic for water than for the non-aqueous solvent. A further feature is that for both solvents as the number of alkyl carbon atoms increases so too does the exothermicity. The increased exothermic change for the transfer to water is easily understood in terms of hydrogen-bonding between the amide groups and water but it seems clear that even in the tetrachloromethane there are quite strong

Table 10. *Standard state (infinite dilution) partial molar volumes and heat capacities of primary amino acids in water at 25 °C*

| Amino acid | $V^{\infty}$ (cm$^3$ mol$^{-1}$) | $C_p^{\infty}$ (J K$^{-1}$ mol$^{-1}$) |
|---|---|---|
| ala | 60.45 | 141.4 |
| arg | 123.7 | 279 |
| asn | 77.18 | 125 |
| asp | 71.79 | 127 |
| cys | 73.62 | 188.0 |
| glu | 89.36 | 177 |
| gln | 94.36 | 187 |
| gly | 43.25 | 39.2 |
| his | 99.14 | 241 |
| ile | 105.45 | 383.3 |
| leu | 107.57 | 397.7 |
| lys | 108.71 | 267 |
| met | 105.3 | 293 |
| phe | 121.92 | 384 |
| pro | 82.65 | 172.3 |
| ser | 60.62 | 117.4 |
| thr | 76.86 | 210 |
| trp | 144.0 | 420 |
| tyr | 123 | 299 |
| val | 90.79 | 301.9 |

interactions occurring between the amides and the solvent. These seem to be at least as strong as the hydrogen-bonding contributions in water. If it is assumed that, as the side chains diminish in size, then one will be left with an increased relative contribution from the peptide groups, then extrapolation in the way shown in Figure 7 leads to the following approximate values for the enthalpies for the solvation of groups in the two solvents:

| | | |
|---|---|---|
| Tetrachloromethane: | primary peptide group | $-35$ kJ mol$^{-1}$ |
| | secondary peptide group | $-34$ kJ mol$^{-1}$ |
| Water: | primary peptide group | $-65$ kJ mol$^{-1}$ |
| | secondary peptide group | $-56$ kJ mol$^{-1}$ |

The indications are therefore that there is little distinction between the two types of peptide groups in tetrachloromethane but the primary group interacts somewhat more attractively with water than does the secondary group. Both of these features are what one would predict, at least qualitatively, from chemical intuition. In many ways the most striking aspect of the information obtained is that notwithstanding the fact that these amides do interact more favourably with water than with tetrachloromethane, the slopes of the lines drawn in Figure 7 are very nearly the same. It would thus seem that whether one considers water or the chlorinated hydrocarbon the contribution to solvation from the apolar residues are not dissimilar. In other words from these data alone there are no indications that

Table 11. *Standard state (infinite dilution) partial molar volumes and partial molar heat capacities of small peptides in water at 25 °C*

| Peptide | $V^{\infty}$ ($cm^3 mol^{-1}$) | $C_p^{\infty}$ ($J K^{-1} mol^{-1}$) |
|---|---|---|
| diglycine | 76.27 [66] | 105.0 [66] |
| triglycine | 111.81 [66] | 185.9 [66] |
| tetraglycine | 149.7 [66] | 283 [66] |
| pentaglycine | 187.1 [66] | 373 [66] |
| dialanine | 110.30 [66] | 330.8 [66] |
| trialanine | 163.80 [66] | 526 [66] |
| tetraalanine | 220.1 [66] | 733.4 [66] |
| diserine | 111.8 [66] | 259.9 [66] |
| triserine | 166.0 [66] | 398 [66] |
| glyala | 92.37 [67] | 216.2 [69] |
| alagly | 94.53 [69] | 203.4 [69] |
| glyasp | 110.11 [68] | 188.1 [68] |
| glythr | 108.50 [68] | 251.0 [68] |
| glyphe | 155.45 [68] | 422.0 [68] |
| glyser | 92.93 [68] | 161.2 [68] |
| glyleu | 139.62 [68] | 473.3 [69] |
| glyval | 121.99 [67] | 352.6 [69] |
| glyalagyl | 129.67 [70] | 289.8 [70] |
| glyvalgly | 160.38 [70] | 443.7 [70] |
| glyleugly | 177.35 [70] | 526.1 [70] |
| glysergly | 131.13 [70] | 263 [70] |
| glyalagy | 129.67 [70] | 289.8 [70] |
| alaglygly | 130.16 [70] | 286.2 [70] |

water behaves any differently towards apolar residues than does tetrachloromethane and that the transfer of a methylene group from the vapour phase to either solvent gives an exothermic change of about $-4.8$ kJ $mol^{-1}$. The indications are therefore that the solvation of such groups is dominated by 'cohesive' i.e., non-directional van der Waals' forces, between the solvents and the group. This is almost certainly an oversimplification and we will return to this matter a little later but the point which should be made is that these particular data can be rationalised rather well using simple and quantifiable ideas without invoking vague and non-quantified solvent structural concepts.

### 3.4. *Partial molar volumes and heat capacities*

A substantial amount, but still far from complete, body of information exists on the partial molar volumes and partial molar heat capacities, both at infinite dilution, of amino acids and small peptides. The data for the amino acids are collected in Table 10 and Table 11 lists the currently available reliable information for small peptides. In a crude sense these generally increase as the molecular weight of the solute increases (see Figure 8) but there is superimposed upon this trend more specific effects which are

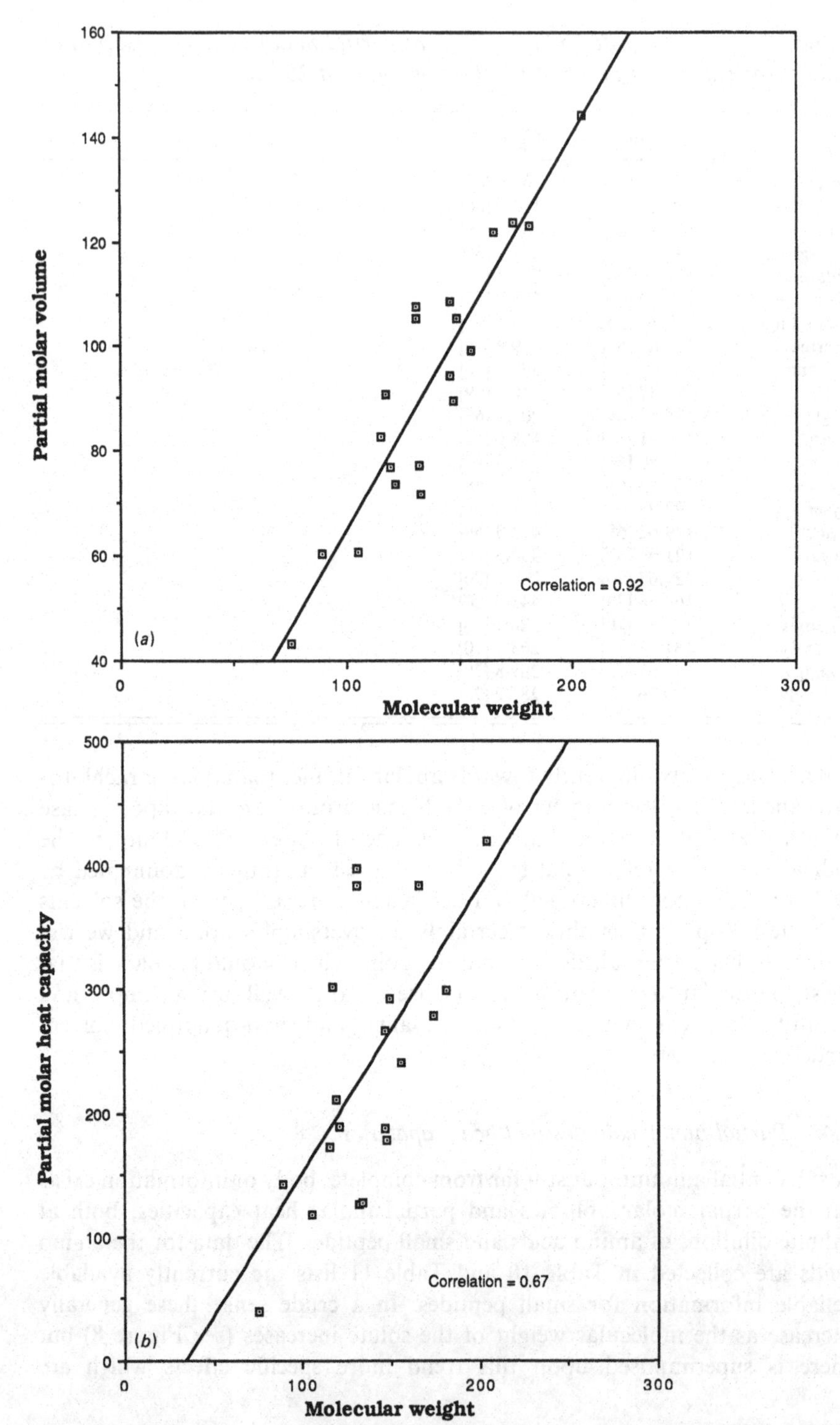


Figure 8. The variation of (*a*) partial molar volume (in $cm^3 mol^{-1}$) and (*b*) partial molar heat capacity (in $J K^{-1} mol^{-1}$) (both at infinite dilution in water at 25 °C) with molecular weight.

apparent from, for example, the difference observed between the isomers leucine and isoleucine.

As mentioned above both sets of compounds are zwitterionic and consequently one would expect there to be a significant contribution to their properties from the presence of the distinct and separate carboxylate and amino charged centres. If it is assumed that these centres are distinct, spherical in form and that the solvent medium is a dielectric continuum then the contribution to macroscopic properties from the charges can be expressed, at least approximately, using the Born model of ion solvation. If we consider the chemical potential then the electrostatic contribution is

$$\mu^{\text{elec}} = (Le^2/4\pi E_0)/E_r\langle r\rangle \qquad (20)$$

In this $L$, $e$, $E_0$, $E_r$ and $\langle r\rangle$ are respectively Avogadro's constant, the electronic charge, the permittivity of a vacuum, the relative permittivity of the medium and the harmonic mean radius of the amino and carboxylate ionic centres. If the experimental value of the relative permittivity for water is used and a mean radius of 100 pm is assumed, the contribution to the chemical potential from this source is 17.7 kJ $\text{mol}^{-1}$ and although this is small compared to covalent bond energies it is nevertheless fairly significant compared to non-bonding energies generally. Appropriate differentiation of Equation (20) gives the following expressions for the contributions from the charges from electrostatic sources for the partial molar volume and heat capacity respectively.

$$V^{\text{elec}} = -[(Le^2/4\pi E_0)/E_r\langle r\rangle(\text{d}\ln E_r/\text{d}P) \qquad (21)$$

$$C_p^{\text{elec}} = [(Le^2/4\pi E_0)/E_r\langle r\rangle[(\text{d}^2\ln E_r/\text{d}T^2)-(\text{d}\ln E_r/\text{d}T)^2] \qquad (22)$$

For water at 25 °C these take the numerical forms

$$V^{\text{elec}}(\text{m}^3\,\text{mol}^{-1}) = -8.33\times10^{-16}/\langle r\rangle \qquad (23)$$

$$C_p^{\text{elec}}(\text{J K}^{-1}\,\text{mol}^{-1}) = -10.82\times10^{-9}/\langle r\rangle \qquad (24)$$

Using the same radius as above the volumetric and heat capacity contributions for a zwitterion are $-8.3\ \text{cm}^3\,\text{mol}^{-1}$ and $-108\ \text{J K}^{-1}\,\text{mol}^{-1}$ respectively. A direct check on the actual contributions from this source for the amino acids is not possible but one can get an indication of the effect by considering the changes in the thermodynamic properties for condensation reactions for the formation of peptides. If we consider homopeptides the condensation process can be formally written as

$$n[\text{H}_3\text{N}^+\text{-CH(R)-CO}_2^-] = \text{H}_3\text{N}^+\text{-[-CH(R)CONH-]}_{n-1}\text{-CH(R)-CO}_2^- + (n-1)\text{H}_2\text{O} \qquad (25)$$

The standard state volume change for this process is

$$\Delta V^\circ = V^\infty_{\text{peptide}} + (n-1)\,V^\circ_{\text{water}} - n\,V^\infty_{\text{aminoacid}} \qquad (26)$$

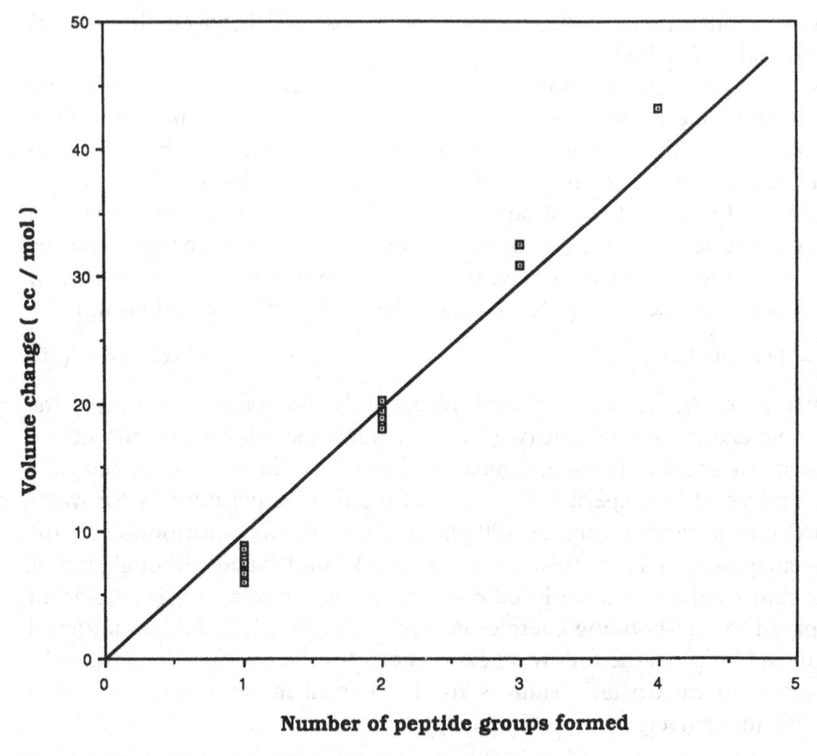


Figure 9. Plot of Equation (27) for the formation of peptides.

and if electrostatic effects are the only ones which need be considered then using Equation (23) we get

$$\Delta V^{\circ}(\mathrm{cm}^3\,\mathrm{mol}^{-1}) = 8.33 \times 10^{-10}(n-1)/\langle r\rangle \tag{27}$$

The analogous expression for the standard state heat capacity change is at 25 °C

$$\Delta C_p^{\circ}(\mathrm{J\,K}^{-1}\,\mathrm{mol}^{-1}) = 108.2(n-1)/\langle r\rangle \tag{28}$$

The available data [**66–70**] for both homo- and heteropeptides are shown plotted according to these equations in Figures 9 and 10. It is apparent from both of these that reasonable linear relationships do exist as predicted and the volume and heat capacity changes which result for the formation of a peptide are respectively about 9.4 $\mathrm{cm}^3\,\mathrm{mol}^{-1}$ and 113 $\mathrm{J\,K}^{-1}\,\mathrm{mol}^{-1}$. Substitution of these into Equations (27) and (28) leads to values for the harmonic mean radius of the two charged groups of 89 and 95 pm. Although these do not agree exactly they are about what one would expect and so the conclusion that can be drawn is that most if not all of the volume and heat

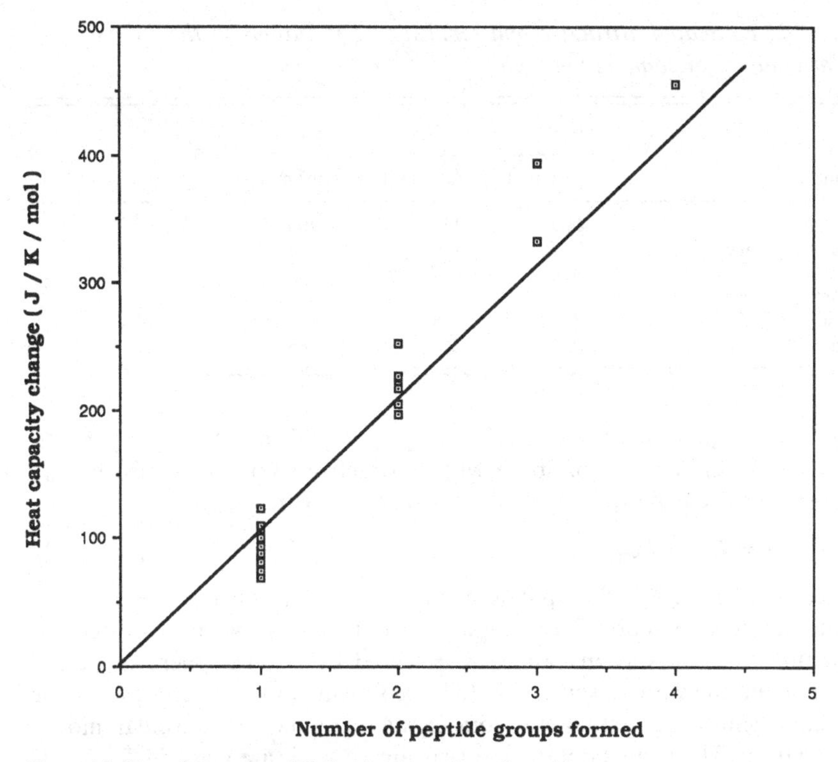


Figure 10. Plot of Equation (28) for the formation of peptides.

capacity changes stem from electrostatic sources and that specific or structural effects are relatively minor although possibly significant.

A correct relationship between molecular events and partial molar volumes was established a good many years ago [**73–5**] but has been little used by workers in this area (but see refs 44 and 76). It was shown [**73**] that there is a direct link between the partial molar volume of a solute at infinite dilution, the isothermal compressibility of the pure solvent ($\kappa$) and the potential of mean force $\{w_{sw}(r)\}$ between one solute molecule and one solvent molecule when separated by a distance $r$ *viz.*

$$V^{\infty} - RT\kappa = 4\pi L \int_0^{\infty} \langle 1 - \exp[-w_{sw}(r)/kT] \rangle r^2 \, dr \tag{29}$$

As things stand at the moment the evaluation of the functional dependence of the potential of mean force is not possible, but it is possible to adopt approximate procedures. This was done by Herrington and her coworkers for sucrose solutions [**77, 78**] and was recently used [**76**] to investigate solute–solvent interactions of some amino acids. The solute–solvent cluster

Table 12. *Estimated attractive and repulsive contributions to the partial molar volumes of some amino acids*

| Amino acid | $V^\infty$ (cm³ mol⁻¹) | $L\phi_R$ (cm³ mol⁻¹) | $L\phi_A$ (cm³ mol⁻¹) |
|---|---|---|---|
| Glycine | 43.25 | 150 | −108 |
| *N*-Methylglycine | 62.79 | 170 | −108 |
| Alanine | 60.45 | 204 | −145 |
| *N*-Methylalanine | 79.74 | 230 | −151 |
| Serine | 60.62 | 235 | −176 |
| *N*-Methylserine | 79.77 | 265 | −186 |

integral in Equation (29) was arbitrarily split [79] into two components representing the regions of space where attractive ($\phi_A$) and repulsive ($\phi_R$) interactions occur between the solute and the solvent.

$$V^\infty - RT\kappa = L\phi_A + L\phi_R \tag{30}$$

Hence in this regard, the split is analogous to that used earlier [47] for solute–solute interactions. The repulsive contributions were estimated by assuming that the solvent can be represented by a hard sphere of radius 0.14 nm and the various solutes could be represented by hard ellipsoids with long and short axes estimated from crystallographic or molecular model information. The results obtained for six amino acids are given in Table 12. The repulsive contributions do, as a necessary consequence of the hard-body model which was used, increase on methyl substitution and the indications are that methyl substitution seems to lead to more attractive contributions as does the incorporation of a hydroxyl group. Estimates such as these probably have some validity but since so many features of the potential energy surfaces are necessarily neglected it is difficult to have too much confidence in any conclusions drawn from them.

There has been a considerable amount of effort directed towards finding predictive schemes to represent the molar volumes of polyfunctional solutes in water and these usually but not invariably take the form

$$V^\infty = \sum n_i V_i^\infty$$

where $n_i$ is the number of defined groups of type $i$ on the molecule and $V_i^\infty$ is the group molar volume. The simplest scheme which could be adopted would be to assume that each molar volume is atom additive. There are several pertinent pieces of information available in the literature to test such an hypothesis. Using some of the recent data [76] obtained from the study mentioned earlier it has been found for the partial molar volumes of alanine and its isomer *N*-methylglycine that the former has a volume of 60.45 cm³ mol⁻¹ and the latter a volume of 62.79 cm³ mol⁻¹. Similarly the

Table 13. *Partial molar volumes at infinite dilution of three isomeric amino acid amides in water at 25 °C*

$$CH_3CON(R^{(1)})-CH(R^{(2)})-CONH(R^{(3)})$$

| $R^{(1)}$ | $R^{(2)}$ | $R^{(3)}$ | $V^{\infty}$ ($cm^3\ mol^{-1}$) |
|---|---|---|---|
| H | H | $CH_3$ | 108.93 |
| H | $CH_3$ | H | 108.06 |
| $CH_3$ | H | H | 107.14 |

molar volumes in water of the three isomeric acid amides of general structure

$$CH_3CON(R^{(1)})-CH(R^{(2)})-CONH(R^{(3)})$$

are given in Table 13. It is apparent from this that there are relatively large differences in the results obtained for these three solutes and this consequently negates the possibility of atom additivity applying.

For compounds of the types addressed in the present chapter it seems reasonable and chemically sensible to define the groups to be used as being the amino acid residues. However, using even these relatively large structural units it is apparent that group volume additivity schemes are strictly inapplicable. There are several examples which could be chosen to illustrate this. If one considers the information [**80**] on substituted amino acid amides then the difference in the molar volumes of *N*-acetylglycinamide and *N*-acetyl-L-alaninamide (17.50 $cm^3\ mol^{-1}$) is by no means the same as the difference between the volumes of *N*-acetyl-*N*-methylglycinamide and *N*-acetyl-*N'*-methyl-L-alaninamide (13.94 $cm^3\ mol^{-1}$). Other examples from sequence isomeric peptidic compounds are presented in Table 14.

The non-additivity of groups is also evidenced from information [**66**] on a series of homopeptides. Figure 11 shows the incremental changes in molar volumes of adjacent isomers for three amino acid residues and it is apparent from this that each of the series shows a similar behaviour in that the volume increases and then tends to level off. Presumably a major factor in the trends is the influence of the ionic head groups and their solvation regions on the hydration of the amino acid and peptide residues. It is difficult to be categorical here because of the relative paucity of data but it does seem that there is a need for about four residues in the chain before constant values are approached.

Notwithstanding the above difficulties it is possible to construct crude additivity schemes that will give *approximate* values for the group

Table 14. *Partial molar volumes at infinite dilution of some isomeric peptides and some peptides derivatives in water at 25 °C*

| Solute[a] | $V^{\infty}$ (cm³ mol⁻¹) | Difference (cm³ mol⁻¹) |
|---|---|---|
| GLYPRO | 161.77 [80] | 2.46 |
| PROGLY | 164.23 [80] | |
| GLYALA | 144.41 [69] | 1.02 |
| ALAGLY | 145.43 [69] | |
| PRO-L-ALA | 180.67 [80] | 0.04 |
| PRO-D-ALA | 180.63 [80] | |
| glyala | 92.37 [69] | 2.16 |
| alagly | 94.53 [69] | |
| glyglyala | 128.79 [70] | |
| glyalagly | 129.67 [70] | 1.37 |
| alaglygly | 130.16 [70] | |

[a] The first six entries are the *N*-acetyl amides. Notice that for entries 5 and 6 the chirality of the alanyl residue is stressed. As expected within experimental error no effect of chirality is observed.

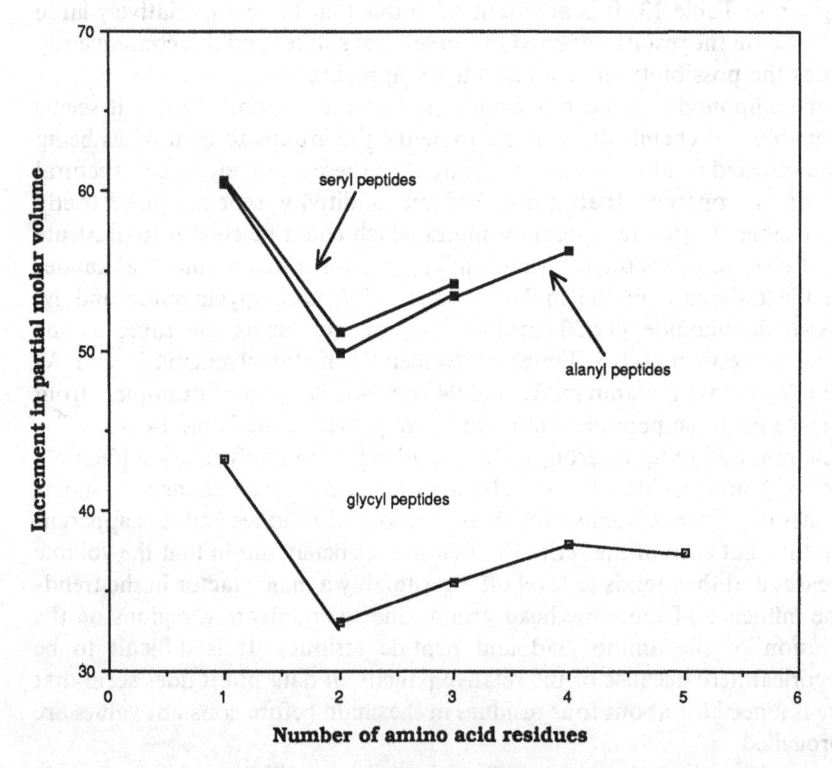


Figure 11. Incremental changes in the partial molar volumes (in cm³ mol⁻¹), of glycyl, alanyl and seryl peptides.

Table 15. *Approximate group contributions to partial molar volumes at infinite dilution in water at 25 °C*

| | $V^{\infty}$ ($cm^3\ mol^{-1}$) | |
|---|---|---|
| Group[a] | Ref. 80 | Ref. 81 |
| gly | 36.8 | 35.7 |
| ala | 53.7 | 51.2 |
| val | 85.1 | 82.6 |
| leu | 104.1 | 98.4 |
| pro | 73.1 | 70.3 |

[a] This denotes, in this instance, the amino acid side chain plus its associated peptide group e.g., gly = $-NH-CH_2-CO-$

contributions and some results obtained using this approach are given in Table 15.

The usual molecular interpretation of the information obtained on partial molar volumes has tended to consider the volume to arise from several sources. One model **[81]** assumes that the volume is comprised of two major contributions, one coming from the 'intrinsic' volume of the acid ($V_{\text{intrinsic}}$) and the other from electrostatic sources ($V_{\text{elec}}$). (The latter is essentially the contribution which was referred to earlier when partial molar volumes and heat capacities were discussed.) The intrinsic contribution is imagined to be composed of two terms, the first from the volumes occupied by the atoms (the van der Waals' volume – $V_w$) and the second from packing effects of the molecule within the solvent ($V_p$). The partial molar volume for any amino acid is represented therefore by

$$V_{\text{amino acid}} = V_{\text{intrinsic}} + V_{\text{elec}} = V_w + V_p + V_{\text{elec}} \tag{31}$$

It is certainly possible to estimate the first term on the rhs tolerably well but the other terms must have considerable uncertainties associated with them. This is illustrated in Table VII of ref. 81 for some of the amino acids. It is transparently obvious from this table that, although the results obtained for any residue are about the same, it would be difficult to draw any molecular conclusions from these. One must also question the assumption that the charged groups will make the same electrostatic contribution for all amino acids. Even neglecting any solvent structural features, the effective dielectric medium adjacent to say the carboxylate group in alanine would be expected to be somewhat different to that for serine.

Some semi-quantitative statements regarding the solvation of the peptide groups *vis à vis* apolar residues can be made **[82]** from a comparison of the difference between the partial molar volume of say diglycine and that for 5-aminopentanoic acid (87.58 $cm^3\ mol^{-1}$ **[71]**). The volume difference in this case corresponds approximately to the difference in volumes of a primary peptide group (–CONH–) and of two methylene groups ($-CH_2CH_2-$).

$$V^{\infty}(-CH_2CH_2-) - V^{\infty}(-CONH-) = 11.31 \text{ cm}^3 \text{ mol}^{-1}$$

The corresponding result on comparing the data on triglycine and 8-amino octanoic acid ($V^{\infty} = 136.03 \text{ cm}^3 \text{ mol}^{-1}$ **[72]**), where two peptide groups are now concerned is 24.3 $cm^3 \, mol^{-1}$. Using these a mean value of 11.9 $cm^3 \, mol^{-1}$ results for the 'transformation' of a $-CH_2CH_2-$ unit to a –CONH– unit. The volume of an isolated methylene group is close to 16 $cm^3 \, mol^{-1}$ **[72, 79, 83, 84]** and so one can conclude that the partial molar volume of the primary peptide group is about 20 $cm^3 \, mol^{-1}$ and this agrees tolerably well with some other estimates. **[66, 85]** The qualitative conclusion one can draw is that in spite of the fact that the intrinsic volume of two methylene groups is somewhat less than that of the peptide group, the latter has a smaller molar volume in water. One imagines that this arises from the relatively strong interaction between the hydrogen-bond donor and acceptor sites on the peptide interacting with water and causing a contraction in volume.

The overall conclusions that can be drawn from partial molar volumes and particularly group approaches are somewhat disappointing. Notwithstanding the fact that group approaches do have some utility in estimating partial molar volumes of amino acids and peptides they yield little molecular information and are virtually useless when considering the subtleties of peptide–water interactions. Clear examples of this come from the uncharged peptide derivatives for which electrostatic contributions are absent *viz.* *N*-acetylglycyl-L-prolinamide (GLYPRO) and *N*-acetyl-L-prolylglycinamide (PROGLY), and *N*-acetylglycyl-L-alaninamide (GLYALA) and *N*-acetyl-L-alanylglycinamide (ALAGLY) (see Table 14). For the former pair the partial molar volume of GLYPRO is smaller ($\sim 2.5 \text{ cm}^3 \text{ mol}^{-1}$) than that for PROGLY. There are two possible reasons for this. The GLYPRO could be interacting more strongly with the solvent than its isomer and this would lead to a volume contraction or it could have a more compact configuration, i.e., is more 'folded'. The indications are, from an nmr study **[86]** of amide proton exchange rates, that the latter is the case. Interestingly enough most workers on the solution chemistry of peptides have tended to ignore the possibility that folding can occur and have implicitly assumed that small peptides have a completely open, solvent accessible structure. This need not necessarily be so although it seems unlikely that any one configuration will be dominant.

The relative insensitivity of molar volumes to the solvent environment is also apparent from information **[87]** which is available on some simple amides and some uncharged amino acid derivatives in both water and *N*-methylacetamide (Table 16). Although the observed differences are experimentally significant the variations observed from one solvent to the other are difficult to explain in anything other than very qualitative terms. The very small volume differences exhibited by these amidic solutes in water and in a

Table 16. *Partial molar volumes of some simple amides and some amino acid amides in water at 25 °C and in N-methylacetamide at 32.6 °C*

| | $V^{\infty}$ (cm$^3$ mol$^{-1}$) | |
|---|---|---|
| Amide | Water | *N*-Methylacetamide |
| F | 38.51 | 38.88 |
| NMF | 56.87 | 59.32 |
| DMF | 74.50 | 77.45 |
| A | 55.60 | 55.67 |
| DMA | 89.65 | 93.24 |
| GLY | 90.56 | 90.65 |
| ALA | 108.06 | 107.57 |
| VAL | 139.00 | 138.40 |
| PRO | 126.51 | 124.93 |
| PHE | | 170.11 |

The abbreviations used are: F = formamide; NMF = *N*-methylformamide; DMF = *N,N*-dimethylformamide; A = acetamide; DMA = *N,N*-dimethylacetamide; GLY = *N*-acetylglycinamide; ALA = *N*-acetyl-L-alaninamide; VAL = *N*-acetyl-L-valinamide; PRO = *N*-acetyl-L-prolinamide; PHE = *N*-acetyl-L-phenylalaninamide.

solvent which to some extent mimics protein interiors seem to parallel the small differences in molar volumes of native and denatured proteins.

### 3.5. *Partial molar heat capacities*

It is well known that the partial molar heat capacities of solutes in water depend very markedly on the nature of the solute and for the amino acids unlike the molar volumes there is a very poor correlation with molar mass. The heat capacities of the primary amino acids and some peptides are given in Tables 10 and 11.

Generally for partial molar heat capacities in water it would be expected that there will be contributions from skeletal motions such as translations, rotations and vibrations of the amino acid. The rotational and translational contributions would be expected to be quite small and considerably less than those which would be present if the molecule was in the gas phase and certainly intramolecular vibrational contributions should be very small since the energy needed to excite transitions would be inadequate at ambient temperatures. Consequently most of the partial molar heat capacity of the solute when present in water will have a solvational origin. In this regard the heat capacity is a much more sensitive probe than is the partial molar volume for which the main or at least a major contribution comes from intramolecular sources.

In a qualitative way, if the zwitterionic head group of amino acids and peptides is considered, it might be expected that there would be a loss of

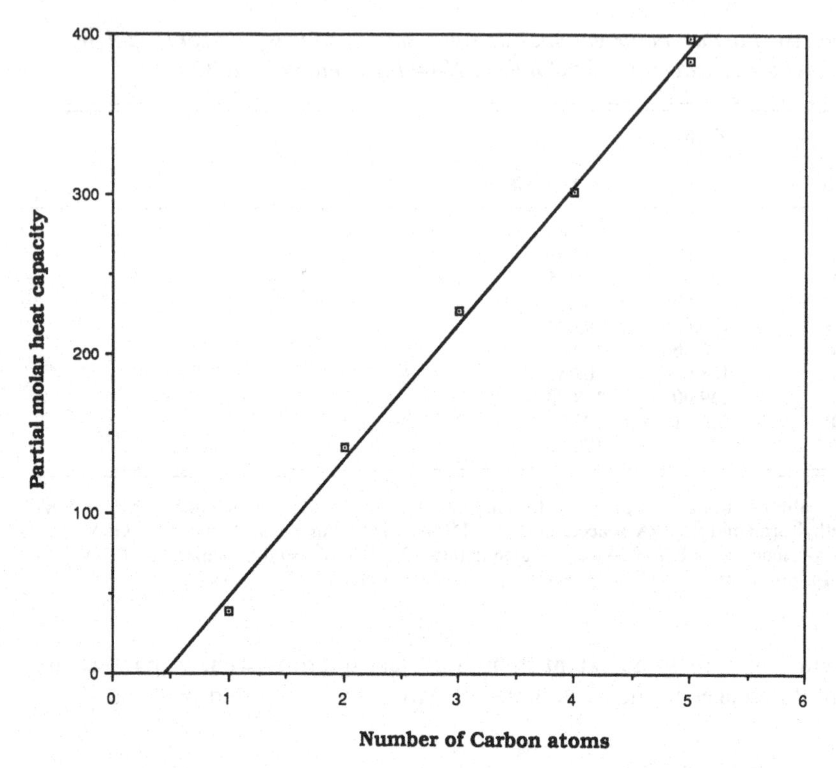


Figure 12. Plot of the partial molar heat capacity (in J $K^{-1}$ $mol^{-1}$) of $\alpha$-amino acids containing aliphatic side chains against the number of carbon atoms in the chain.

'motion' from those water molecules which are involved in direct solvation of the ionic residues and this in turn would give rise to a low or possibly negative (as is found for most electrolytes [**88**]) contribution to the partial molar heat capacity. The indications are from trends in the heat capacities of $\alpha$-amino acids containing only aliphatic side chains that this is the situation (see Figure 12) and the partial molar heat capacity of the ionic head group is of the order of $-30$–40 J $K^{-1}$ $mol^{-1}$. There is a considerable body of evidence [**89**] which indicates that the partial molar heat capacity of an 'isolated' methylene group in a hydrocarbon chain is, in water at 25 °C, 90 J $K^{-1}$ $mol^{-1}$. Presumably apart from a small intramolecular skeletal contribution most of this will arise from the structural organisation, of a transient clathrate type, of solvent about the apolar residue. There will be some diminution of the heat capacity because of the loss of translational motion in these structured solvent molecules but, since the structures are held together by relatively weak hydrogen-bonds and these will have low energy vibrations associated with them, it will be these easily excitable vibrations

Table 17. *Partial molar heat capacities at infinite dilution of some amino acids at 25 °C*

| Amino acid | $C_p^\infty$ (J K$^{-1}$ mol$^{-1}$) | Difference (J K$^{-1}$ mol$^{-1}$) |
| --- | --- | --- |
| glycine | 39.2[a] | |
| | | 102.2 |
| α-Alanine | 141.4[a] | |
| | | 85.8 |
| α-Aminobutyric acid | 227.2[b] | |
| Glycine | 39.2[a] | |
| | | 34.0 |
| β-Alanine | 73.2[b] | |
| | | 63.0 |
| γ-Aminobutyric acid | 136.2[b] | |
| α-Aminobutyric acid | 227.2 | |
| β-Aminobutyric acid | 177.6 | |
| γ-Aminobutyric acid | 136.2 | |

[a] See Table 10.
[b] Ref. 90.

which will dominate the partial molar heat capacity. The indications are that the solvation about apolar residues is rather fragile and easily disrupted by the presence of other groups. As an example we consider the heat capacities of the first few members of the α- and α,ω-amino acid series (see Table 17). The differences between the heat capacities of adjacent members of the α-acid series are not too different from the value referred to above for isolated methylene groups but this is not the situation for the α,ω-acids where the heat capacities are considerably lower than those for the α-isomers suggesting disruption of the apolar solvation regions by the closeness of the $NH_3^+$ and $CO_2^-$ groups and their solvation regions. This point is further illustrated by the three isomers of amino-butyric acid (Table 17) where as the distance between the positively charged and negatively charged groups is increased, the heat capacity decreases. It would seem from the information presented in Figure 13 that some destruction of the solvation about methylene groups by the ionic groups is still present when the charges are separated by four methylene groups. (This is probably a lower estimate since the data [**90**] on the longer α,ω-acids are somewhat uncertain because of the experimental method used to determine the heat capacities.)

Similar effects to these are found for homopeptides. The information available is rather limited but in Figure 14 some of the results presented in Tables 10 and 11 are shown in incremental form. Now, of course, there are peptide groups present and if anything these seem to worsen the situation and the peptide chain really has to be rather long before anything approaching a constant value is attained. The values do seem eventually to

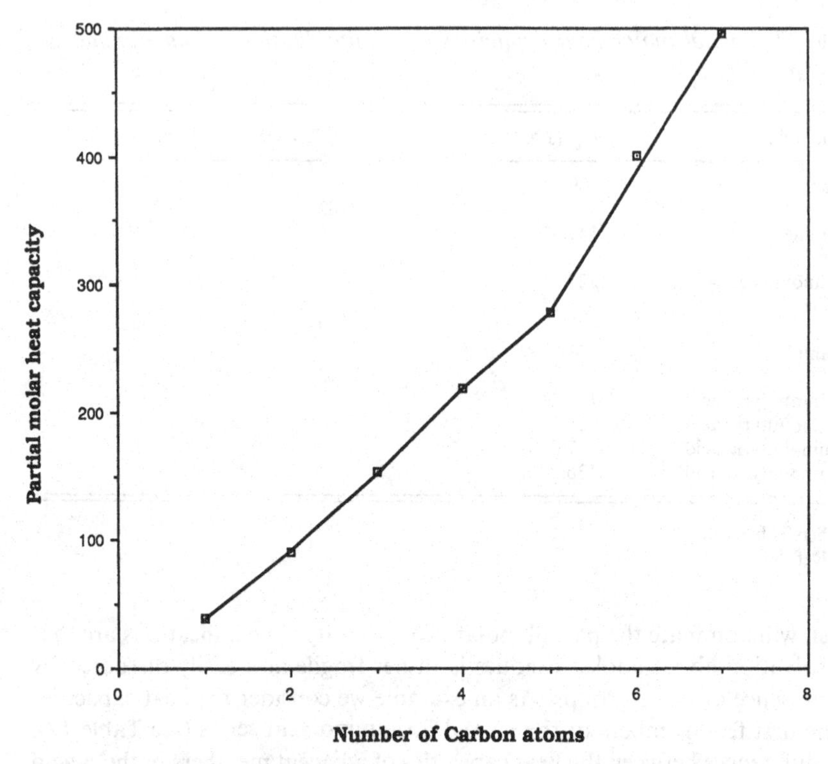


Figure 13. Plot of the partial molar heat capacity (in J $K^{-1}$ $mol^{-1}$) of $\alpha,\omega$-amino acids containing aliphatic side chains against the number of carbon atoms in the chain.

level off to give approximate values for the glycyl ($-NH-CH_2-CO-$) and alanyl ($-NH-CH(CH_3)-CO-$) residues of 95 and 205 J $K^{-1}$ $mol^{-1}$ respectively. Clear indications of the perturbation effects caused by the presence of adjacent groups and not only the charged residues, are apparent if the differences between the heat capacities of glycyldipeptides (gly-X) are compared with those of the amino acids (X). From the information given in Tables 10 and 11 these differences range from about 38 to about 75 J $K^{-1}$ $mol^{-1}$. There seems to be no obvious link with the general nature (i.e., hydrophilic, hydrophobic etc.) of the amino acid residue. Similar conclusions can be drawn from the diglycyl-tripeptides and the sequence isomers containing two glycyl and one alanyl residue (see Table 11).

A study has recently been completed **[91]** in which the partial molar heat capacities of some terminally substituted amino acids and peptides have been obtained. These are summarised in Table 18. It is apparent from this that even when charges are absent, perturbation by proximate groups still occurs to some extent which is evidenced by the different values obtained for the two glycyl, alanyl containing compounds. The data set is relatively limited but

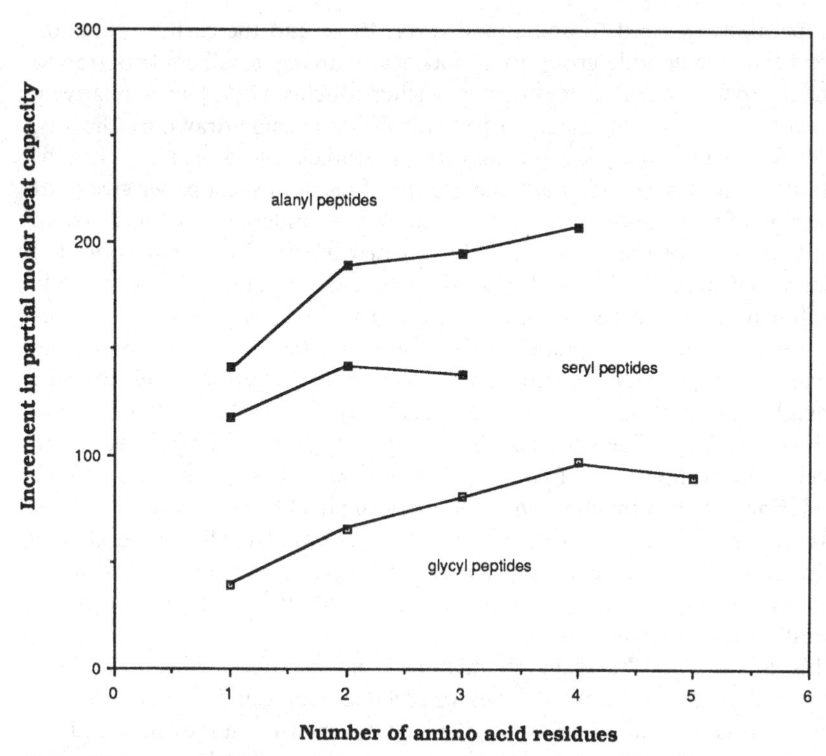


Figure 14. Incremental changes in the heat capacities (in J $K^{-1}$ $mol^{-1}$) of glycyl, alanyl and seryl peptides.

Table 18. *Partial molar heat capacities of some N-acetyl amino acid and peptide amides in water at* 25 °C

| Amide | $C_p^{\infty}$ (J $K^{-1}$ $mol^{-1}$) |
|---|---|
| *N*-Acetylglycinamide | 238.4 |
| *N*-Acetyl-L-alaninamide | 346.4 |
| *N*-Acetylglycyglycinamide | 322.1 |
| *N*-Acetylglycyl-L-alaninamide | 422 |
| *N*-Acetyl-L-alanylglycinamide | 430.9 |
| *N*-Acetyl-L-alanyl-L-alaninamide | 537.5 |

assuming that each group acts independently and using also the data [**92**] for acetamide, the following group contributions are found:

$-NH-CH_2-CO-$ 79.6 J $K^{-1}$ $mol^{-1}$
$-NH-CH(CH_3)-CO-$ 187.4 J $K^{-1}$ $mol^{-1}$

These differ somewhat from those inferred from the glycyl and alanyl homopeptides referred to above. However, these and the earlier results do suggest that the peptide group itself makes a relatively small contribution to the heat capacity which agrees with an earlier conclusion [**93**] using relatively uncertain data and is in disagreement with the conclusion drawn by Nichols, Skold Suurkuust and Wadso [**89**] from studies on aqueous solutions containing amides. (Incidentally the group values in this last paper give poor predictions for the substituted amino acid and peptide amides and it would seem that some of the values given are incorrect or at least inappropriate.)

The point made earlier with regard to volumes certainly seems to apply also, but more so, to heat capacities. Group values can give, in favourable situations, reasonable estimates of the heat capacities of complex molecules but can never give any detailed information since the perturbations induced by neighbouring atoms and groups, particularly when charges are present, can be rather large. The effects of charge can be to some undetermined extent cancelled by comparison of two amino acids [**65**] and there is a very rough correlation between the differences between the heat capacities of the amino acids and glycine and the 'hydrophobic' area, calculated [**94**] in a similar way to that used for a protein's surface accessibility, of the side chains in the acids. This is not too unexpected given the contribution that apolar residues generally make to heat capacities.

This is perhaps the place to mention some studies which have been performed on amino acids, peptides and related compounds in aqueous urea solution. It is not the intention to give a comprehensive survey here and the earlier work was discussed at length some years ago. [**95**] It is apparent for the amino acids that the partial molar volumes in 8 M urea at 25 °C [**65**] are all about 4 $cm^3\ mol^{-1}$ greater than the corresponding values in water and it seems likely [**44**] that most if not all of this volume change arises from the different dielectric behaviour of the mixed solvent. This is not to indicate that structural features are absent in such solvent systems but rather that the presence of the charged groups tends to mask other effects. Such relatively simple and consistent behaviour is not apparent for the partial molar heat capacities. It is clear that the zwitterionic charge does play a role [**44**] but the more hydrophilic amino acids (arg, asp) have heat capacities which are about 100 $J\ K^{-1}\ mol^{-1}$ greater in urea solutions than in water whereas the more hydrophobic acids (leu, ileu) have heat capacities which are considerably ($\sim 30\ J\ K^{-1}\ mol^{-1}$) less in the urea containing systems. Consequently for any amino acid there are at least two opposing factors involved in determining the experimental quantity. There are fairly strong indications[**42, 96, 97**] that urea does interact relatively favourably with polar residues and for the amino acids, presumably superimposed on any bulk solvent electrostatic effect, there will be a contribution to the heat capacity arising from the fact that intermolecular 'soft' vibrations between urea and the amino and carboxylate groups will be different to the vibrations arising from water–ionic group interactions. It is not possible to quantify these differences, however, but it seems that similar effects are found for small peptides. [**98, 99**] It was mentioned above that there is a correlation between one measure of the hydrophobicity of amino acid side chains and partial molar heat capacities in water. This correlation is absent when urea–water is the solvent and this seems consistent with the idea that the presence of urea inhibits the formation of structured regions about apolar residues. [**95**]

### 3.6. *Partial molar compressibilities and expansibilities*

There have been some reports on the pressure and temperature variations of partial molar volumes. Most of the experimental information on compressibilities has been obtained via velocity of sound measurements [**81–3**] and consequently adiabatic or more properly isoentropic, partial molar compressibilities are obtained. Given appropriate ancillary data these can easily be transformed to isothermal compressibilities but generally the corrections involved are relatively small and usually approximately within the reproducibility of work from different laboratories. Table 19 lists the data which have been obtained for amino acids and peptides.

An attempt was made [**81**] to assess the solvation numbers of the amino acids and an average value of 4.1 was arrived at for the solvation number of the $H_3N^+$–CH–$COO^-$ entity. Given the various approximations which were used and the range of values found in obtaining this average it is difficult to imagine that it has more than a notional link with molecular reality.

It seems to be fairly clear from studies on electrolytes [**100**] that the negative values found for the compressibilities of those zwitterionic species arise principally from the charged groups and in particular from the carboxylate group. There does seem to be a slight tendency to more negative values as the amino acid side chain becomes more apolar and this is consistent with information from studies on a wide range of non-electrolytes in water [**101–7**]. A similar trend is found also for the more hydrophilic species and so little discrimination is apparent. If one looks at the $\alpha,\omega$-acids there does seem to be a trend towards more negative values as the series is ascended. Whether this arises from the apolar residues, for which it has been estimated [**107**] that each methylene group contributes about $-1.9 \times 10^{-4}$ cm$^3$ mol$^{-1}$ bar$^{-1}$ or from the charged groups becoming individually more effective with increase in their separation it is difficult to say. It does seem that a similar but considerably enhanced effect is seen for glycyl peptides and from differences between these peptides and $\alpha,\omega$-acids with the same number of atoms in the backbone, a value of $-12.4$ cm$^3$ mol$^{-1}$ bar$^{-1}$ was derived [**82**] for the isoentropic compressibility of the peptide group. Incidentally the strictly linear increase in compressibility of the glycyl peptides as the number of residues is increased is quite astonishing given the fact that, for example, no such linear increase is found for say, volumes or heat capacities. Using an approach analogous to that mentioned above the

Table 19. *Isoentropic ($\kappa_S^\circ$) and isothermal ($\kappa_T^\circ$) partial molar compressibilities and partial molar expansibilities ($E^\circ$) of some amino acids and small peptides, in water at 25 °C*

| Compound | $-10^4\ \kappa_S^{\circ a}$ ($cm^3\ mol^{-1}\ bar^{-1}$) | $-10^4\ \kappa_T^\circ$ ($cm^3\ mol^{-1}\ bar^{-1}$) | $10^2\ E^{\circ a,a}$ ($cm^3\ mol^{-1}\ K^{-1}$) |
|---|---|---|---|
| gly | 26.0[b] | 22.4 | 10.0 |
| ala | 24.4[b] | 21.3 | 8.8 |
| val | 27.0[b] | 24.0 | 8.0 |
| leu | 31.8 | | |
| ser | | | 8.0 |
| phe | 34.5[c] | | |
| thr | | | 7.9 |
| asn | 26.6[c] | | |
| asp | 33.1[c] | | |
| cys | 32.8[c] | | |
| glu | 36.2[c] | | |
| his | 31.8[c] | | |
| met | 31.2[c] | | |
| trp | 30.2[c] | | |
| pro | 23.3[c] | | |
| α-aminobutyric acid | 21.8[d] | 19.2 | 7.6 |
| β-alanine | 21.5[d] | 18.3 | 8.9 |
| β-aminobutyric acid | 18.7[d] | 16.8 | 5.8 |
| γ-aminobutyric acid | 27.0[d] | 24.5 | 7.3 |
| 5-aminopentanoic acid | 27.3[d] | 25.2 | 6.4 |
| 6-aminohexanoic acid | 29.2[d] | 26.1 | 9.2 |
| glygly | 35.7[e] | | 10.3 |
| triglycine | 44.4[f] | | 13.9 |
| tetraglycine | 53.1[f] | | |
| glyala | | | 10.2 |

[a] Typical errors are about ±1 in the given units.
[b] Mean of the values given in refs 81 and 83. In some instances the discrepancies in values are quite appreciable.
[c] Ref. 81.
[d] Ref. 83.
[e] Mean of the values given in refs 82 and 83.
[f] Ref. 82.

same workers obtained a value of $-8.5\ cm^3\ mol^{-1}\ bar^{-1}$ for the contribution of a glycyl residue.

It is difficult to make any quantitative comments about the partial molar expansibilities; trends may well exist in various series but the indications are that several effects are contributing to the values found and there is not even an indication of a major effect of charges. [83]

### 3.7. *Enthalpy changes*

As was mentioned earlier partial molar enthalpies are experimentally inaccessible quantities and only differences can be obtained. For the amino

Table 20. *Enthalpies of transfer of some amino acids and triglycine from water to 8 M urea solutions at 25 °C. Data at other urea molarities are given in ref.* 108

| Solute | $\Delta_{trans}H^{\infty}$ (kJ mol$^{-1}$) |
|---|---|
| gly | −4.54 |
| ala | −2.32 |
| leu | −0.82 |
| phe | −4.42 |
| trp | −8.86 |
| met | −5.21 |
| thr | −3.97 |
| tyr | −5.05 |
| his | −6.99 |
| asp | −8.97 |
| gln | −7.64 |
| triglycine | −9.52 |

acids, in view of the presence of the charged centres and the fact that these must themselves have a marked influence on enthalpies of transfer, simple relationships with solvation are not to be expected. As an example if we consider the partial molar enthalpies of transfer [**108**] of some amino acids and one peptide from water to urea–water mixtures (see Table 20) it is apparent that all of the solutes considered have exothermic enthalpies of transfer, some of which probably arises from the presence of the charged groups. The amino acids bearing aliphatic side chains do show a tendency towards more positive enthalpies of transfer as the apolarity increases but, in contrast, the relatively hydrophobic acid phenylalanine has an enthalpy of transfer about the same as that for glycine, which is usually considered to be fairly hydrophilic. The tripeptide certainly has a quite exothermic transfer enthalpy but whether this stems from the peptide group or the disposition of the charges is a matter of conjecture.

It is possible, by making the assumption that the ionic head group makes a constant contribution to the enthalpies of transfer, to get relative measures of these. Whether such an assumption can be justified in view of some comments which were made earlier is a matter of opinion, but it does seem that there is little correlation between the results obtained and, say, the heat capacities or free energies of transfer referred to earlier. An enthalpy of transfer study has recently been conducted [**42**] on some of the uncharged amino acid derivatives mentioned earlier and some simple amides (see refs 109 and 110 for some related investigations). The solvent systems studied included urea–water mixtures and the amidic solvent *N*,*N*-dimethylformamide. Some of the results obtained are given in Table 21. There are some similarities in the overall trends of the results in 8 M urea with those referred to above [**108**] but it is apparent by comparing the glycyl, alanyl and

Table 21. *Enthalpies of transfer of some N-acetyl and N-acetyl-N′-methyl amino acid amides from water to 8 M urea and to N,N-dimethylformamide (DMF) at 25 °C*

| | $\Delta_{trans}H^{\infty}$ (kJ mol$^{-1}$) | |
|---|---|---|
| Amide[a] | 8 M urea | DMF |
| GLYMe | −2.03 | 12.97 |
| ALAMe | 0.87 | 19.87 |
| LEUMe | 4.75 | 22.16 |
| PROMe | −1.08 | 26.43 |
| GLY | −4.73 | 0.25 |
| ALA | −2.18 | 6.68 |
| LEU | 1.06 | 9.33 |
| PRO | −4.02 | 13.07 |

[a] The abbreviations are the same as those used in Table 16. The suffix 'Me' denotes the *N*-methyl compound.

leucyl compounds in each of the three sets of data that the intramolecular environment does seem to have a significant effect. The most striking feature of the information given in Table 21 is the wide variation in the enthalpies of transfer from water to DMF. For the solutes considered charged groups are not playing a part and perhaps complicating the issue. The overall picture which emerged **[42]** is that, as in an earlier study, **[111]**, what is being seen is an endothermic effect which increases in magnitude with increase in overall apolarity of the solute and which arises from the breakdown of the hydrophobic solvation regions of the amides. The indications are that the polar parts of the molecules (i.e., the peptide groups) interact thermochemically more favourably with DMF than with water. This is consistent with other work. **[96, 97]**

## 4. Spectroscopic investigations

There is an enormous literature on the use of various spectroscopic probes to investigate the behaviour of amino acids, peptides and many of their derivatives. Most of this is directed towards structural determinations **[112]** and there has been, relatively, only a small amount of work performed on solvent effects. By far the most popular and important technique which has been used is nmr spectroscopy in its various forms and for various nuclei and a great deal of attention has been directed towards the use of nmr to determine conformations in solution (see below). The earlier work using proton resonances **[113]** has been reviewed and this is still a readable and useful introduction to the ideas used. More recent developments in instrumentation have led to an increased use of $^{13}C$ and $^{15}N$ in peptide

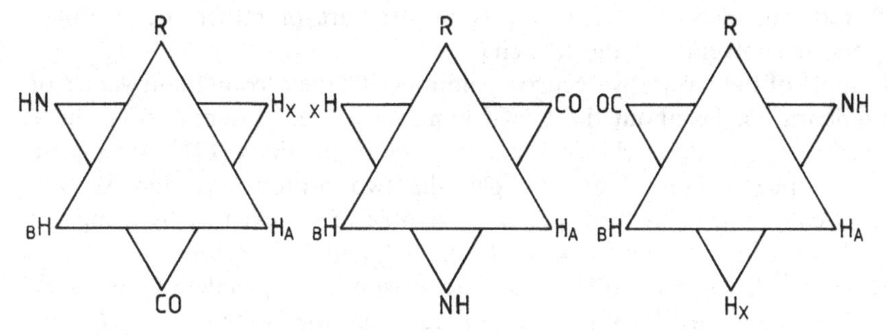


Figure 15. Newman projections along the $C^{\alpha}$—$C^{\beta}$ bond of amino acid residues. The side chain on the $\alpha$-carbon is labelled R.

Table 22. *Rotamer populations of some terminally substituted amino acid and peptide compounds at 22 °C*

| | | Rotamer population | | |
|---|---|---|---|---|
| Solute[a] | Solvent[b] | *a* | *b* | *c* |
| PHEGLY | $D_2O$ | 0.424 | 0.316 | 0.250 |
| GLYPHE | $D_2O$ | 0.171 | 0.481 | 0.348 |
| PHEGLY | $CD_3OD$ | 0.269 | 0.453 | 0.278 |
| PHEGLY | DMSO | 0.138 | 0.558 | 0.304 |
| PHE | $D_2O$ | 0.228 | 0.487 | 0.285 |
| PHE | DMF | 0.230 | 0.404 | 0.366 |

[a] These are all *N*-acetylamides – see Table 16.
[b] The solvents were all deuterated.

containing systems and have considerably enhanced the use of particularly, spin–spin coupling constants **[14]** but other observables **[115, 116]** too, as a conformational probe. There has, however, been little work performed which has tried to link solvation *per se* and nmr information. It is certainly well known, see earlier, that prolyl residues are affected by the solvent environment in that significant variations in *cis–trans* ratios are apparent from one solvent to another. The reasons for such variations are, however, not at all well understood any more than are the reasons why prolyl containing compounds are so soluble in water. The important point can be made that the conformations adopted by even small peptides in condensed liquid phases often bear no relationship to the conformations in the solid state **[117]** and that molecular conformations in crystalline compounds are heavily biased by peptide group–peptide group interactions and by packing considerations. **[118]** In contrast the side chains of most amino acid residues in both amino acids and peptides have the possibility of exhibiting free rotation in solution, although this rotation can be restricted to some extent not only by neighbouring intramolecular structure (i.e., steric effects)

[**119, 120**] but also of course by solvent structure or rather the interplay between the solvent and the side chain.

In most of the naturally occurring amino acids the predominant values of the dihedral angles about the $C^{\alpha}$–$C^{\beta}$ bond have been assumed to be those pertaining to the three classical staggered conformations [**121**] (see Figure 15). For phenylalanine for example, the two protons $H_A$ and $H_B$ are magnetically non-equivalent and it is possible to exploit this by using the vicinal coupling constants between the $C^{\alpha}$ and $C^{\beta}$ protons. The same situation arises for many other amino acids with non-equivalent $C^{\beta}$ protons. [**122**] If it is assumed [**123**] (but see ref. 124) that the individual *gauche* and *trans* vicinal coupling constants are the same in all three rotamers, it is possible to obtain values of the relative rotamer populations of a side chain about the $C^{\alpha}$–$C^{\beta}$ bond from an analysis of the spin–spin coupling pattern.

The coupling constants $J_{AX}$ and $J_{BX}$ are given by the population weighted averages (*a,b* and *c*) of the rotamers and for example

$$J_{AX} = aJ_g + bJ_t + cJ_g \tag{32}$$

where $J_g$ and $J_t$ are the coupling constants of the *gauche* and *trans* forms, with a similar expression for $J_{BX}$. Utilisation of the Karplus relationship [**125**] with appropriate values [**126**] for $J_g$ and $J_t$ allows the rotamer populations to be deduced.

To illustrate that for even small substituted amino acids and peptides rather marked changes in populations are evident and dependent on both sequence and solvent, some information [**87, 127**] is presented in Table 22. There are several things apparent from this table. Firstly in a given solvent the sequence in which the amino acid residue occurs has a marked effect on the rotamer population distribution. This must mean from packing considerations alone that in the solvation of the phenylalanyl the residue must be markedly influenced by its intramolecular neighbours. Secondly, for a given solvent the populations of the rotamers vary quite significantly with the solute and it is not possible to say that the local solvational environment of the phenylalanyl residue is the same for different peptidic species. In other words, the indications are that, even for relatively small molecules, it seems unlikely that one can use information obtained from one species to predict the behaviour of another. Thirdly, and finally, for the same solute in different solvents the populations may or may not vary. The thing one can say, quite categorically, is that all of these results indicate that solvation is clearly important in determining the 'shape' of the molecules in solution. What is needed now is for a link to be established between results such as these and the macroscopic information which has largely been the scope of the present chapter. This will be no easy task since, for most small molecules in water, usually no one molecular conformation is dominant but rather what one has is a large number of accessible states all of which contribute to observed, experimental properties.

## References


1. S. Hunt. In *Chemistry and Biochemistry of the Amino Acids* (ed. G. C. Barrett). Chapman and Hall: London, 1985, Ch. 4.
2. See e.g., J. L. Bada. *Interdiscip. Sci. Rev.* **7**, 30(1982).
3. Y.-C. Tse, M. D. Newton, S. Vishveshwara and J. A. Pople. *J. Amer. Chem. Soc.* **100**, 4329(1978).
4. M. J. Locke, R. L. Hunter and R. T. McIver. *J. Amer. Chem. Soc.* **101**, 272(1979).
5. P. Haberfield. *J. Chem. Educ.* **57**, 346(1980).
6. See e.g., I. D. Kuntz and W. Kauzmann. *Adv. Protein Chem.* **28**, 239(1974).
7. E. R. Stimson, S. S. Zimmerman and H. A. Scheraga. *Macromolecules* **10**, 1049(1977).
8. C. Grathwhol and K. Wuthrich. *Biopolymers* **15**, 2025(1976).
9. J. Wyman. *Chem. Revs.* **19**, 213(1936).
10. L. Onsager. *J. Amer. Chem. Soc.* **58**, 1486(1936).
11. A. D. Buckingham. *Aust. J. Chem.* **6**, 93 323(1953).
12. C. J. F. Bottcher. *Rec. Trav. Chim.* **62**, 119(1943).
13. J. J. Edward, P. G. Farrell and J. L. Lob. *J. Amer. Chem. Soc.* **96**, 902(1974).
14. J. Kirchnerova, P. G. Farrell and J. T. Edward. *J. Phys. Chem.* **80**, 1974(1976).
15. J. A. Walder. *J. Phys. Chem.* **80**, 2777(1976).
16. U. Kaatze, H. Bieler and R. Pattel. *J. Mol. Liquids*, **30**, 101(1985).
17. J. L. Salefran. *Adv. Mol. Relaxation Interaction Processes* **19**, 75(1981).
18. D. A. Horsma and C. P. Nash. *J. Phys. Chem.* **72**, 2351(1968).
19. J. W.-O. Tam and C. P. Nash. *J. Phys. Chem.* **76**, 4033(1972).
20. S. U. Kokpol, P. B. Doungdee, S. V. Hannongbua, B. M. Rode and J. P. Limtrakul. *J. Chem. Soc. Faraday Trans 2.* **84**, 1789(1988).
21. J. Sadlej. *Semi-Empirical Methods of Quantum Chemistry.* Ellis Horwood: Chichester, 1985.
22. A. Pullman and B. Pullman. *Quarterly Rev. Biophys.* **7**, 4(1975).
23. P. Claverie, J. P. Daudey, J. Langlet, B. Pullman, D. Piazzola and M. J. Huron. *J. Phys. Chem.* **82**, 405(1978).
24. L. Carozzo, G. Corongiu, C. Petrongolo and E. Clementi. *J. Chem. Phys.* **68**, 787(1978).
25. E. Clementi. *J. Chem. Phys.* **46**, 3851(1967).
26. N. R. Kestner. *J. Chem. Phys.* **48**, 252(1968).
27. S. F. Boys and F. Bernardi. *Mol. Phys.* **37**, 1529(1979).
28. J. H. van Lentre, T. van Dam, F. B. van Duijneveldt and L. M. Kroon-Battenberg. *Faraday Symp. Chem. Soc.* **19**, 125(1984).
29. F. T. Marchese, P. K. Mehrotra and D. L. Beveridge. *J. Phys. Chem.* **88**, 5692(1984).
30. W. L. Jorgensen and C. J. Swenson. *J. Amer. Chem. Soc.* **107**, 14899(1985).
31. P. J. Rossky, M. Karplus and A. Rahman. *Biopolymers* **18**, 825(1979).
32. P. J. Rossky and M. Karplas. *J. Amer. Chem. Soc.* **101**, 1913(1979).
33. M. Karplus and P. J. Rossky. In *Water in Polymers* (ed. S. P. Rowland). Amer. Chem. Soc.: Washington DC, 1980, p. 25.
34. M. Mezei, P. K. Mehrotra and D. L. Beveridge. *J. Biomolec. Struct. Dyn.* **2**, 1(1984).

**35.** S. Romano and E. Clementi. *Int. J. Quantum Chem.* **XIV**, 839(1978).
**36.** O. Matsuoka, E. Clementi and M. Yoshimine. *J. Chem. Phys.* **64**, 1351(1976).
**37.** D. L. Beveridge, M. Mezei, P. K. Mehrotra, F. T. Marchere, G. Ravi-Shankar, T. R. Vasu and S. Swaminathan. In *Molecular Based Study of Fluids* (eds J. M. Haile and G. A. Mansoori). Amer. Chem. Soc.: Washington DC, 1983.
**38.** S. Swaminathan, S. W. Harrison and D. L. Beveridge. *J. Amer. Chem. Soc.* **100**, 5701(1978).
**39.** For a review see F. H. Stillinger. *Adv. Chem. Phys.* **31**, 1(1975).
**40.** G. Ferro, G. Della Gatta and G. Barone. In press.
**41.** G. Barone, G. Della Gatta and T. H. Lilley, unpublished observations.
**42.** A. H. Sijpkes, A. A. C. M. Oudhuis, G. Somsen and T. H. Lilley. *J. Chemical Therm.* in press.
**43.** H. L. Friedman. *Ionic Solution Theory*, Interscience, New York, 1962.
**44.** T. H. Lilley. In *Biochemical Thermodynamics*, 2nd edition (ed. M. N. Jones). Elsevier: Amsterdam, 1988, Ch. 1. This gives reference to earlier work.
**45.** W. G. McMillan and J. E. Mayer. *J. Chem. Phys.* **13**, 276(1945).
**46.** H. L. Friedman. *J. Solution Chem.* **1**, 387, 413, 419(1972).
**47.** J. J. Kozak, W. S. Knight and W. Kauzmann. *J. Chem. Phys.* **48**, 675(1968).
**48.** R. H. Wood, T. H. Lilley and P. T. Thompson. *J. Chem. Soc. Faraday Trans. 1*, **74**, 1990(1978).
**49.** G. R. Hedwig, J. F. Reading and T. H. Lilley. In preparation.
**50.** A. Ben-Naim. *J. Phys. Chem.* **82**, 792(1978).
**51.** J. Fendler, F. Nome and J. Nagyvary. *J. Mol. Evol.* **6**, 215(1975).
**52.** S. Damodaran and K. B. Song. *J. Biol. Chem.* **261**, 7220(1986).
**53.** Y. Nozaki and C. Tanford. *J. Biol. Chem.* **246**, 2211(1971).
**54.** R. Wolfenden, L. Andersson, P. M. Cullis and C. C. B. Southgate. *Biochemistry* **20**, 849(1981).
**55.** J. A. V. Butler. *Trans. Faraday Soc.* **33**, 229(1937).
**56.** J. Hine and P. K. Mookerjee. *J. Org. Chem.* **40**, 292(1975).
**57.** E. Wilhelm, R. Battino and R. J. Wilcock. *Chem. Revs.* **77**, 219(1977).
**58.** H. B. Bull and K. Breese. *Arch. Biochem. Biophys.* **161**, 665(1974).
**59.** C. Chothia. *J. Mol. Biol.* **105**, 1(1976).
**60.** B. Lee and F. M. Richards. *J. Mol. Biol.* **55**, 379(1971).
**61.** J. Janin. *Nature(London)* **277**, 491(1979).
**62.** R. Wolfenden. *Biochemistry* **17**, 201(1978).
**63.** G. Barone, G. Castronuovo, G. Della Gatta, V. Elia and A. Iannone. *Fluid Phase Equilibria* **21**, 157(1985).
**64.** G. Della Gatta, G. Barone and V. Elia. *J. Solution Chem.* **15**, 157(1986).
**65.** C. Jolicoeur, B. Riedl, D. Desrochers, L. L. Lemelin, R. Zamojska and O. Enea. *J. Solution Chem.* **15**, 109(1986). This paper also gives references to earlier work.
**66.** C. Jolicoeur and J. Boileau. *Canad. J. Chem.* **56**, 2707(1978).
**67.** A. K. Mishra and J. C. Ahluwalia. *J. Phys. Chem.* **88**, 86(1984).
**68.** J. F. Reading and G. R. Hedwig. *J. Solution Chem.* **18**, 159(1989).
**69.** R. Bhat and J. C. Ahluwalia. *J. Phys. Chem.* **89**, 1099(1985).
**70.** G. R. Hedwig, private communication.
**71.** J. C. Ahluwalia, C. Ostiguy, G. Perron and J. E. Desnoyers. *Canad. J. Chem.* **55**, 3364(1977).

**72.** F. Shahida and P. G. Farrell. *J. Chem. Soc. Faraday Trans. 1* **74**, 858(1978).
**73.** J. G. Kirkwood and F. P. Buff. *J. Chem. Phys.* **19**, 774(1951).
**74.** J. E. Garrod and T. M. Herrington. *J. Phys. Chem.* **73**, 1877(1969).
**75.** A. Ben-Naim. *Water and Aqueous Solutions*. Plenum: New York, 1974, pp. 137–69.
**76.** J. F. Reading, P. A. Carlisle, G. R. Hedwig and I. D. Watson. *J. Solution Chem.* **18**, 131(1989).
**77.** T. M. Herrington and E. L. Mole. *J. Chem. Soc. Faraday Trans. 1* **78**, 213(1982).
**78.** T. M. Herrington, A. D. Pethybridge, B. A. Larkin and M. G. Roffey. *Chem. Soc. Faraday Trans. 1* **79**, 845(1983).
**79.** T. H. Lilley. In *Chemistry and Biochemistry of the Amino Acids* (ed. G. C. Barrett). Chapman and Hall: London, 1985, Ch. 21.
**80.** T. E. Leslie and T. H. Lilley. *Biopolymers* **24**, 695(1985).
**81.** F. J. Millero, A. LoSurdo and C. Shin. *J. Phys. Chem.* **82**, 784(1978).
**82.** M. Iqbal and R. E. Verrall. *J. Phys. Chem.* **91**, 967(1987).
**83.** S. Cabani, G. Conti, E. Matteoli and M. R. Tine. *J. Chem. Soc. Faraday Trans. 1* **77**, 2377(1981).
**84.** C. Jolicoeur and G. Lacroix. *Canad. J. Chem.* **54**, 624(1976).
**85.** E. J. Cohn and J. T. Edsall. *Proteins, Amino Acids and Peptides*, Amer. Chem. Soc. Monographs. Hafner: New York, 1965.
**86.** See ref. 44.
**87.** T. E. Leslie and T. H. Lilley, unpublished work.
**88.** J. E. Desnoyers, C. De Visser, G. Perron and P. Picker. *J. Solution Chem.* **5**, 605(1976).
**89.** See e.g. N. Nichols, R. Skold, J. Suurkuusk and I. Wadso. *J. Chem. Therm.* **8**, 1081(1976).
**90.** K. P. Prasad and J. C. Ahluwalia. *J. Solution Chem.* **5**, 491(1976).
**91.** J. F. Reading, G. R. Hedwig and T. H. Lilley. In preparation.
**92.** G. Roux, G. Perron and J. E. Desnoyers. *Canad. J. Chem.* **56**, 2808(1978).
**93.** S. Cabani, G. Conti and E. Matteoli. *Biopolymers* **16**, 465(1977).
**94.** R. B. Hermann. *Proc. Nat. Aca. Sci. USA* **74**, 4144(1977).
**95.** F. Franks and D. Eagland. *CRC Critical Revs in Biochem.* **3**, 165(1975).
**96.** P. J. Cheek and T. H. Lilley. *J. Chem. Soc. Faraday Trans. 1* **84**, 1927(1988).
**97.** G. Barone, G. Castronuovo, P. Del Vecchio and C. Giancola. *J. Chem. Soc. Faraday Trans. 1* **84**, 1919(1988).
**98.** O. Enea and C. Jolicoeur. *J. Phys. Chem.* **86**, 3870(1982).
**99.** B. Riedl and C. Jolicoeur. *J. Phys. Chem.* **88**, 3348(1984).
**100.** J. G. Mathieson and B. E. Conway. *J. Solution Chem.* **3**, 455(1974).
**101.** F. Franks, J. R. Ravenhill and D. S. Reid. *J. Solution Chem.* **1**, 3(1972).
**102.** T. Nakajima, T. Komatsu and T. Nakagawa. *Bull. Chem. Soc. Japan* **48**, 788(1975).
**103.** H. Hoiland and E. Vikingstad. *Acta Chem. Scand. Ser. A.* **30**, 692(1976).
**104.** A. LoSurdo, C. Shin and F. J. Millero. *J. Chem. Eng. Data* **23**, 197(1978).
**105.** S. Harada, T. Nakajima, T. Komatsu and T. Nakagawa. *J. Solution Chem.* **7**, 463(1978).
**106.** M. V. Kaulgud and K. S. M. Rao. *J. Chem. Soc. Faraday Trans. 1.* **75**, 2237(1979).

**107.** S. Cabani, G. Conti and E. Matteoli. *J. Solution Chem.* **8**, 11(1979).
**108.** M. Abu-Hamdiyyah and A. Shehabuddin. *J. Chem. Eng. Data* **27**, 74(1982).
**109.** E. R. Stimson and E. E. Schrier. *Biopolymers* **14**, 487(1975).
**110.** P. K. Nandi and D. R. Robinson. *Biochemistry* **23**, 6661(1984).
**111.** A. C. Rouw and G. Somsen. *J. Chem. Soc. Faraday Trans. 1* **78**, 3397(1982).
**112.** See e.g., several of the chapters in *Chemistry and Biochemistry of the Amino Acids* (ed. G. C. Barrett). Chapman and Hall: London, 1985.
**113.** F. A. Bovey, A. I. Brewster, D. J. Patel, A. E. Tonelli and D. A. Torchia. *Acc. Chem. Res.* **5**, 193(1972).
**114.** V. F. Bystrov. *Prog. NMR Spectros.* **10**, 41(1976).
**115.** R. Deslauriers and I. C. P. Smith. In *Topics in $^{13}C$ NMR Spectroscopy* (ed. G. C. Levy). Wiley: New York, p. 1.
**116.** O. W. Howarth and D. M. J. Lilley. *Prog. NMR Spectros* **12**, 1(1978).
**117.** H. Kessler, G. Zimmerman, H. Forster, J. Engel, G. Oepen and W. S. Sheldrick. *Angew. Chem. Int. Edn.* **20**, 105333(1981).
**118.** M. Avignon and J. Lascombe. *Jerusalem Symposium on Quantum Chemistry* **5**, 97(1973).
**119.** A. Nakamura and O. Jardetsky. *Biochemistry* **7**, 1126(1968).
**120.** V. F. Bystrov, S. L. Portnova, V. I. Tsetlin, V. T. Ivanov and Yu. A. Ochinnikov. *Tetrahedron* **25**, 493(1969).
**121.** J. R. Cavanaugh. *J. Amer. Chem. Soc.* **90**, 4533(1968).
**122.** See e.g., A. J. Fischman, H. R. Wyssbrod, W. C. Agosta, F. H. Field, W. A. Gibbons and D. Cowburn. *J. Amer. Chem. Soc.* **99**, 2953(1977).
**123.** K. G. R. Pachler. *Spectrochim. Acta* **20**, 581(1964).
**124.** J. Feeney. *J. Mag. Resonance* **21**, 473(1976).
**125.** M. Karplus. *J. Chem. Phys.* **30**, 11(1959).
**126.** See e.g., R. B. Martin. *J. Chem. Phys.* **83**, 2404(1979).
**127.** H. E. Kent. Unpublished observations.

# Solution properties of low molecular weight polyhydroxy compounds

FELIX FRANKS

*Biopreservation Division, Pafra Ltd, 150 Cambridge Science Park, Cambridge CB4 4GG, UK*

AND

J. RAUL GRIGERA

*University of La Plata, IFLYSIB, cc 565, 1900 La Plata, Argentina*

## 1. Introduction

### 1.1. *Historical perspective: scope of review*

Polyhydroxy compounds (PHC) in general, and carbohydrates in particular, have until recently been the preserve of the organic chemist. The emphasis has been on synthesis, stereochemistry, derivatization, reaction mechanisms and, more recently, crystal structures. During the past decade biochemical interest in sugars has grown, partly from the realization that oligosaccharide chains (attached to proteins) may have important biological functions. A recent review, entitled 'Glycobiology', [**1**] in which glycosylation mechanisms are discussed in depth, well illustrates the increasing importance of this new subdiscipline.

PHCs are characterized by their polar nature and their ability to participate in hydrogen bonding, both as donors and acceptors. As a class of chemical compounds they therefore display a high affinity for water and other polar solvents. It is to be expected that solute–solvent hydrogen bonding plays some role in determining the structures and conformations of PHCs in solution, and indirectly, also their interactions with one another, perhaps even intramolecular interactions. Surprisingly, until quite recently, little attention was paid to such problems; the solution chemistry of simple carbohydrates was almost non-existent. Suggett, in what is probably the first criticial review of the subject, concluded that throughout the vast literature devoted to polysaccharides, water '...was treated implicitly as the universal inert filler'. [**2**] This situation is almost as true today as it was in 1975 when the observation was made.

The aim of this review is to summarize and analyse the present state of our knowledge regarding the influence of solvation interactions on the structural, equilibrium and dynamic properties of monomeric and oligomeric PHCs, with a few short excursions into the realm of their polymers. Although the discussions will concentrate on sugars and sugar alcohols, some of their more common derivatives will also be considered. Brief mention will also be made of solution interactions of PHCs with third components, in particular ions, proteins and lipids. Emphasis will be placed on more recent experimental, theoretical and computational techniques which are now being increasingly applied to studies of PHC conformations and interactions. While it has long been realized that the molecular shapes of polysaccharides are largely responsible for the wide range of physico-chemical properties displayed by chemically very similar molecules, **[3]** any possible solvent involvment in determining, or even modifying, these shapes has been almost universally ignored.

In view of their ubiquity and varied functions, it is surprising that PHCs should have remained the Cinderellas of physical chemistry for so long and that so little attention has been paid to the influence of the solvent, usually water, on their stereochemistry and, hence, their interactions with one another and with other types of molecules. The available information, meagre though it is, has been reviewed, first by Suggett, **[2]** and later by one of us. **[4, 5]** Cesaro, in his summary review of the solution thermodynamics of PHCs, may have correctly identified the reason for their long neglect by physical chemists. **[6]** He quotes statements made by two previous workers, judging the state of our knowledge of the conformational interconversions and the predictive ability of current theoretical treatments.

In 1979, one of us took the pessimistic (realistic?) view that 'there is, as yet, no unified treatment which is capable of accounting for internal energies of monosaccharides, let alone free energies in solution' and 'attempts to incorporate solvation corrections into the conformational energy calculations cannot be regarded as anything better that arm-waving.' **[4]** A more sanguine view was expressed three years later: **[7]** '... the calculations are now refined to a point at which an accurate prediction can usually be made about the shapes and configurations that will predominate in sugar solutions, as well as the proportions of each form.' The interested reader may judge, if he stays the course of this narrative, which of these assessments corresponds more closely to the state of affairs now, a decade later.

### 1.2. *The place of PHCs in chemistry, biochemistry and chemical technology*

Along with amino acids, purine and pyrimidine bases and lipids, sugars and their derivatives are building blocks in the chemistry of life processes. They play a central role in metabolic reaction cycles, while their polymers fulfil biological functions, almost as diverse as those of proteins. Carbo-

Table 1. *Occurrence and application of PHCs in biology and technology*

| Function | Example | Nature of PHCs |
|---|---|---|
| Genetic control | Nucleotides | Ribose, deoxyribose |
| Energy source | Starch | Glucose ($\alpha$1,4-linked) |
| Load bearing fibres | Cellulose | Glucose ($\beta$1,4-linked) |
| Metabolic intermediates | Glycolytic products | Sugar phosphates |
| Sweeteners | Food additives | Sucrose |
| Cell–cell recognition | Blood group determinants | Mannose oligomers |
| Connective tissue | Hyaluronic acid | Glucuronic acid, *N*-Acetyl glucosamine |
| Natural antifreezes | Antarctic fish Anti-freeze protein | Galactose |
| Antigens | *B. dysenteriae* | Uronic acids, amino sugars |
| Osmoregulators | Cryoprotectants | Sorbitol, trehalose |
| Oil drilling fluid | Xanthan gum | Glucose, mannose, Glucuronic acid |
| Food gelling agents | Alginates | Mannuronic acid, guluronic acid |

hydrates are also used extensively in many branches of chemical technology. Thus, 95% of all water soluble polymers used by industry are carbohydrate-based, mostly chemically modified in various ways.

Some important biological and technological functions of PHCs and their derivatives are summarized in Table 1. Sugars also occur covalently linked to other molecules where their functions have not yet been clearly identified. For instance, it is believed that a majority of proteins *in vivo* are in fact glycoproteins, since they contain short oligosaccharide chains which, during isolation, separation and purification processes, are removed, perhaps inadvertently, by hydrolysis. The influence of such short sugar chains on the folding, stability, activity or transport of the protein is a subject of lively current investigation. [**8**]

Sugars are also encountered as polar head groups in membrane lipids. Indeed, some organisms seasonally exchange phospholipids for glycolipids, usually in response to some physiological stress, such as chill, desiccation of salinity. [**9**] The significance of such modifications, as they affect the chemical or mechanical properties of membranes, is not fully understood.

Many of the technological applications of carbohydrates rely on the unique colloidal and rheological properties of their dilute solutions. The various types of flow behaviour encountered encompass thixotropy, dilatancy and ion-, pH- or temperature-induced gelling. [**10**] It is hard to see how a true understanding of these properties can be achieved without due consideration of the solvent contribution to chain–chain interactions, especially, since most of the exotic rheological properties appear to be confined to *aqueous* solutions.

Of very recent realization is the significance of sugar–water interactions in low moisture carbohydrate-based systems. [**11**] Such mixtures resemble the better-studied synthetic polymer/plasticizer systems, their observed thermomechanical properties being governed by glass–rubber transitions. An increased acceptance of these developments will have a profound effect on several processing industries, especially foods and pharmaceuticals, where one still finds equilibrium concepts, such as water activity or sorption isotherms, being applied in situations where process and product attributes are determined by very slow kinetics, i.e., stationary states, rather than by equilibrium properties.

The possibility of glassy carbohydrate/water states is now also being discussed in connection with mechanisms of cold resistance in plants and insects, [**12**] whilst cryobiologists are speculating on the possibility of utilizing such aqueous vitreous states towards the achievement of the long-term preservation of tissues and organs. [**13**]

## 2. Stereochemistry and nomenclature

### 2.1. *Configurational complexity and chemical heterogeneity*

To the student of hydration effects, PHCs present a unique challenge because, unlike lipids and peptides, PHCs are stereochemically labile and offer a large variety of chemically very similar, but configurationally quite dissimilar molecules. Furthermore, in some cases free energy differences between various isomers are small so that, under suitable conditions, two or more structurally distinct species can coexist in equilibrium. The nature of the solvent medium can then significantly affect the positions of such equilibria.

The chemical heterogeneity of PHCs in solution gives rise to problems in the interpretation of experimental results in terms of stereochemical detail and also reduces the credibility of some theoretical approaches to the study of sugar–solvent interactions, because in real life, unique conformational species rarely exist in solution. The various modes of intramolecular conversions are treated in more detail in later sections. The problems posed by compositional heterogeneity can of course be overcome by chemical modifications, e.g., through the blocking of anomeric —OH groups by esterification, but the hydration interactions of sugars are so sensitive to stereochemical detail that such chemical interference may lead to unacceptably large changes in these very interactions (and also in solute–solute interactions), [**14**] so that even minor chemical modifications can turn out to be self-defeating, where the parent sugar is to be the object of study.

Figure 1. Envelope (E) and twist (T) forms of furanoid rings. Altogether ten possible E and T conformers can exist.

## 2.2. *Nomenclature*

The simple generic formula of carbohydrates, $(C.H_2O)_n$, belies their extreme stereochemical complexities. It has therefore been necessary to construct a framework of nomenclature to describe the various isomeric forms. The procedures to be used are embodied in 35 Rules of Carbohydrate Terminology which are lucidly described and explained in Shallenberger's classic, *Advanced Sugar Chemistry*, **[15]** and will only be summarized here insofar as they affect the main topic of this review. We shall therefore dispense with a discussion of the problems relating to chirality, despite scattered reports claiming distinguishable differences in the interactions between D- and L- sugars in solution. **[16]** Hereafter it will be assumed that all references are to D-enantiomers, unless otherwise stated.

The numbering of atoms in this review is as follows: the anomeric carbon atom is denoted by C(1) and the pyranose ring oxygen by O(5). Hydrogen atoms attached to carbons bear the number of the carbon atom e.g. H(1) is the (ring) hydrogen on C(1) and OH(1) is the anomeric —OH group. In the polyol or sugar methylene groups, hydrogens are shown as H(*n*) and H*n*′).

*Furanoses* The furanoses are a family of five-membered rings, composed of four carbon atoms and one oxygen atom, based on the tetrahydrofuran (I) structure. Examples of common sugars are $\beta$-glucofuranose (II), $\beta$-ribofuranose (III) and $\beta$-fructofuranose (IV).

$CH_2OH$ OH $CH_2OH$ OH $HOCH_2$ HO
O HO O O $CH_2$ O OH
HO OH OH
OH OH $CH_2OH$

I II III IV

The five-membered ring can exist in two basic conformations: the envelope (E) and the twist (T), shown in figure 1. In the E-form, four atoms are coplanar, while in the T-form the plane is defined by three atoms. In principle, each form can exist in ten distinguishable conformations. In solution, the energy barriers separating the various conformations are believed to be low, as are also the differences in free energies of the various species. Furanoses are therefore believed to exist as rapidly interconverting equilibrium mixtures rather than as unique conformers. The —OH group orientations cannot thus be exactly defined, nor can neighbouring —OH groups be descirbed exactly as *trans* or *gauche* to one another. We shall return later to the computational problems caused by furanoid ring flexibility.

*Pyranoses* The pyranoses are a family of six-membered rings, composed of five carbon atoms and one oxygen atom and based on the tetrahydropyran (V) structure which resembles that of cyclohexane. Common examples are β-glucopyranose (VI), β-ribopyranose (VII) and β-fructopyranose (VIII). Eight strainless pyranoid rings can exist, two chair forms and six boat forms. In each case the plane is defined by the four atoms C(2), C(3), C(5) and O. The two chair forms are written respectively as $^4C_1$ and $^1C_4$, as shown in IX and X. Thus, the superscript denotes the atom which lies above the plane and the subscript the atom below the plane. The respective abbreviated notations which will be used from here on are C1 and 1C.

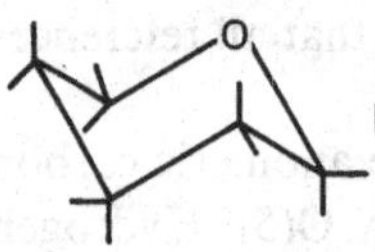

V 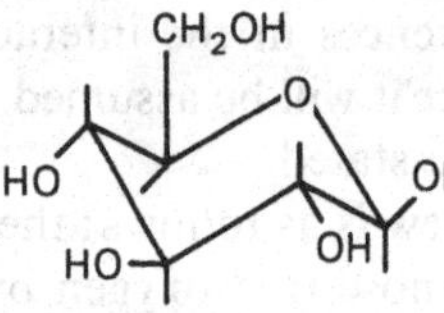 VI 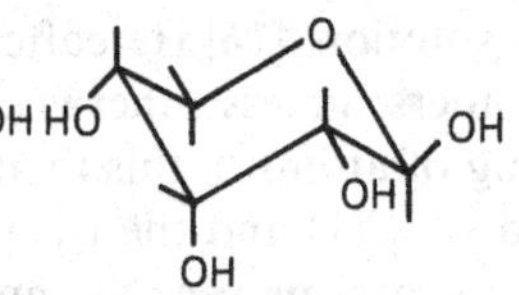 VII

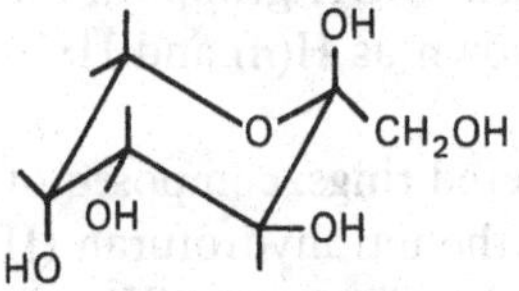

VIII 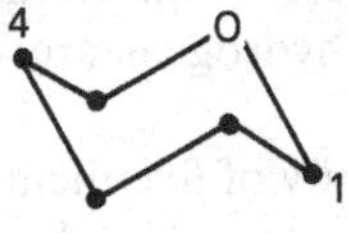 IX 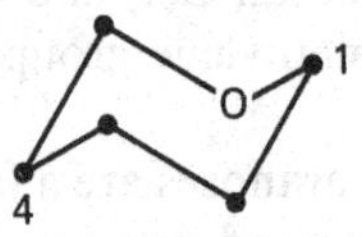 X

In addition to the strainless conformers, 'skewed boat', 'half-chairs' (H) and 'sofa' (S) structures can be envisaged, presumably as possible transition states during the interconversion of energetically more stable forms. Figure 2 shows the various conversion pathways of α-pyranoses.

*Tautomers and tautomerism* Tautomers are easily interconvertible configurational isomers which, in the case of sugars, often coexist in equilibrium, e.g., pyranose C1 ⇌ pyranose 1C or pyranose ⇌ furanose.

*Anomers and anomerization (mutarotation)* This is the name given to configurational isomers at the C(1) atom of the ring form; in α- and β-pyranose sugars the —OH group on C(1) is in the axial and equatorial

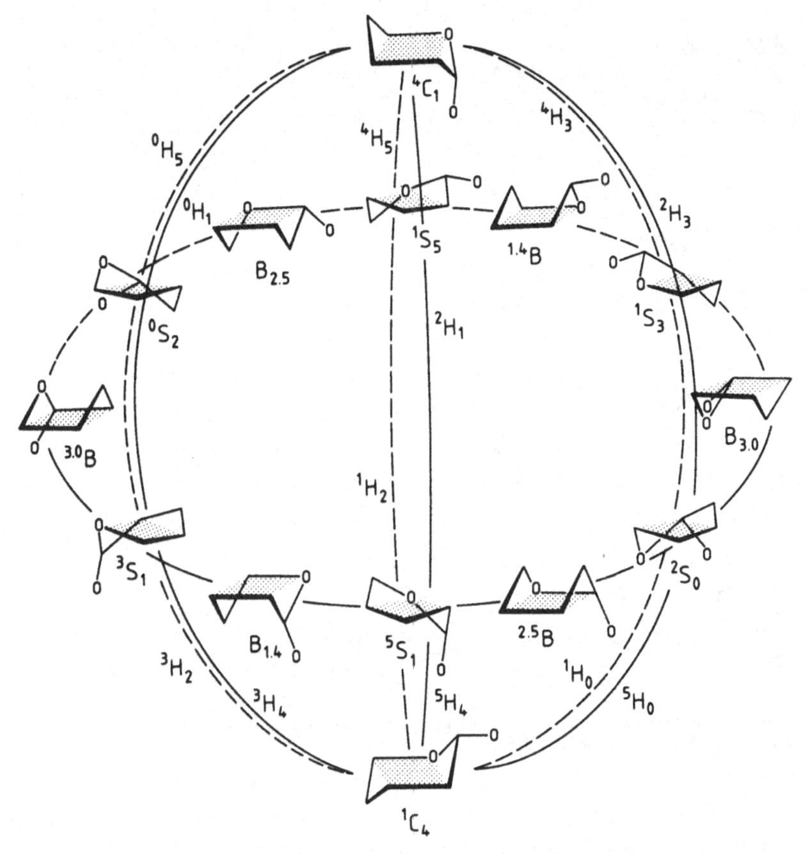


Figure 2. Ring conformations of α-pyranoses and their positions on the conformational sphere. The twelve half-chair conformations and the interconversion pathways are also indicated. Reproduced, with permission, from ref. 6.

position respectively. During anomerization the ring opens to produce an acyclic intermediate. The ring is then reformed, but different ways of closing the ring are possible, as shown in Figure 3, some involving tautomeric changes (pyranose–furanose conversion). In a given sugar solution a number of different anomeric forms can therefore coexist in equilibrium. Anomerization is described as simple or complex, depending on whether the process is of a two-state nature or whether multiple steps and several equilibria are involved.

*Diastereoisomerism* This is the name given to configurational isomers containing two or more asymmetric carbon atoms. Thus, mannose (XI) and galactose (XII) are 2,4-diastereoisomers.

*Epimerism* This is the simple case of diastereoisomerism in which a

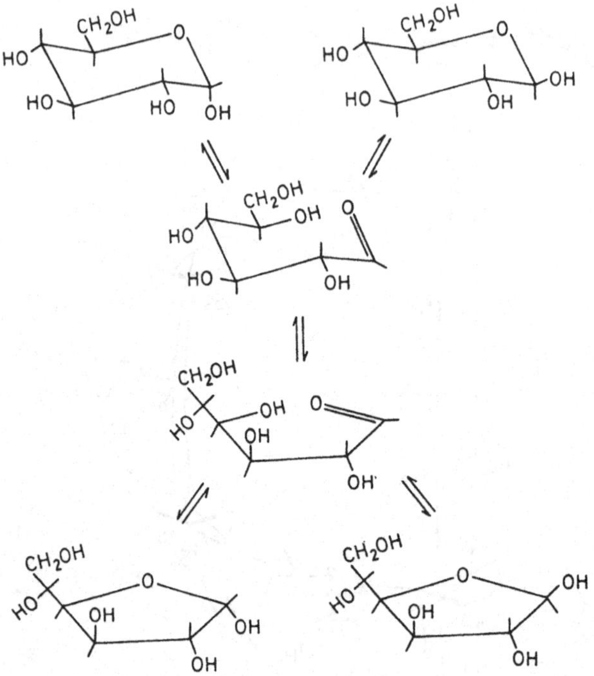

Figure 3. Anomeric and tautomeric forms of glucose which can be formed by ring opening and closure; only C1 pyranoses are shown.

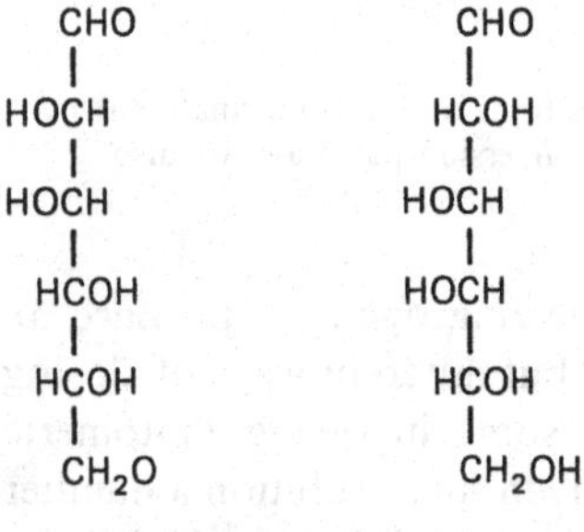

| XI | XII |
|---|---|
| CHO | CHO |
| HOCH | HCOH |
| HOCH | HOCH |
| HCOH | HOCH |
| HCOH | HCOH |
| $CH_2O$ | $CH_2OH$ |

transposition at only one carbon atom, usually C(2), is involved. Glucose and mannose are epimers: in glucopyranose C1 the —OH group on C(2) is equatorial, while in the corresponding mannose tautomer it is axially placed.

*Acyclic sugars and sugar alcohols* Open chain (acyclic) sugars and sugar alcohols possess a degree of chain flexibility. It is often assumed that for a complete rotation about a C—C bond there exist three stable positions. The two *gauche* positions are equivalent and are expected to be less stable than

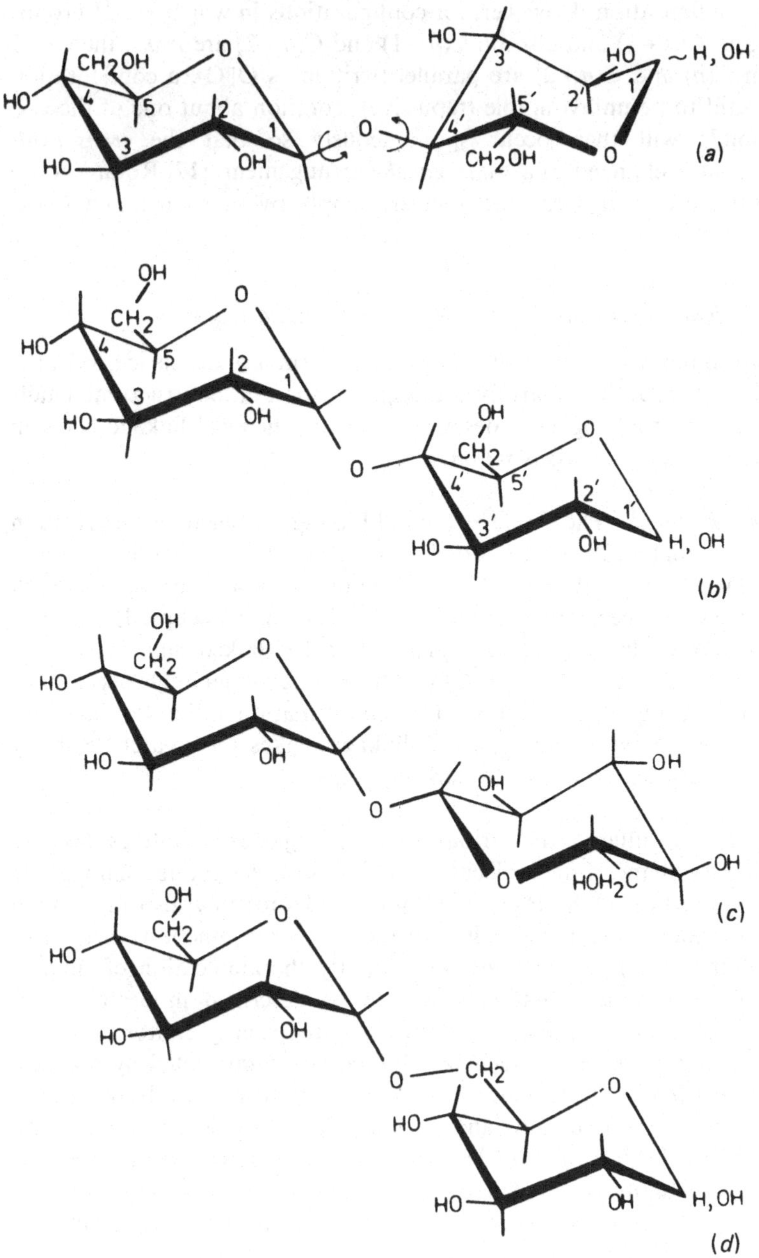

Figure 4. Diglucoses with different glucosidic linkages: (*a*) cellobiose ($\beta$ 1→4 linkage); (*b*) maltose ($\alpha$ 1→4 linkage); (*c*) $\alpha,\alpha$ trehalose ($\alpha$, 1→1 linkage); (*d*) isomaltose ($\alpha$, 1→6 linkage).

the *trans* configuration. However, for configurations in which —OH groups on C($n$) and C($n+1$) and also on C($n+1$) and C($n+2$) are *trans*, the C—O bonds on C($n$) and C($n+2$) are parallel (written as O‖O), a configuration which is said to be unfavourable (repulsive); rotation about one of the two C—C bonds will then occur spontaneously so that the *trans–trans* configuration is changed to a *trans–gauche* arrangement. [**17**] Rotamers are therefore isomers which can interconvert simply by rotation about C—C bonds.

### 2.3. *Additional nomenclature for di- and oligosaccharides*

When two monosaccharides are condensed to form a disaccharide, with the elimination of $H_2O$, then apart from the specification of the structural details of the two indivdual sugars, a description of the chemical linkage between them (the glycosidic bond) is required.

*The glycosidic bond* The C—O—C′ bond linking the sugars is described in terms of two angles of rotation. For 1,4′-linked sugars the significant angles are $\phi$ (C(1)—O) and $\psi$ (C(4′)—O), as shown in Figure 4 for cellobiose. Each angle is zero when the relevant C—O and C—H bonds are eclipsed, i.e., when H, C, O and C′ lie in the same plane. For 1,6′-linked sugars, such as isomaltose, also shown in Figure 4, in additional torsional angle needs to be specified for a complete description of the conformation of the two rings with respect to each other. Clearly the 1,6′-linkage makes for greater flexibility than direct linkages between two carbon atoms.

*$\phi$, $\psi$ maps* The values of the various torsional angles are of interest because they define the shape of the molecule which, in turn, determines many of the physical properties of the PHC in solution and probably also in its solid amorphous states. The actual values of the angles are functions of the free energy of the molecule in its environment. By the application of suitable theoretical or computational methods, to be described in Section 4, it would, in principle, be possible to construct free energy contour maps in terms of $\phi$ and $\psi$. Actually, such calculations are complicated by practical problems relating to the estimation of entropic contributions. In practice, it is usually the potential energy, and not the free energy, which is calculated and related to the torsional angles; an example is shown in Figure 5. The simplifying assumption is then made that the difference between the potential energy and the free energy is insignificant or, at least, constant and independent of the environment (i.e. vacuum or solution).

A particularly interesting application of torsional angle measurements is in comparisons of PHC conformations in the crystal, in the isolated molecule and in solution. [**4, 18**] Recent results of such studies will be discussed in Section 10.

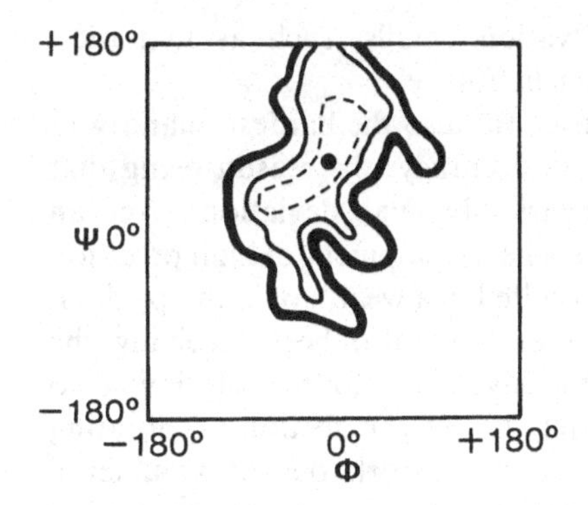


Figure 5. The $\phi,\psi$ conformational potential energy map of sucrose; $\phi$ is defined by H(1)—C(1)—O(1)—C′(2) and $\psi$ by C(1)—O(1)—C′(2)—C′(1). The area outside the bold line is forbidden, since it corresponds to conformations with at least one non-bonding atom–atom contact distance less than half the sum of the van der Waals radii. A Buckingham potential (Equation (15)) was used for the energy minimisation. The dot corresponds to the minimum energy conformation and the intermediate contours are 42 and 210 kJ mol$^{-1}$, respectively.

## 3. Experimental techniques

This section consists of a summary of the techniques that are commonly employed in studies of PHC solutions. For convenience they are subdivided into methods which provide macroscopic information (thermodynamics, transport) and those which probe the shapes of molecules and specific interactions (optical activity, spectroscopy). Strictly speaking, computer-based methods could be included here, but we have chosen to introduce such techniques under the general heading of theoretical approaches (Section 4).

### 3.1. *Thermodynamics*

The value of thermodynamics measurements as applied to PHCs in solution has recently been reviewed. [6] The significant experimental methods fall into three groups: (1) measurements of chemical potential (water activity), i.e., freezing point depression, osmotic pressure, vapour pressure, (2) calorimetery, for the determination of enthalpies and heat capacities, and (3) measurements of density and its temperature and pressure derivatives (expansibility and compressibility).

Two distinct types of information are of importance: (1) the limiting (infinite dilution) values of the various thermodynamic quantities, and (2) the concentration dependences of such quantities. The former measurements provide information about the solvation of isolated solute molecules, whereas concentration dependence reflects solute–solute effects, albeit modified by the solvent environment. It must be stressed, however, that measurements of the concentration dependence of any property cannot provide explicit information about solvation details, such as hydration numbers, despite such claims in the older literature.

The chemical potential is the most basic, but also the hardest quantity to measure with a sufficiently high degree of accuracy, the reason being that most aqueous PHC solution exhibit remarkably small deviations from an apparently ideal behaviour (Henry's law). The development of high precision microcalorimeters during the 1970s has resulted in a wealth of enthalpy data, and, more recently, partial heat capacities have also been receiving the attention of experimentalists. A comprehensive survey, although limited to pentoses and hexoses, can be found in ref. 19. The perfection of vibrating tube densimeters has taken much of the hard work out of volumetric measurements, whilst the availability of automated sound velocity measuring devices permits the rapid determination of isentropic compressibilities. [20] The most serious source of experimental error now no longer lies in the measurement of the particular thermodynamic property but in the measurement of concentration. Surprisingly, none of the above techniques has yet been applied to the study of PHCs in non-aqueous solvents. The peculiarities (if any) of water as a solvent have therefore not yet been put to the test against PHC solution properties in 'normal' solvents.

*Limiting partial molar properties* A partial molar property of a component $i$ in a mixture in the limit of infinite dilution, $X_i^\circ$, is defined by

$$\Delta X_i^\circ = X_i^\circ - X_i^* \tag{1}$$

where $X_i^*$ is the property of component $i$ in some reference state. Where the aim is to investigate the effects of minor structural differences between similar solute species on $X_i^\circ$, the adoption of a common reference state is desirable, e.g., the ideal gas. For PHCs this is not possible because of their low volatilities; nor is it possible to adopt a reference state based on their identical atomic compositions, because of a lack of high-precision heat of formation data. In any case, experimental errors in such measurements are likely to be of the same order as are differences between $H_i^\circ$ values for a series of stereoisomers.

In practice, the PHC crystal is usually adopted as the reference state, although there must be doubt about the value of this procedure for comparisons of $X_i^\circ$ data of a series of isomers. Crystal structures and lattice energies can and do vary significantly within such a series. Possibly the amorphous PHC at its glass–rubber transition temperature might serve as a better common reference state (see Section 12).

*Concentration dependence* Cesaro has described the value of such measurements. [6] The value is based on the applicability of the McMillan–Mayer theory of solutions, [21] according to which the excess thermodynamic properties can be expressed in terms of virial expansions of the molal concentration $m_i$:

$$\Delta X^{\mathrm{E}} = x_{ii} m_i + 2x_{iii} m_i^2 + \ldots \quad (2)$$

where the homotactic (self-)coefficients $x_{ii}$ etc. are formally related to interactions between pairs, triplets, etc. of solute molecules in solution. For ternary solutions, solutes $i$ and $j$ in solution, the expression for $\Delta X^{\mathrm{E}}$ also includes heterotactic (cross-)interactions, characterized by the coefficients $x_{ij}$, $x_{iij}$, $x_{ijj}$ etc. The excess property $\Delta X_i^{\mathrm{E}}$ is defined by

$$\Delta X^{\mathrm{E}} = X_{\mathrm{expt}} - X_{\mathrm{ideal}} \quad (3)$$

where $X_{\mathrm{ideal}}$ is usually, but necessarily, related to Henry's law, and $X_{\mathrm{expt}}$ is the measured value.

*Thermodynamics and molecular detail* Thermodynamic measurements, as such, are blunt instruments; they cannot provide direct information about structure and weak interactions at the molecular level. However, statistical thermodynamics provides a bridge between structure and energetics. It is for this reason that thermodynamic data are significant, because the more direct structural techniques, e.g., diffraction, cannot be applied to solutions of molecules with as many scattering centres as PHCs.

The relationship between thermodynamic quantities and structural detail, as applied to PHCs, has been discussed in a previous review. [4] The formalism of the McMillan–Mayer theory is used, with modifications due to Friedman. [22] The second virial coefficient $x_{ii}$ in Equation (2) can be related to certain integrals of the molecular pair distribution function $g(\mathbf{r},\Omega)$ and the potential of mean force $W(\mathbf{r},\Omega)$ between the molecules in solution; $\mathbf{r}$ is the distance of separation between the solute molecules and $\Omega$ are the mutual orientations, expressed by the Euler angles. For dilute solutions we write

$$g(\mathbf{r},\Omega) \approx \exp[-W(\mathbf{r},\Omega)/kT] \quad (4)$$

The effect of the solvent is averaged out and is included implicitly in $W$. In principle, the thermodynamic excess functions can be expressed as functions of $W(\mathbf{r},\Omega)$. In practice, experimental measurements are not sensitive enough for the evaluation of $W(\mathbf{r},\Omega)$ which contains many unknown parameters and can adopt complicated, oscillating forms. It is common practice to simplify the treatment by averaging over all possible orientations, in which case the radial distribution function $\langle g(r)\rangle$ is obtained. The averaging process is equivalent to treating the PHC molecule as a sphere or an ellipsoid of revolution, and one must fear that the very feature which makes diastereoisomers distinctive, i.e., the —OH group topology, is averaged out. It would be expected that, at constant $r$, $W(\Omega)$ of two PHC molecules can adopt a complex shape. Even for a pair of 'simple' molecules, such as the water dimer, the interaction energy depends sensitively on the hydrogen bond donor and acceptor angles, whatever the values chosen for the atom–atom potentials. [**23**]

Probably at the present time, computer simulation is the only means of probing $g(\mathbf{r},\Omega)$ of two PHC molecules, whether in vacuum or in a 'computer' solvent. Nevertheless, the thermodynamic functions of dilution and mixing serve to highlight differences in the virial coefficients which must originate from subtle differences in the interaction energies of chemically very similar molecules. Comparisons of the virial coefficients in different solvents might also provide a better insight into the details of solute–solute interactions, but such data are completely lacking for PHCs.

The limiting partial thermodynamic functions are of interest because they reflect solute–water effects which would formally be related to $g_{sw}(\mathbf{r},\Omega)$, where the subscripts refer to solute and water, respectively. Here again, the replacement of water by another solvent should provide useful information which could be compared with results from spectroscopic measurements and computer simulation.

### 3.2. *Transport properties*

*Viscosity* Because of their propensity for hydrogen bonding, solutions of PHCs can exhibit complex rheological behaviour; this is especially true for their polymers. [**10**] Many polysaccharides, whether from animal, plant or microbial sources, find application (also *in vivo*) as thickeners, stabilizers, gelling agents, emulsifiers, wetting agents, etc. Even oligomers often yield highly viscous solutions and, with a few exceptions, PHCs are quite reluctant to crystallize from aqueous solution.

Although viscosities are important in chemical technology, such data provide only limited information regarding interactions of PHC molecules with one another and with the solvent. When there is a possibility of non-Newtonian flow, information about shear rates is required, but this is not generally available in the technical literature. The viscosity of a PHC solution is usually related to molecular detail via the simple Einstein equation or one of its many extensions, e.g., the Vand equation:

$$\log \eta_{rel} = 2.5cV_e/(1-K_c V_e) \tag{5}$$

where $\eta_{rel}$ is the relative viscosity, $K_c$ is a crowding parameter, $c$ is the concentration and $V_e$ is the effective hydrodynamic volume associated with the solute at infinite dilution. On the assumption of sphericity, some workers have drawn conclusions about hydration numbers, based on a comparison of $V_e$ with the intrinsic volume. We doubt the credibility of calculations based on Equation (5), according to which the solute molecule is regarded as a sphere moving through a structureless continuum, a model which hardly corresponds to a PHC molecule in water.

*Diffusion* Few data exist on diffusion rates of PHCs in solution. The reasons are not clear, because diffusion is of prime importance in several biologically and technologically important processes, e.g., the tranport of cryoprotectants across cell membranes, and the crystallization of sugars. Diffusion coefficients can be obtained most directly by radiotracer methods or, more indirectly, by nmr, but neither method has been extensively applied to PHCs.

### 3.3. *Optical activity*

Until the advent of advanced nmr techniques, optical rotation was the preferred experimental method in studies of PHC conformations. In recent years its place has been taken by nmr methods of various types (see below), because of their ability to identify particular species in complex mixtures, provided that interconversion rates are slow on the nmr time scale. Optical rotation measurements (or associated techniques, such as optical dichroism and optical rotatory dispersion) provide only an average result for all the constituents in a mixture. This is often still adequate as a quality control criterion, and optical rotation is therefore still widely used in analytical laboratories. Compared to the more refined nmr techniques (see Section 3.5.), optical rotation measurements are simple, rapid and inexpensive.

Results are usually expressed as the specific rotation, measured at the wavelength of the sodium D-line:

$$[\alpha]_D = \alpha/(g\,ml)(dl) \qquad (6)$$

where $\alpha$ is the measured rotation and is divided by the solute concentration and the decimetre-length of the light path. To obtain the molecular rotation $[M]$, the specific rotation is multiplied by $10^{-3}$(molecular weight).

*Monosaccharides* Many different relationships have been proposed between $[M]$ and molecular configuration of a monosaccharide. For a lucid discussion, see ref. 15. Even now, all calculations of $[M]$, based on molecular structure, are semi-empirical, with only moderate agreement between experiment and calculated values. In studies of solvent effects on sugar conformation, an additional element of uncertainty is introduced: specific solute–solvent interactions give rise to changes in $[M]$ of each individual conformer, over and above any solvent-induced changes in the equilibrium composition of the anomer/tautomer/conformer mixture; this is shown in Table 2 for glucose in water and pyridine. Therefore, unless $[M]$ values for all possible coexisting species are available for a given solvent, changes in $[M]$ cannot be interpreted with a high degree of reliability.

*The optical rotatory power of the glycosidic linkage* It became apparent early on [**24**] that in oligomers the glycosidic linkage contributes to the

Table 2. *Solvent effect on the molar optical rotatory power of glucose anomers*

| | $[\alpha]_D$ (deg) | |
|---|---|---|
| | $\alpha$-glucose | $\beta$-glucose |
| Water | +122 | +16 |
| Pyridine | +152 | +11 |

observed optical rotatory power. With increasing degree of polymerization, the dextrorotatory power of the pyranoid ring decreases, due to the partial screw pattern of the glycosidic linkages which introduces an axis of chirality. It is thus not possible to calculate the optical rotation of a di- or trisaccharide from the $[M]$ values of the constituent sugars alone. In any case, since $[M]$ is sensitive to the ring conformation, it is reasonable to expect that the stereochemistry of the glycosidic linkage should also affect the optical rotation; in other words, $[M]$ contains a contribution which can be expressed in terms of the torsional angles $\phi$ and $\psi$.

The quantitative treatment of the linkage rotation $[\Gamma]$ ws developed by Rees and coworkers who derived the following expression: **[25, 26]**

$$[\Gamma] = [M_{nr}] - ([M_{me,n}] - [M_r]) \tag{7}$$

where $[M_{nr}]$ is the molar rotation of a disaccharide which contains a non-reducing (n) and a reducing (r) group. $[M_{me,n}]$ is the molar rotation of the methyl glycoside of the non-reducing sugar, in the same anomeric configuration as it exists in the disaccharide. The treatment is based on two assumptions: (1) the anomeric ratios of r and nr are similar, and both hexose residues are in the C1 configuration. To test Equation (7), $[\Gamma]$ has been expressed in terms of $\phi$ and $\psi$, values for which were taken from crystal structure data. For cellobiose and lactose, the agreement with experiment was close, but not for $\beta$-methyl maltoside, where $[\Gamma]_{calc} = 107°$, compared to the experimental value in aqueous solution of $+52°$. Possible reasons for such discrepancies are discussed later.

## 3.4. *Optical spectroscopy*

PHCs lack the chromophores which are required for uv and visible absorption spectroscopy in the conveniently accessible frequency range, i.e., $> 200$ nm. On the other hand, their ir and Raman spectra are so complex that the available normal mode analytical data are limited to a few common sugars, mainly in their crystalline states. Solution spectra are particularly complicated because of the coexistence of anomers, tautomers, etc., added to which are the intermolecular solute–solvent hydrogen bond modes. Nonetheless, comparisons of PHCs in the crystalline and amorphous states, anhydrous or hydrated, might prove informative.

### 3.5. *Dielectric properties*

Measurements of dielectric permittivity $\epsilon$ as a function of concentration and frequency have proved to be useful in solution studies. In some situations the measurements are subject to technical or interpretative complications arising from the conductivity of samples, but this is not the case with neutral PHCs. Suggett and his colleagues used such measurements in conjunction with nmr relaxation, to probe the solvation details of some simple sugars. **[27, 28]** The results and their interpretations are disucssed in Section 6.

The quantities of greatest interest include the molecular dipole moment $\mu$, the real and imaginary parts ($\epsilon'$ and $\epsilon''$) of the measured permittivity $\epsilon(i\omega)$ and their amplitudes and the dielectric relaxation times $\tau_i$ of the $i$ components in a mixture. Measurements can be made in the frequency domain, i.e., at fixed frequencies, or in the time domain which involves measurements of reflection and/or transmission coefficients, typically in the frequency range of $10^8$–$10^{10}$ Hz. **[29]**

Frequency domain data are often represented in the form of Cole–Cole (complex plane) diagrams of $\epsilon'(\epsilon'')$. For an 'ideal' Debye relaxation process (single relaxation time) this takes the form of a semi-circle. Deviations from Debye behaviour are then accounted for in terms of a distribution of relaxation processes. Time domain data can be analysed by fitting to an equation of the form

$$\epsilon(i\omega) = \epsilon_\infty + \sum_m A_j / [1 + (i\omega\tau_j)^{1-h_j}] \tag{8}$$

where $m$ is the number of distinct relaxation processes with amplitudes $A_j$, decay times $\tau_j$ and distribution parameters $h_j$. For a pure Debye relaxation, $h_j = 0$.

### 3.6. *Nmr spectroscopy*

Without doubt, recent technical developments have turned nmr spectroscopy into the most powerful tool for studies of PHC structure and conformation. Such developments include, in particular, the construction of stable, high-field, superconducting magnets (> 600 MHz), able to produce very high resolution spectra, multinuclear methods ($^1$H, $^2$H, $^{13}$C, $^{15}$N, $^{17}$O, $^{31}$P,, etc.) with possibilities of spin decoupling, paramagnetic shift methods, magic angle spinning techniques for solid state high-resolution spectra (MA-CAS), a wide choice of pulse sequences for measurements of nuclear relaxation rates, e.g., correlated spectroscopy (COSY, HECTOR), heteronuclear multiple quantum spectroscopy (HMQ) and cross-relaxation and chemical exchange

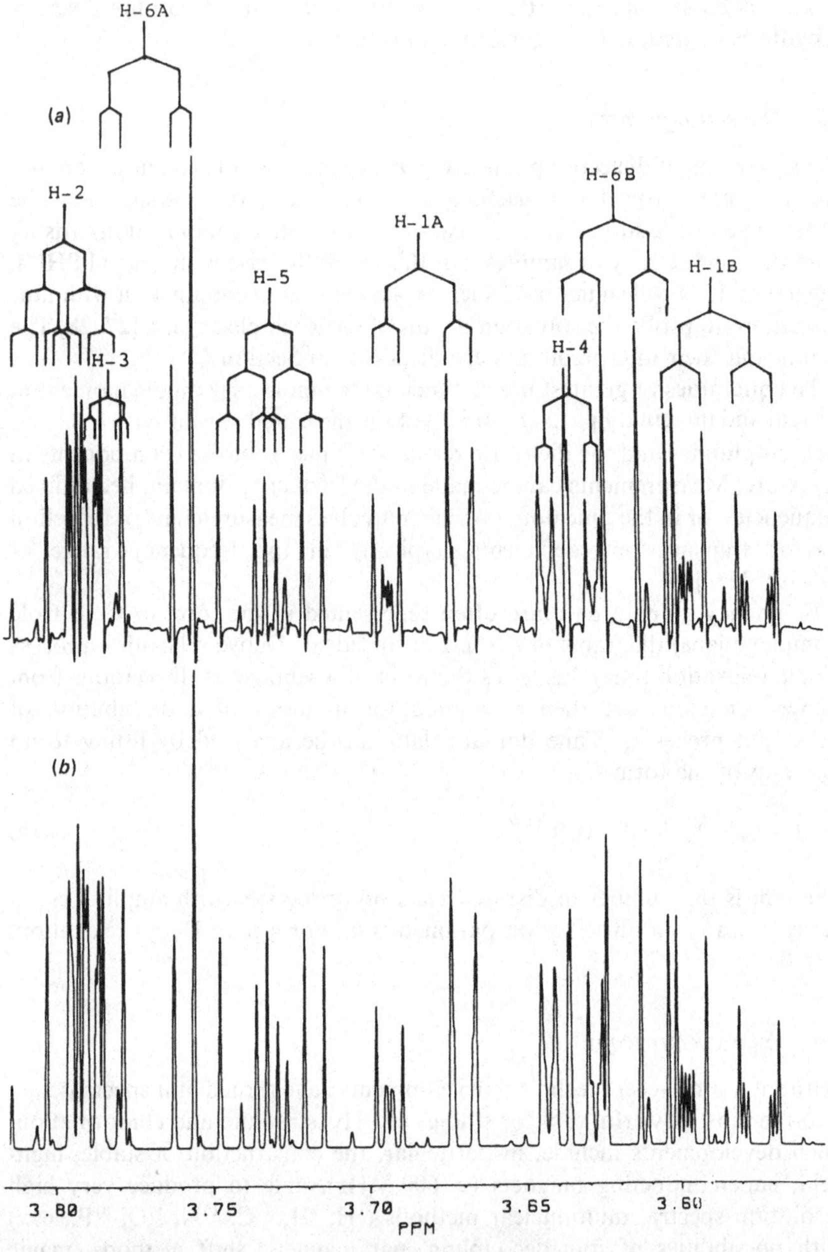

Figure 6. 500 MHz proton nmr spectrum of sorbitol in $D_2O$: (*a*) experimental spectrum with assignments and (*b*) computer simulated spectrum. Reproduced, with permission, from ref. 31.

(NOESY). Last, but not least, powerful computers are being increasingly applied for purposes of spectrum simulation.

The full armoury of nmr techniques has for several years been brought to bear on oligopeptides and small proteins, [30] but so far, relatively few studies of oligosaccharide solution structures are on record. Nmr would appear to be of particular value because of the remarkably complex nature of PHC spectra. Most of the molecules in question contain only three nuclear types, and many of the nuclei may exist in very similar environments. For instance, the $^1$H spectrum of a simple hexitol with eight non-exchangeable protons can exhibit 40–60 individual signals within a total range of 0.3 ppm. Such spectra had defied resolution until the advent of stable, high-field magnets. The 500 MHz spectrum of sorbitol, in Figure 6, serves as an example. [31] Preston and Hall drew attention to the potential of detailed nuclear magnetic relaxation rate measurements, as applied to sugars as early as 1974, [32] but such measurements are only now coming into their own. Thus, it is now not only possible to assign all chemical shifts in oligosaccharides, but the shapes of small and intermediate size PHCs in solution can be determined with a high degree of precision and compared with the molecular shape in the solid state (crystal or glass) or in the 'vacuum' state, as determined from theoretical calculations.

Nmr techniques are being increasingly used in the analysis of oligosaccharide residues in glycoproteins, e.g., the blood group determinants. [33] Where experimental techniques cannot yet provide unambiguous conformational details, the results can be combined with information derived from calculations or computer simulations (see Section 11) and areas of agreement established.

## 4. Theoretical approaches to the study of PHC conformation

Conformation is of major importance in PHC chemistry. Since all monosaccharides belong to a few families of diastereoisomers, they differ, within each family, only in their steric arrangements. The theoretical evaluation of conformational factors allows for the prediction of different properties of PHCs. Even when an accurate theoretical evaluation cannot be achieved, approximate results will help with the prediction of some properties. Alternatively, such theoretical results may provide explanations for some observed facts that are often presented without proper understanding.

It is necessary to consider carefully the theoretical treatments and always to place them in a correct framework. For PHCs, a considerable degree of complexity is unavoidable. To solve the problem, many approximations and assumptions are needed, and any analysis of results must pay regard to such shortcomings.

Several methods can be, and have been used to predict PHC conformations. In all cases, two decisions have to be taken before any calculation is attempted: the method itself and, in most cases, the form of the interaction potential to be used. Both aspects become important at the time of the discussion of results. Rather than make a listing of all the reported studies, we prefer to present the different methods, the source of the interaction potentials, and then discuss some examples to illustrate the various techniques.

The final goal of the computation is to account for, or even predict, the molecular conformation in solution. *In situ* predictions would be highly desirable but are probably unattainable with the present state-of-the-art. Different methods will provide either an average conformation for a given solvent or the percentage of the more stable conformers coexisting in equilibrium in the solution. We can roughly classify the methods in terms of different groups, each having its range of applicability and possibilities. The selection depends not only on the specific problem but also on the facilities available.

(1) Phenomenological methods. This group includes diverse methods that make use of empirical information and adjustable parameters to obtain the energy of the molecule.

(2) Strictly *ab initio* methods, such as SCF-LCAO-MO calculations. No empirical parameters are required but the calculations make a heavy demand on computing time. For that reason, only few results exist for carbohydrates.

(3) Semi-empirical quantum mechanical methods. They require the knowledge or assumption of a few empirical parameters, including the molecular geometry. Methods in this group achieve different degrees of refinement. They include EHT, CNDO, MINDO, NDDO, PCILO and others.

(4) Computer simulation. These methods require the knowledge of interatomic potentials; the number of empirical parameters varies. Often bond lengths and bond angles form part of the initial input. The simulation itself does not use adjustable parameters. However, it is possible to readjust the parameters used in one simulation for future use elsewhere. We clearly distinguish between two methods: Monte Carlo (MC) and Molecular Dynamics (MD).

We shall review the basic features of these methods, starting with the phenomenological approaches which are the simplest to perform and were the first to be applied to PHCs. It is not our aim to give a full account of the methods themselves but to quote their main features and limitations as regards their application to PHCs.

### 4.1. *Phenomenological methods*

One can start conformational studies with an attempt to obtain the relative populations of different conformers which, according to the experimental information, coexist in solution. To this end, some interaction potentials between all, or the relevant atoms, have to be assigned and a geometry stated for each conformer. With this information, energies are calculated. The conformation ratio of the conformers is then obtained. This scheme has many variants, but relies on several drastic assumptions.

The relative conformer populations in solution are determined by their *free* energies, while the calculation from interaction potentials yields the *internal* energies. This serious drawback is, unfortunately, often overlooked. One frequently finds references to free energies but, since the calculations lack the proper evaluation of all possible entropic contributions, the results are, in fact, potential energies.

Since, by and large, we deal with free energy differences, then, provided that the entropy of the two states is identical, it need not be evaluated. Two situations are of interest in this respect. One case is the calculation of the free energies of different conformers in the same solvent. The other case relates to the computation of free energy differences of the same substance in different solvents. The latter is relevant when changes in relative populations of conformers resulting from a change in the solvent medium are to be computed. It is often considered that a change of solvent does not affect the entropy of the system or, what is less restrictive, that the medium does not affect the *differences* between the entropies of the conformers. **[34]** These assumptions are hard to justify in the light of recent dynamical computations, **[35, 36]** as will be described presently.

The pioneering work was due to Angyal who performed an analysis of the conformations of aldopyranoses in aqueous solution. **[37]** He assumed that free energy differences are due only to steric factors and can be calculated through the interaction energies between non-bonded atoms in an additive way. However, he included an extra contribution to the conformational free energy, the so-called 'anomeric effect'; see Section 4.6. An additional assumption was that the pyranose ring has the same geometry as cyclohexane.

Calculations of the Angyal type are referred to as *hard sphere calculations* (HSC), because they are based only on spherical interaction potentials. Actually this description is misleading, since a true hard sphere system only has a hard core repulsion and no attractive part in the potential energy function.

More elaborate procedures make use of minimization techniques. After the definition of the various interaction terms, the molecular geometry is varied in such a way that the total energy decreases to a minimum. Much has been written about the difficulties inherent in locating the potential energy minimum in proteins but, although we may here be dealing with molecules

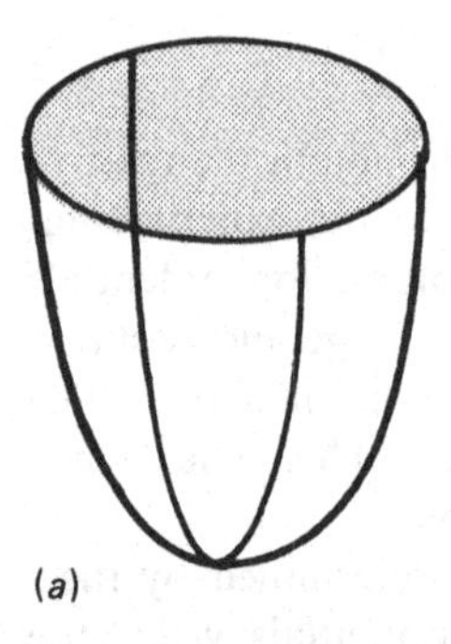


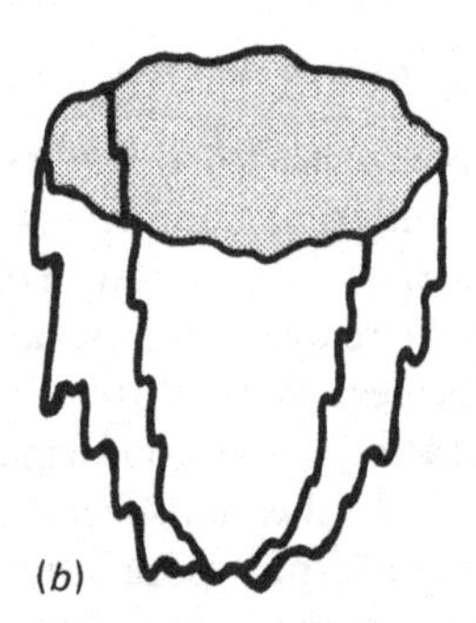


Figure 7. Potential energy surfaces with (*a*) a single minimum and (*b*) multiple minima.

very much smaller than proteins, the situation is anything but simple. Figure 7 shows two types of limiting potential surfaces. In the first type (Figure 7(*a*)), any method in which the conformational changes are driven by the potential energy gradient, will eventually lead to the minimum, although this process may take a long time if the number of degrees of freedom of the system is large. In the second case (Figure 7(*b*)), many substates exist. The presence of *local minima* makes it more difficult to identify the *global minimum*. The global potential may well be shallow, so that several local minima near the bottom may differ only slightly in their energies. Many substates may be occupied at room temperature, and the average conformation will not correspond to any one of the allowed conformations.

When bond lengths and bond angles are allowed to relax, the degrees of freedom increase in number, and searching for a global minimum demands much computation. Usually, especially for large molecules, only part of the *conformational map* is explored, taking as starting conformations those previously obtained by simpler methods and/or from experimental information.

When working with phenomenological methods, a good interaction potential (*force field*) is required. We recall that the calculations do not yield free energies and that the entropy contribution *may* be important. Most of the calculations contain no provision for the solvent, but even then, a static picture is obtained. The probability of trapping the system in a local energy minimum, rather than the true free energy minimum, is high.

For the case of disaccharides, a useful representation is provided by the energy contours in a $\phi$, $\psi$ plane; for 6-linked sugars, like gentiobiose or isomaltose, another angle $\omega$ is required. One can chose between global minimization, with the consequent demands on computing time, or a sampling of the $\phi$, $\psi$ ($\omega$) space. In the latter case, isoenergy curves are obtained. This kind of representation is common in protein conformation studies and is referred to as a Ramachandran map (see Figure 5). Often,

Table 3. *Total energy for the $H_2O$ molecule (in au) without polarization. The first column X/Y gives the number X of s functions and the number Y of p functions for the oxygen. The numbers in the first row give the number of s functions for each hydrogen; after E. Clementi, Determination of Liquid Water Structure. Coordination Numbers for Ions and Solvation for Biological Molecules. Springer Verlag: Berlin,* 1976.

| | H | | | | | |
|---|---|---|---|---|---|---|
| O | 2 | 3 | 4 | 5 | 6 | 7 |
| 4/2 | −75.15053 | −75.17413 | −75.17890 | | | |
| 5/2 | −75.42926 | −75.45301 | −75.45778 | | | |
| 6/2 | −75.50143 | −75.52124 | −75.52585 | | | |
| 5/3 | −75.74707 | −75.77026 | −75.77419 | | | |
| 6/3 | −75.80813 | −75.82896 | −75.83298 | | | |
| 8/3 | | | −75.90618 | −75.91008 | −75.91171 | −75.91212 |
| 7/4 | | | −75.97129 | −75.97500 | −75.97614 | −75.97603 |
| 8/4 | | | −75.98710 | −75.99073 | −75.99187 | −75.99175 |
| 9/4 | | | −75.99126 | −75.99489 | −75.99603 | −75.99592 |
| 9/5 | | | −76.01153 | −76.01467 | −76.01567 | −76.01545 |
| 10/5 | | | −76.01419 | −76.01721 | −76.01818 | −76.01796 |
| 10/6 | | | −76.01824 | −76.02118 | −76.02199 | −76.02178 |
| 11/6 | | | −76.01946 | −76.02234 | −76.02313 | −76.02292 |
| 12/6 | | | −76.01982 | −76.02267 | −76.02345 | −78.02325 |
| 11/7 | | | −76.02088 | −76.02381 | −76.02457 | −76.02435 |
| 12/7 | | | −76.02120 | −76.02412 | −76.02486 | −76.02465 |

instead of the full energy contours, short paths are shown from some starting configuration to the minimum.

### 4.2. *Ab initio* methods

The quantum mechanical treatment of relatively large molecules requires some kind of approximation. The SCF-LCAO-MO methods (self-consistent field-linear combination of atomic orbitals) are commonly used. They attempt a description of the molecules in terms of electrons and nuclei of the constituent atoms.

Two approximations are involved in the LCAO-MO method: (1) the description of the molecular orbitals by means of the coordinates of only one electron explicitly, and (2) the expansion of the molecular orbital in terms of the linear combination of atomic orbitals. The wave function for each electron is expressed as

$$u_i = \sum c_{ij} \chi_j \tag{9}$$

where $\chi_j$ were formerly taken as the atomic orbitals but are presently taken to represent one of the terms of the basis set. In order to arrive at the best possible linear combination, the self-consistent field technique is introduced.

The selection of the basis set is one of the critical aspects of these methods. Table 3 shows the total energy for water (without polarization), computed with different basis sets. For accuracies of 0.001 Å, 6s functions for the hydrogen and 11s+7p functions for the oxygen are required. For an accuracy of 0.01 Å, a set with 5s for hydrogen and 9s, 5p for oxygen will suffice. However, there is no physical reason for chosing a set 11s, 7p, 6s rather than 12s, 6p, 6s or 11s, 6p, 6s. The final choice is dictated by the computation efficiency. The primary choice of the basis set depends on what particular feature is to be computed with the highest accuracy. A particular basis set may give low accuracy in the total energy but good accuracy in the valence electrons or, conversely, high accuracy for the inner shells and low accuracy in the valence electrons.

Obviously, the use of *ab initio* methods requires considerable expertise, coupled with a corresponding computer power which, although essential, becomes a secondary factor for the success of the procedure.

### 4.3. *Semi-empirical methods*

Quantum mechanical methods become increasingly complex with increasing molecular size. New approximations can be introduced, generating a new family of methods which can be described as 'semi-empirical'. In these methods the Coulomb integral is expressed in a way that separates out the electron repulsion terms and involves some $\gamma$-terms from the so-called core integrals which are the sums of the true one-electron operators. One example is the CNDO (complete neglect of differential overlap) approximation in which many electron repulsion terms are ignored. The main problem lies in the choice of parameters, especially the $\gamma$-values.

In the field of PHCs, one of the semi-empirical methods used is the so-called PCILO (perburbative configuration interaction using localized orbitals). [**38**] The method is based on the use of bond orbitals and perturbation theory for the calculation of the ground state energy of the molecule.

### 4.4. *Simulation methods*

Two major simulation methods now find application in many research fields, ranging from solid state physics to galaxy simulation: Monte Carlo (MC) and Molecular Dynamics (MD). The main difference is the capability of the latter method to provide true dynamical information. In both cases the main input is the interaction potential. The geometry may be included as an aid to checking different physical properties. It is possible to fix the molecular topology and allow the bond lengths and angles to be adjusted in accordance with the respective potentials. It is also possible to include, apart from the topology, the bond lengths and angles as fixed values.

Full descriptions of MC and MD techniques are out of place here. There are, however, some facets which need to be stressed for the sake of clarity. Both MC and MD techniques yield a collection of states of the system under study, but the difference between these states is important. With MC, all states are accessible for the randomly obtained equilibrium. With MD, one obtains the *evolution of the system* as a function of time which means that the collection of states also corresponds to accessible states at equilibrium, but they are ordered in time. The states obtained by MC provide information to be used in computations of equilibrium properties. With MD, time-dependent properties, e.g., diffusion coefficients, can additionally be obtained.

Simulations are often termed *computer experiments*. This designation is based on the belief that, for a given model, the results of the simulation are exact. Probably *computational theories* is a better term, even on the assumption of exact results for a given model, but quite apart from the name, we need to be aware of certain limitations. To avoid surface effects, *periodic boundary conditions* are applied in which a 'simulation box' and its images are treated. Since the number of atoms that can be handled by the computer is limited, computations are usually limited to relatively few molecules, and this can lead to physically quite unacceptable situations. For simulations of molecules in solution, two extreme cases are considered: *infinite dilution*, when the simulation is limited to one solute molecule, and very concentrated solutions when more than one solute molecule is included in the simulation. This abrupt change, even by just passing from one to two molecules, poses a major problem. However, all other theoretical methods also deal with isolated molecules *in vacuo* or at infinite dilution in a solvent; see Section 5.

A further, curious paradox needs to be stressed: a solution containing one solute molecule per 220 solvent molecules would normally be regarded as a 0.25 molal solution. However, in a simulation it qualifies as an infinitely dilute solution, because only a single solute molecule (and 220 individual solvent molecules) is included in the calculation. The total free energy of the system therefore contains no contribution from solute–solute interactions which is the definition of an infinitely dilute solution; see Section 3.1.

### 4.5. *The interaction potential*

For most of the methods described, the selection of the interaction potential is of overriding importance. Any procedure used for the calculation will be a computation of the mutual interaction between atoms. Consequently, a good result can hardly be expected, if the basic definition of such an interaction is incorrect.

For small molecules one can list interaction pairs and assign an energy contribution to each pair. This procedure is specific to the actual molecule under consideration and was used in the early stages of the development of the field. [37] The most general and useful method is to define the potential as a function of distance and/or angles with an appropriate set of parameters. The relationships between such a function and the experimentally accessible thermodynamic excess functions have been discussed in Section 3.1. Although any number of parameters can be introduced in a potential function, it is sensible to use the minimum compatible with an adequate description of the potential shape. This norm is dictated not only by a need to keep the computations as simple as possible, but also by the limited available experimental information which is available for the calculation of the parameters.

For a limited range of closely-related molecular structures it is not difficult to devise a set of parameters on an *ad hoc* basis. However, this set will be of such restricted utility that it would not differ too much from the single-application potentials already referred to. The challenge is to find a set of parameters that can adequately describe *any* compound, i.e., that allows comparisons of molecules of different chemical types, having the same functional groups in different chemical environments, and the computation of their energies.

For a rigid bond assumption one needs to specify only the non-bonded potential functions, but it is also possible to define a potential for bonded atoms. In this case, many more degrees of freedom are available, giving more scope for molecular arrangements but, at the same time, increasing the scope of the computation, in some cases beyond the practical possibilities. In general, the potential energy function can incorporate some or all of the following contributions:

A harmonic potential can describe the bonds as

$$U_b = \tfrac{1}{2}K_b(b-b_0)^2 \tag{10}$$

where $K_b$ is the force constant, $b$ is the bond length and $b_0$ is the bond length at the minimum potential energy (at equilibrium).

If the force constant is too large, any distance other than $b_0$ will yield a large contribution to the total potential energy. This result is equivalent to the fixing of the bond length which may be a useful approach. However, if the fixed bond length is preferred, several other, more economical methods are available to apply bond length constraints.

*Bond angles* For a set geometry, not only bond lengths but also bond angles are fixed. However, one may allow for a relaxation of the bond angles; a corresponding potential energy function then needs to be defined. This alternative not only gives more flexibility to the molecule, permitting the detection of more stable conformations, but it also better describes the dynamics of the systems in the case of MD simulations. In such studies a finite time step is defined to solve the equations of motion. This period is often selected to be as long as possible, in order to simulate over longer times with lower computation needs. Assuming that intermolecular vibrations are not of interest, this period ranges mostly between $10^{-15}$ and $10^{-14}$ s. On this time scale the intermolecular vibrations are indeed fast and can be neglected. Bond angle librations, however, are much slower and it is pertinent to consider such motions, for instance, by means of the function

$$U_\theta = \tfrac{1}{2}K_\theta(\theta-\theta_0)^2 \tag{11}$$

where $K_\theta$ is the force constant for bond angle deformation and $\theta$ and $\theta_0$ are the bond angle and the angle at minimum energy.

*Torsional (dihedral) angles* Normally, rotation about a bond is not free because of steric hindrance which gives rise to distinctive equilibrium values. Since the geometries of different conformers of an organic molecule differ considerably in their torsional angles, a conformational analysis places a great deal of emphasis on this aspect.

The nature of torsional angles, as commonly defined for PHCs, has been described in Section 2.3. and Figure 4. Many of the force fields in use include explicit potential energy functions (torsional potentials) for these angles. A common approximation is

$$U_\phi = K_\phi[1+\cos(n\phi-\phi_0)]^2 \tag{12}$$

where the symbols have the same meaning as before, with $n$ being the multiplicity (degeneracy) of the torsional angle $\phi$.

Since the conformation of the molecule is so sensitive to the torsional angles, it follows that the quality of the final result will depend sensitively on the choice of the parameters in Equation (12). Some authors consider that the torsional potential function should follow as a direct consequence of the atom–atom interactions between neighbouring atoms, [**39, 40**] a view which we share.

*Improper torsional angles* To preserve the tetrahedral (or planar) conformation of, for instance, four atoms attached to a carbon atom, it is possible to define a class of *improper torsional angles* (out-of-plane, out-of-tetrahedral). Such an angle is defined as the dihedral between two planes which are fixed by three out of five atoms. Figure 8 shows the improper torsional angles required to maintain the tetrahedral geometry of a set of five atoms. Atoms 1, 2 and 3 determine one plane (*z*–*x* plane) and atoms 3, 4 and 5 the other (parallel to the *x*–*y* plane). A suitable potential function reads

$$U_\pi = \tfrac{1}{2}K_\pi(\pi-\pi_0)^2 \tag{13}$$

the symbols being defined as earlier.

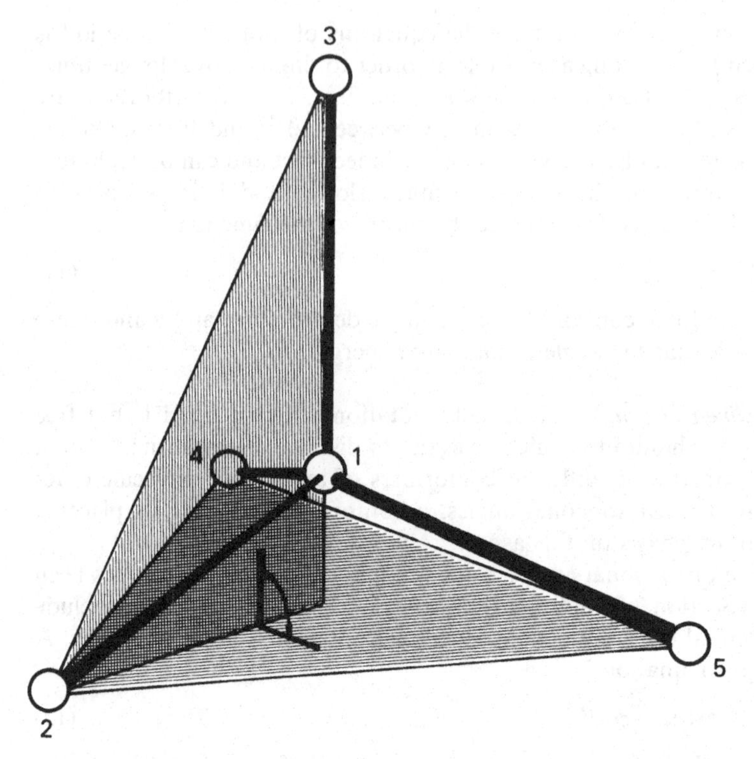


Figure 8. Improper torsional angle definition for a tetrahedral arrangement.

*Non-bonded interactions* Non-bonded interactions are the most relevant in determining the behaviour of the system. First neighbours correspond to bonded atoms and second neighbours are geometrically related by bond angles. Therefore the third neighbours (1,4-interactions) and beyond, fall within the category of non-bonded. It is often advisable to distinguish between third neighours and the remainder. However, the difference lies in the values of the potential parameters and not in the shape of the potential. This difference is relavant to the inclusion (or not) of torsion potentials. When explicit torsion potentials are used, the 1,4-interactions can be set equal to zero. However, as already mentioned, in order to predict molecular properties reliably, it is better to include the 1,4-interactions rather than torsion potentials.

Two different types of interactions are considered for non-bonded atoms: dispersion/repulsion and Coulombic interactions. There are several ways of writing the dispersion/repulsion term. A familiar expression is the Lennard–Jones potential function:

$$U_{ij} = C_n r_{ij}^{-n} + C_m r_{ij}^{-m} \tag{14}$$

where $n$ and $m$ are integers, usually 12 and 6, and $r$ is the atom–atom distance. The first term represents the repulsion and is sometimes replaced by a three-parameter exponential form, known as a Buckingham potential, so that the complete dispersion/repulsion expression reads

$$U_{ij} = A \exp(-a r_{ij}) - C_m r_{ij}^{-n} \tag{15}$$

The exponential form has a sounder theoretical basis. However, the $C_n r_{ij}^{-n}$ term, with the appropriate selction of $n$ (usually 12) and $C_n$, can often reproduce the shape of the repulsion part of the potential in a satisfactory manner. It has the advantage of requiring only two parameters.

*The parameters* We have mentioned that the interaction potential should reflect the properties of almost any molecule without the need for new assumptions. Apart from the shape, about which there should be little doubt, a key to the success of the potential lies in the selection of the parameters. The molecule under study should not be used in the fitting procedure, but related compounds can be used as a source of the parameters. It is not advisable to have different sets of parameters, to be used for different families of molecules, but rather to rely on *only one* universal set.

Different parameter sets have been developed independently by different workers and, as a consequence, we are far from the ideal situation of one universal set for any situation. Each set has certain advantages, and the decision to use one or another is based on the personal preference of the researcher and his acquaintance with the set (or with the author of the set). A series of parameters for the force field has been devised by Rasmussen and coworkers, [**39**] specifically for calculations on carbohydrates. The different force fields derived, lead to a continuous improvement of the parameters which includes the elimination of the torsion potential; this is achieved by a large reduction in the parameter $K_\phi$ in Equation (12).

More recently, Marsden, Robson and Thompson have derived a parameter set for sugars, [**40**] to be compatible with existing amino acid empirical interatomic parameters. This attempt to unify the parameters for different classes of compounds is of great importance, not only for the particular case of glycopeptides in which compatible parameters are mandatory, but as an effort to derive a universal force field. Apart from the particular numerical values obtained, it is important to note that only small differences were found between the 'pure' peptide parameters and the best-fitted common ones.

It is beyond the scope of this review to list all the available parameters used for PHCs, and the reader is referred to the primary sources. While most of the available sets have been derived by matching the properties of similar compounds in aqueous solution, it is possible to find pure 'vacuum' potential parameter sets. Such sets need the explicit addition of the solvent contributions, as described in refs 36 and 41; the procedure makes use of parameter sets derived for proteins and nucleic acids.

When water is included explicitly, a potential function with its respective parameters has to be provided for such molecules. As far as we know, only MD simulations have been used in such procedures. [**35, 36, 41**] 'Computer water' of the types SPC [**42**] and SPC/E [**43**] was employed. Finney has provided a detailed discussion of the available water model potentials which now exceed 25 in number. [**23**]

*Partial charges* Even neutral groups of atoms can be appropriately represented by atoms bearing partial charges. If a reasonable source of information for the assignment of atom charges is available, the results are more realistic. In order explicitly to introduce electrostatic interactions, it is not necessary to employ extra terms lacking in theoretical significance. The electrostatic interaction is described by the familiar Coulombic term

$$U_{ij} = q_i q_j / (4\pi\epsilon_0 r_{ij}) \qquad (16)$$

where $q_i$, $q_j$ are the atomic charges, $r_{ij}$ the atom–atom distance and $\epsilon_0$ is the dielectric permittivity of the medium (vacuum).

When the solvent is explicitly included elsewhere in the calculations, there is no doubt about the use of the vacuum permittivity, but when it is intended to mimic the solvent through its charge screening effect, then a different choice for the permittivity is indicated. This choice presents problems. The use of the bulk solvent permittivity is not fully justified for short or medium range values of $r_{ij}$. A reduction in its value, although common practice, turns $\epsilon$ into an adjustable parameter.

The reader may be surprised at the uncertainties in the value to be assigned to $\epsilon$ in aqueous solutions. However, several subtle points have to be clarified, such as the dipolar shielding at short distances and the charge screening in the case of electrolyte solutions. For short distances, the only reliable procedure, up to date, is the explicit inclusion of the solvent.

Possible existing dipoles in the molecules are described by the atom charges. This represents an advantage over the inclusion of dipoles themselves, because the treatment of dipole–dipole interactions requires the specification not only of distances but of angles as well. For radially interacting charges, only $r_{ij}$ enters into the calculations, making the energy computation easy.

Atomic charges are generally expressed as fractions of the electronic charge. The determination of these *partial charges* is of importance in a proper decription of the electric interaction. It can be achieved by the use of *ab initio* methods for similar compounds or simpler molecules that contain the chemical groups of interest. The literature contains many data for partial charges, a proof of the extensive search for better descriptions of electrostatic interactions. One of the most used methods to compute partial charges is due to Del Re. [**44**] Allinger *et al.* [**45, 46**] have gone further, directing their work to the representation of partial charges in molecular mechanics calculations.

## 4.6. *The anomeric effect*

Despite several efforts over the past two decades to calculate conformational free energies of sugars, it is still not obvious why the $\alpha:\beta$ ratio for glucose is 36:64, whereas for mannose it is 67:33. Edward first postulated the existence of an interaction to account for differences in anomeric ratios. [**47**] The effect, which came to be known as the 'anomeric' effect, has been discussed at length in the carbohydrate literature. [**48, 49**] Basically it is postulated that an equatorial —OH group at the anomeric site produces a repulsive dipole–dipole interaction with the ring oxygen orbitals; hence the $\alpha$-anomer is favoured. The effect is of an electrostatic nature, so that it is assumed to vary inversely with the dielectric permittivity of the solvent. Interpreted in this way, however, the anomeric effect is not only affected by changes in the solvent but also by the nature and configuration of substituents in other parts of the pyranose ring.

The evaluation of the contribution of the anomeric effect to the total free energy of the molecule becomes rather uncertain. The values are obtained from a comparison of the experimental and calculated free energies of the anomers. Angyal's data will served as an example. [**49**] For glucopyranose, the free energy difference between the anomers is calculated as 3.8 kJ mol$^{-1}$. The experimental value in aqueous solution is 1.5 kJ mol$^{-1}$. The difference, i.e., 2.3 kJ mol$^{-1}$, is ascribed to the anomeric effect. This method of calculating the contribution of the anomeric effect to the total free energy lacks rigour and is of little predictive value.

Angyal expressed his doubts as follows: 'The evaluation of the anomeric effect is the least satisfactory part of the calculations of the free energies of sugars, because the effect varies considerably not only with the nature of the solvent and of the anomeric group but it is affected even by the presence and the configuration of substituents in other parts of the pyranose molecule.' In fact, the anomeric effect has been used (at times quite indiscriminately) as an adjustable quantity to account for discrepancies between calculated and experimental conformational free energies. For instance, to reconcile the conformational free energies of the mannose and deoxyarabinose anomers it has been assumed that the magnitude of the anomeric effect depends on the presence and configuration of the —OH group on C(2) and also on the presence or absence of an —OH group on C(6), 'and probably in other positions'. [**49**]

The anomeric effect is, without doubt, an important feature in the determination of molecular conformation. With the correct choice of atom–atom interactions in the energy calculations, the anomeric, as well as any other effect should naturally contribute to the total energy. The use of partial charges gives rise to effective dipoles when two neighbouring atoms have opposite charge signs. The dipoles will interact, according to their separations and orientations. The contribution to the energies of different anomers will appear as a *consequence* of the atom–atom interaction without the need to postulate a special effect in the form of a 'correction potential'. In summary, therefore, the anomeric effect does exist as an observable fact but should never be included as a special component in the calculated interaction potential.

### 4.7. *The hard-sphere-exoanomeric method (HSEA)*

One of the commonest methods for the prediction of the preferred conformations of di- and oligosaccharides is the so-called hard-sphere-exoanomeric-effect calculation. The primary source for the computation of the energies of conformers is the hard-sphere approximation in which every non-bonded atom pair of the molecule, with a given geometry, interacts according to some prescribed potential. As an example, Bock **[50]** uses the potential proposed by Kitaygorodsky: **[51, 52]**

$$U = 14.6(-0.04/z^6 + 8.5\times 10^3 e^{-13z}) \text{ kJ mol}^{-1} \qquad (17)$$

where $z = r_{ij}/r_0$, $r_0$ being the equilibrium distance between atoms $i$ and $j$.

The interaction energies are calculated for a given set of torsional angles, defining the conformation. The sum corresponds to the total non-bonded interactions for that conformation. However, in the HSEA method, this potential is further modified by a term thal allows for the exoanomeric effect which is said to arise from the anomeric effect and which produces a reorientation of an aglycon with respect to the sugar ring, such that the preferred torsion angles C(aglycon)—O(1)—C(1)—O(5) are $+60°$ for $\alpha$-, and $-70°$ for $\beta$-glycosides. **[53]** This device is equivalent to the introduction of a biased torsional potential which is written as

$$U_{\text{exo}-\alpha} = 6.6(1-\cos\phi) - 3.1(1-\cos 2\phi) - 2.9(1-\cos 3\phi) + 7.2 \text{ kJ mol}^{-1}$$

$$U_{\text{exo}-\beta} = 11.0(1-\cos\phi) - 5.1(1-\cos 2\phi) - 5.0(1-\cos 3\phi) + 12.1 \text{ kJ mol}^{-1}$$

For each $\phi$ value, this energy contribution is added to the non-bonded energy computed with the hard-sphere potential. Obviously, the introduction of this torsional potential, designed to produce a particular result, will bias the calculation in a predetermined direction. Lemieux *et al.* justify this dubious procedure because of savings achieved in computing effort. **[53]** Nowadays, many force field calculations can be performed with a personal computer, and computing facilities allowing the use of rather sophisticated methods are available to most workers. It is hard to justify an approach that neglects dipole–dipole interactions (either explicitly or through the use of effective charges on atoms) and solvent effects. However, the most serious criticism of the HSEA method is that it strongly biases the trend of the results.

## 5. Theoretical treatments of solvent effects

Few would deny the relevance of solvent effects on the conformation of PHC molecules in solution. However, the term 'solvent effect' embodies a large collection of unknowns, myths and prejudices. Most of the theoretical work on PHCs relates to isolated molecules but frequently some provision for the influence of the solvent has been included.

The explicit inclusion of solvent molecules is possible in simulation methods, although the necessary computational effort then increases considerably. Phenomenological methods for energy minimization also allow for the inclusion of a limited number of solvent molecules; it appears that this type of approach has only rarely been considered in the field of carbohydrates. The commonest *explicit* device to account for solvent contributions involves the incorporation of some extra terms in the energy calculations and/or an adjustment to the dielectric permittivity, as already discussed above. There is, however, another, implicit, and not always clearly stated method for considering the effect of the solvent, namely via the potential parameters. In the discussion of the interaction parameters we referred to sets designed to work *in vacuo*, and others which are obtained from solution properties. Many of the phenomenological conformational analyses make use of the potential parameters obtained for aqueous solutions, i.e., some 'correction' is applied to the potential, in order to obtain the aqueous solution properties.

This approach is not as good as it may appear at first glance. On the one hand, it disregards any possible specific solute–solvent interaction, and on the other, the resulting force field will not reflect the *in vacuo* properties and will be restricted to one particular solvent, without even properly describing that solvent.

One way in which the solvent effect has been considered in the carbohydrate literature is through the addition of a solvation contribution to the computed solute energy. Assuming that free energies can be rigorously calculated, then

$$G = G_u + G_{solv} \quad (18)$$

where $G_u$ represents the free energy of the solute and $G_{solv}$ the contribution from the solute–solvent interaction free energy. This latter term can then be further composed:

$$G_{solv} = G_{cav} + G_{es} + G_{disp} + G_{spec} \quad (19)$$

The term $G_{cav}$ accounts for the free energy required to create a cavity of the appropriate size to accommodate the solute molecule, $G_{es}$ represents the electrostatic contribution, $G_{disp}$ is the free energy due to the dispersion interaction between solute and solvent, and $G_{spec}$ accounts for specific interactions.

The assignments of proper expressions to each of the four terms is fraught with problems. Usually, the continuum model is assumed and, within that model, the specific solute–solvent interaction is taken to be independent of conformation. As regards the other contributions, the cavity term has been estimated [54] as

$$G_{\text{cav}} = \left\{ -\ln(1-y) + \left(\frac{3y}{1-y}\right) R_a + \left[\frac{3y}{1-y} + \frac{9}{2}\left(\frac{y}{1-y}\right)\right] R_a^2 + \frac{yP}{kT} \right\} RT \qquad (20)$$

In this expression which is based on the Scaled Particle Theory, $y = 4\pi d a_{\text{v}}^3$ is the reduced number density, $d$ being the density; $R_a = a_{\text{u}}/a_v$ is the ratio of the radii of the hard sphere solute molecule $a_{\text{u}}$ and the solvent molecule $a_{\text{v}}$. The cavity radius is then $a = (a_{\text{u}} + a_{\text{v}})$ while $R$ and $k$ are the gas and the Boltzmann constants.

The electrostatic contribution $G_{\text{es}}$ depends more directly on the nature of the solute. A given solute has a charge distribution which interacts with the solvent dielectric and through the reaction field which it induces. To consider the solvent effect only through its macroscopic dielectric constant is not entirely appropriate. Near the solute, the solvent may experience a dielectric saturation effect, with a lower effective dielectric permittivity. This effect is somewhat indeterminate but can be approximated by, for example, the Onsager reaction field theory. [55] The following expression [54] takes into account all terms up to, and including, the quadrupole term:

$$G_{\text{es}} = kx/(1-lx) + 3hx/(5-x) + bf[1 - \exp(-bf/16RT)] \qquad (21)$$

where $k = \mu_{\text{u}}^2/a^3$, $h = q_{\text{u}}^2/a^5$ and $l = 2(n_{\text{u}}^2 - 1)/(n_{\text{u}}^2 + 2)$, $\mu_{\text{u}}$, $q_{\text{u}}$ and $n_{\text{u}}$ being, respectively, the solute dipole moment, quadrupole moment and refractive index. The dielectric permittivity appears in Equation (21) in $x = (\epsilon - 1)/(2\epsilon + 1)$ and in $f = [(\epsilon - 2)(\beta + 1)/\epsilon]^{1/2}$.

The direct van der Waals interaction has to be included in the overall solute–solvent effect; it contains the non-bonded, non-electrostatic attractive and repulsion terms. In the continuum approximation, the contribution of dispersion forces to the free energy results from the interaction of the solute molecule within its cavity with an average distribution of all the solvent molecules. It may be obtained from the integration of an effective intermolecular (solute–solvent) potential. The following expression has been used for saccharides:

$$G_{\text{disp}} = -(1-z)\, K_{\text{r}}(3/2)\, N_{\text{v}}\, \alpha_{\text{u}}\, \alpha_{\text{v}} \left(\frac{I_{\text{u}} I_{\text{v}}}{I_{\text{u}} + I_{\text{v}}}\right) r_{\text{uv}}^{-6} \qquad (22)$$

where $\alpha$ and $I$ are the molecular polarizability and the ionization potential respectively, $0.2 < K_{\text{r}} < 0.7$ is a semi-empirical constant, and $(1-z)$ is a proportionality constant to convert from energy to free energy. $N_{\text{v}}$ is the number of nearest neighbour solvent molecules surrounding a given solute conformer. Expressing the effective radius of the solvent molecule in terms of the molar volume $V_v$, and the Avogadro number $N_A$, as $a_{eff} = 3V_v/4\pi N_A$, then the number of nearest neighbours becomes

$$N_v = [(a_v + 2a)^3 - a_v^3]/a_{eff} \tag{23}$$

The careful theoretical reasoning that leads to the above equations is not reflected in our short exposition. However, it must be stressed that the disregard of specific interactions, which could hardly be evaluated by this approach, adversely affects the results. For instance, there is reliable evidence that the non-radial symmetry of the water interactions is one of the sources of the peculiar behaviour of water as a solvent. Additionally, any realistic treatment of PHCs should also be applicable to aqueous solutions. MD simulations on mannitol have shown differences in the conformation of this polyol in an argon-like solvent, an (imaginary) argon-like solvent of high dielectric permittivity and a model water solvent. Neither the molecular discreteness of the solvent, nor a high dielectric permittivity was sufficient to mimic the real water–solute interaction.

The explicit inclusion of the solvent as a finite number of discrete molecules, able to interact with the solute, can be performed in energy minimizations and in MC or MD simulations. This approach should, in principle, yield the most realistic results, were it not for problems associated with the correct modelling of the solvent. Although many attempts are on record to model all physical properties of water in all its states of aggregation, the matter is far from resolved. [**23**] The inclusion of a polarizability contribution in the water potential seems to promise a major advance but, here again, success has been elusive. So-called 'effective' potentials are claimed to allow for the polarization in the liquid through an increase in the dipole moment, but the introduction of such a 'polarized dipole moment' in addition to the permanent dipole moment implies an addition to the energy of the molecule. This contribution must be accounted for when the physical properties of water are used to parameterize the model. To cite one example, the latent heat of evaporation should be affected in a model that does not account for the loss of the polarization energy when the molecule passes from the condensed phase to the gas phase, since in the 'real world' the dipole moment is sensitive to the electrical environment experienced by the molecule. This has only recently been realized, [**43**]after a decade of 'successful' water modelling. It is by no means implied that all water models are grossly inadequate, but the hidden problems associated with theoretical predictions need stressing. It is to be hoped that the reliability of treatments of the solvent effect will improve with time, but there is still need for critical assessment. However valuable theoretical treatments might be, it is important not to overestimate their reliability.

## 6. Solvent effects in the chemistry of small PHCs: history and current position

Carbohydrate chemistry, like other branches of chemistry, is subject to sets of rules which have been framed to explain the observed molecular configurations, their relative stabilities and the reactivities of sugars in general. Most of the rules were formulated by crystallographers and organic chemists and are therefore based on the conformations of the molecules found in the crystalline state. The possible influence of solvents on the configurational and macroscopic properties of PHCs has seldom been taken into account in any but the most superficial manner. This may appear an extravagant statement, but it can be validated by the following example.

### 6.1. *The simple equilibrium hydration model*

Early vapour pressure measurements on aqueous solutions of some common sugars led to the realization that deviations from ideal solution behaviour (Henry's law) were marginal and could be accommodated by simple hydration models. **[56]** Hydration was expressed in terms of 'hydration numbers' $n_h$, i.e., the number of bound water molecules per sugar molecule. The thermodynamic properties of aqueous solutions of a sugar S could then be expressed by a series of simple equilibria of the type

$$S_{i-1} + H_2O \overset{K_{i-1}}{\rightleftharpoons} S_i \qquad i = 1, 2, \ldots, n_h \tag{24}$$

Assuming all hydration sites (and therefore all equilibrium constants $K_{i-1}$) to be equivalent, an average hydration number could be derived which depended only on the water chemical potential (water activity). Remarkably, such a simple model adequately accounts for the water activity of sugar solutions at 25 °C up to high concentrations. For instance, glucose obeys Equation (24) up to saturation ($K_n = 0.789$, $n_h = 6$) and sucrose up to 6 M ($K_n = 0.994$, $n_h = 11$).

The weakness of such simple hydration models was pointed out by Kozak, Knight and Kauzmann **[57]** who demonstrated that they incorrectly predicted the sign of the enthalpy of dilution. The prediction was that, with increasing temperature, the equilibrium would be driven to the right, a result which is inconsistent with experiment and with physical reality. The real reason for the apparent success of the model lies elsewhere, as shown in Figure 9 for mannose and glucose. **[58, 59]** The apparent ideal behaviour ($\Delta G^E \approx 0$) has its origin in the almost complete cancellation of $\Delta H^E$ by $T\Delta S^E$, neither of which has the value required by ideality.

As a rough-and-ready correlation, Equation (24) may have its uses for the prediction of some technological or biological attributes of complex, carbohydrate-containing systems. As a credible physical model, the concept

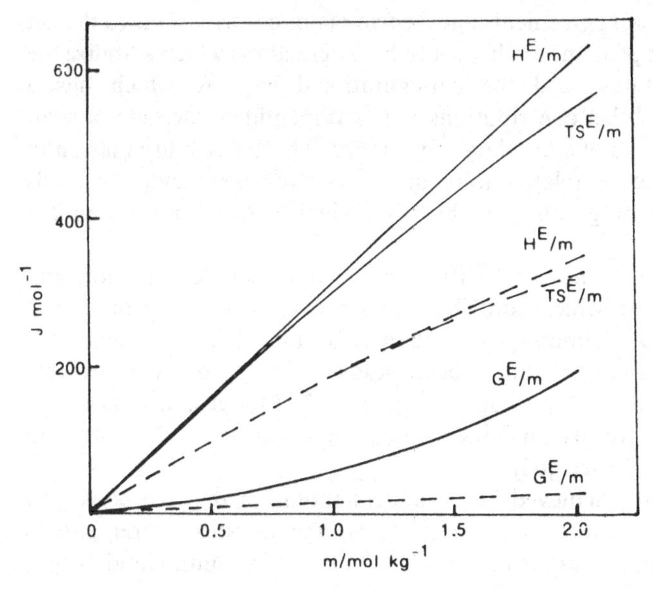


Figure 9. Excess thermodynamic quantities, at 25 °C, of aqueous glucose (full lines) and mannose (dashed lines) solutions; see Equation (3). **[58, 59]**

of simple hydration, based only on the number of —OH groups, should now be discarded. It cannot account for the solution properties of aqueous solutions, let alone for the behaviour of PHCs in mixed solvents. The pretence that the solution properties of PHCs can be so simply accounted for has delayed more penetrating studies into their physical chemistry for many years.

## 6.2. *Hydration as a determinant of protein stability: implications for PHCs*

During the past 15 years the realization of the importance of hydration interactions in determining native state stabilities of proteins has grown rapidly. The phenomena of hydrophobic hydration and hydrophobic interactions were first discussed by biochemists; **[60]** their molecular origins have subsequently been studied by physical chemists, and there now exists a consensus on the structural and energetic features of hydrophobic aggregation processes in aqueous solution. **[61]** In particular, it is realized that the main driving force for non-polar group aggregation derives from the reorganization and reorientation of water molecules in their effort to maintain, despite the presence of the 'foreign' residues, the integrity of the hydrogen-bonded network which is a characteristic feature of the pure liquid. In other words, the *apparent* net attraction between alkyl residues in aqueous solution is in reality the sum of several water–solute repulsions.

There is now general agreement that the two major contributions to the net stability of globular proteins are hydrophobic interactions which stabilize the uniquely folded states, and the configurational entropy which has a destabilizing effect. Other contributions, e.g., intrapeptide or peptide–solvent hydrogen bonding, are of secondary importance. The *net* stability margin of a native (active) protein, relative to its many inactive states, seldom exceeds 50 kJ $mol^{-1}$, a free energy margin which is equivalent to no more than 2–3 hydrogen bonds.

In recent years attention has shifted to the more subtle solvation and configurational effects which contribute to the overall *in vivo* protein and peptide structures and functions (see chapter by Lilley). In particular, the *cis–trans* isomerization of proline is being studied, because of its critical role in protein unfolding and refolding kinetics. [**62**] Model calculations on substituted amides are also helping with the elucidation of the solvation details of peptide linkages. [**63**]

In view of progress achieved in the understanding of the role played by weak, non-covalent interactions in proteins, the question surely arises whether similar approaches could be applied to PHCs in solution and *in vivo*. The problems are somewhat different, in that the amino acid residues in proteins are joined by *identical* links (peptide bonds) which are all in the *trans*-planar configuration, but any given peptide chain can contain $> 20$ different monomer types, with an enormously large number of possible sequence isomers. PHCs, on the other hand, derive their complexity mainly from the different linkage types and chain branchings, while being composed of relatively few monomer types. In some cases the physical diversity can be achieved completely by differences in the linkages, e.g., amylose and cellulose are both homopolymers of glucose but differ markedly in their properties and biological functions. These differences are due solely to the steric nature of the glucosidic bonds, i.e. $\alpha 1,4$- and $\beta 1,4$-, respectively.

It is difficult to decide whether proteins or PHCs present more problems to experimentalists and theoreticians. On balance we favour the view that PHCs are more complex, because (1) they lack the chromophores which would allow uv and visible spectrophotometry to be employed in solution studies; (2) sugar rings are flexible and labile; (3) PHCs in crystals and in solution exhibit complex patterns of intra- and intermolecular hydrogen bonds; and (4) such solute–solute hydrogen bonding closely resembles solute–water hydrogen bonds, with neither type easy to characterize in any but a qualitative manner.

Presumably hydrophobic interactions which play an important role in protein stability are not of such great importance in PHCs, although opinions differ on this point. [**64**] There is no definite indication that biologically active oligosaccharide residues (linked to proteins or lipids) derive their functional specificity from 'native' conformations with limited *in vitro* stabilities, and if so, what are the major energetic factors which determine such function-specific stereochemistry (see Section 4). The limited experimental and theoretical information on oligosaccharides has not yet been able to provide many answers to the questions raised increasingly by 'glycobiology'. **[1]**

### 6.3. *Direct measurements of hydration*

Several attempts have been made to gain a more fundamental insight into the molecular details of sugar solvation. Water is of greatest interest, not only because of its ubiquity or its biological and technological importance, but because aqueous solutions of PHCs seldom conform to the behaviour predicted by theory. The main features that set water apart from other liquids are its spatial and orientational intermolecular correlations, dominated by labile hydrogen bond interactions, and its time-averaged tetrahedral geometry. PHCs contain the same chemical groups (—OH) and are able to interact with water such that, from an energetic point of view, differences between solute–water and water–water interactions are expected to be marginal, unless cooperative effects (about which little is known) act so as to favour water–water interactions. Configurationally the situation is complex because the mutual spacings and orientations of the —OH vectors in different solute molecules can match those in water to varying degrees. If a maximum degree of hydrogen bonding is to be maintained, then in the neighbourhood of the solute the water molecules will suffer a disturbance. Bearing in mind the wide range of hydrogen bond lengths encountered in different systems and the ease with which the HOH angle can accommodate minor distortions, such hydration disturbances are likely to be of short-range nature.

*The specific hydration model* Kabayama and Patterson first drew attention to the spatial compatibility of the —OH topology in water with that of equatorial —OH groups on pyranose sugars. **[65]** In the sugars the spacings between equatorially linked oxygen atoms on next-nearest carbon neighbours are 0.485 nm which is also the distance between next-nearest neighbour oxygen atoms in liquid water. **[66]** Figure 10 illustrates how the glucose anomers could interact with an idealized, unperturbed water lattice, without causing strain in the overall hydrogen bonding network.

*Hydration dynamics* Suggett and his colleagues pioneered direct measurements of sugar hydration, by a combination of dielectric and nmr relaxation techniques. **[27, 28, 67, 68]** It was noted that the Cole–Cole plots of the dielectric permittivities for sugar solutions were not semi-circular (see Section 3.5 or ref. 28 for details). All data could be fitted by two discrete Debye-type relaxation processes, with amplitudes and relaxation times as shown in Table 4. At higher (2.8 m) sugar concentrations, significantly better fits (at a 99%

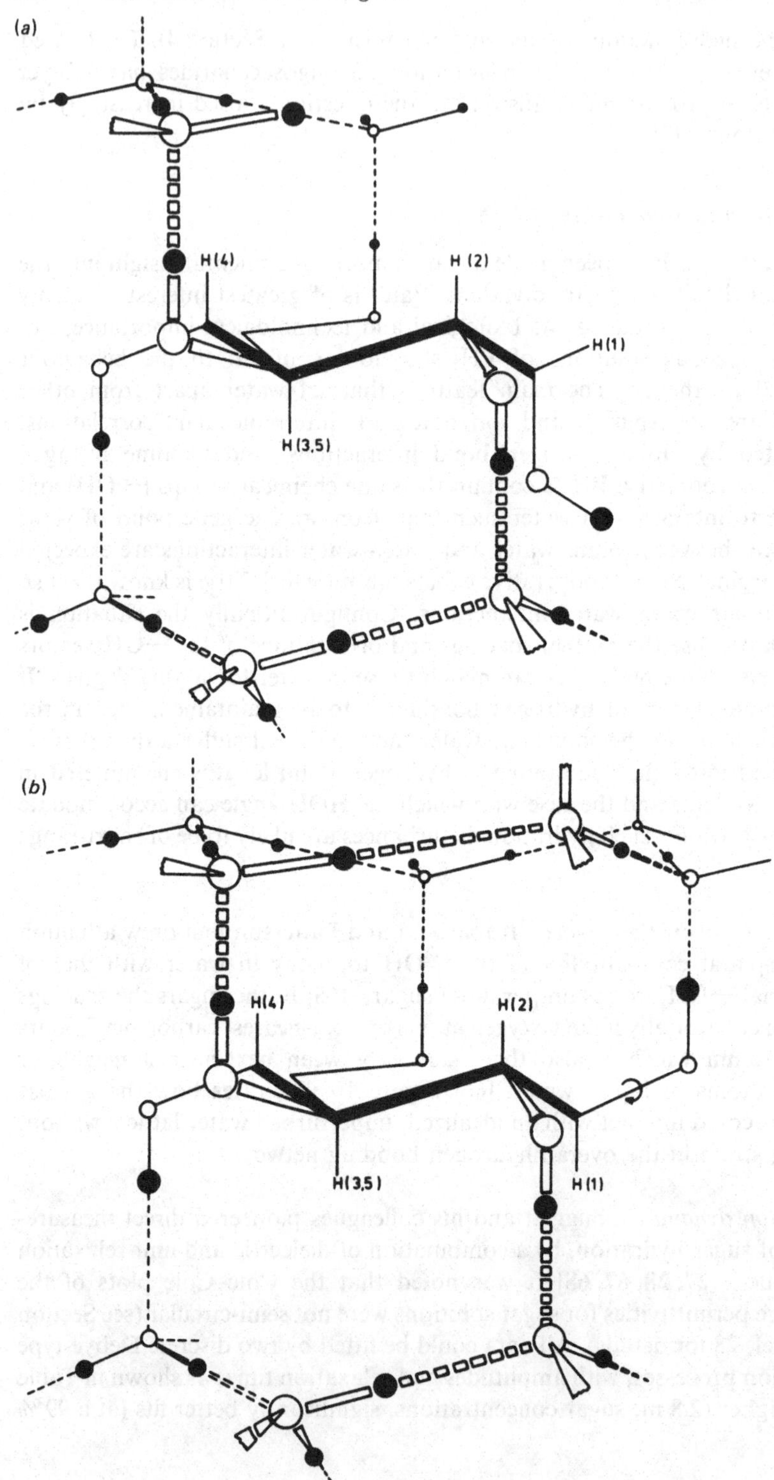
(a)
H(4)
H(2)
H(1)
H(3,5)
(b)
H(4)
H(2)
H(3,5)
H(1)

Table 4. *Two-Debye-process refinement of dielectric permittivity data, according to Equation* (8)

| Sugar | Concn (m) | $\epsilon_\infty$ | Process I $A_1$ | Process I $\tau_1$ (ps) | Process II $A_2$ | Process II $\tau_2$ (ps) | $n_h$ (mol/mol) |
|---|---|---|---|---|---|---|---|
| Glucose | 2.8 | 6.6 | 48.4 | 22.0 | 20.7 | 112 | 3.7 |
| Mannose | 2.8 | 6.6 | 47.7 | 21.0 | 21.0 | 100 | 3.9 |
| Ribose | 2.8 | 6.6 | 53.1 | 20.3 | 18.3 | 92 | 2.5 |
| Maltose | 2.8 | 6.6 | 52.0 | 21.6 | 16.8 | 143 | 5.0 |
| Sucrose | 2.8 | 6.6 | 50.4 | 21.6 | 20.2 | 130 | 6.6 |
| Maltose | 1.85 | 6.3 | 61.7 | 21.1 | 11.3 | 111 | |
| Maltotriose | 1.85 | 6.3 | 61.2 | 21.0 | 12.3 | 272 | |

probability level) were obtained by the addition of a third, lower frequency process in Equation (8). Process I was ascribed to the reorientation of the solvent molecules, although $A_1$, which varied from sugar to sugar, was not large enough to account for all the water in the solutions, and $\tau_1$ exceeded the relaxation time $\tau_0$ of bulk water by 10%. On the other hand, $A_2$, which was interpreted as due to sugar ring tumbling, was larger than expected. The results were consistent with hydration, defined in terms of slow exchange, i.e., $\tau_{ex} > 100$ ps. Interestingly, distinct differences in $\epsilon'(\omega)$ and $\epsilon''(\omega)$ were observed for $H_2O$ and $D_2O$ solutions.

Nmr studies were then performed on both sugar and water molecules, consisting of $^1H$, $^2H$, $^{13}C$ and $^{17}O$ relaxation rate measurements. Although the solvent relaxation was markedly affected by the presence of sugars, relative populations of 'hydration water' could not be resolved, because on the nmr time scale, exchange is too fast. Remarkable differences in reorientational correlation times ($\tau_c$) between axial and equatorial anomeric ring protons were recorded; for glucose these were 440 ps for H(1)$\alpha$ and 360 ps for H(1)$\beta$. The corresponding values for ribopyranose were 130 ps for H(1)$\alpha$ and 150 ps for H(1)$\beta$. The differences had already earlier been found to diminish with rising temperature. [27] Similar differences have sometimes been interpreted in terms of the bulk viscosity, but that begs the question *why* ribose solutions should be less viscous than glucose solutions. The cause must lie in the hydration details.

According to theory, the dielectric relaxation time is related to the nmr reorientational correlation time $\tau_c$ (obtained from the nuclear magnetic

Figure 10. D-glucose hydrogen bonded into a hypothetical tetrahedral water structure, according to the specific hydration model. Water molecules below and above the plane of the sugar ring are shown: (*a*) $\alpha$-anomer, (*b*) $\beta$-anomer. The pyranose ring is indicated by the bold lines. Oxygen and hydrogen atoms are represented by open and filled-in circles; covalent and hydrogen bonds by solid and broken lines. The hydroxymethyl protons (H(6)) are omitted for the sake of clarity. Reproduced, with permission, from ref. 68.

relaxation rate) by $\tau_{\text{diel}} = 3\tau_c$. From his experiments, Suggett found that for the sugars studied, this relationship was not obeyed. He thus concluded that different relaxation processes were measured by the two techniques. Since $\tau_c$ could be shown to monitor the isotropic tumbling of the sugar ring, $\epsilon(\omega)$ clearly reflected some other reorientation process, especially since $\tau_{\text{diel}}$ had also been found to be insensitive to the size of the sugar molecule.

In order to explain the discrepancy, the dielectric spectrum was resolved into discrete processes, according to Equation (8). With the aid of the parameters $A_i$, $\tau_i$, etc., it should then be possible to calculate the $^{17}O$ spin-lattice relaxation time for water and $\tau_c$ for the sugar. The two methods gave good agreement for the tumbling of water in the mixtures, but not for the solute. Two possibilities were investigated: (1) dielectric measurements monitor rotations of —OH groups which are not as fast as is usually assumed, (2) —OH rotations are coupled to ring motions. The experimental evidence favoured the second alternative. The dielectric spectra are now interpreted in terms of a major water relaxation (with an enhanced $\tau_0$) and coupled motions of hydrated side chains and the sugar ring, with slow water exchange. Hydration numbers $n_h$ therefore have a meaning within the limits set by the residence times. From a comparison of the experimental and calculated $A_1$ values, Suggett was able to obtain $n_h$ for several sugars; the results are included in Table 4. The results support the 'special' hydration characteristics of equatorial —OH groups, as earlier proposed on the basis of the specific hydration model. [**27**]

Further analysis of the $^1H$ and $^{13}C$ nmr spectra for the $—CH_2OH$ protons and the anomeric ring proton H(1) clearly showed that for $\beta$-glucopyranose the relaxation behaviour was of a purely intramolecular nature, whereas for the $\alpha$-anomer measurable differences in the C and H relaxation times were observed which decreased with decreasing concentration. The results indicate that H(1)$\alpha$ is susceptible to intermolecular interactions, as had already been suggested earlier by the results of Preston and Hall, [**32**] although these authors did not believe that the measured differences in the $^1H$ and $^{13}C$ relaxation times were significant.

All the results point to stereospecific hydration: in $\beta$-glucose, H(1) is shielded by the motion of the hydrated equatorial —OH group, with $\tau_c <$ 100 ps. In $\alpha$-glucose, on the other hand, H(1) is the only exposed equatorial proton in the molecule, as illustrated in Figure 10.

Suggett concludes that dielectric data, taken in isolation, can be misleading, because they can usually be fitted adequately to some model which may, however, be incorrect. Nmr measurements, while useful for monitoring the motions of the solute, can give quite the wrong picture for the motions of the solvent molecules. Yet, the extensive literature devoted 'bound' water is based almost exclusively on nmr measurements on the aqueous component only; see, for instance, ref. 69. Suggett, whose work should serve as a model for the student of hydration effects, recommends that

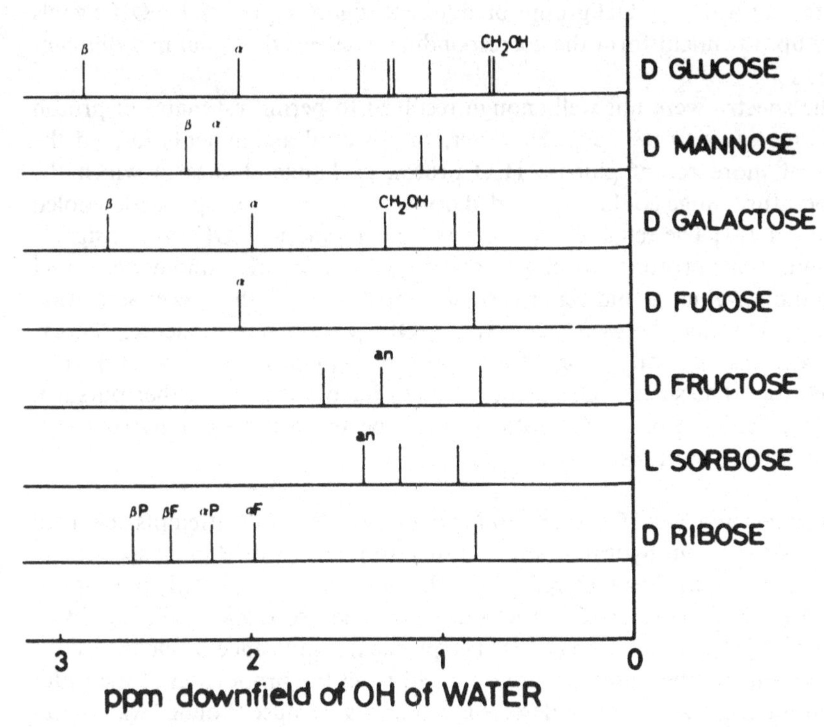


Figure 11. Hydroxyl proton chemical shifts of sugars in aqueous ($H_2O$) solutions at −6 °C, according to data from ref. 70. α and β refer to anomeric —OH protons, an = anomeric, where the assignment is uncertain; P = pyranose, F = furanose. Unmarked signals refer to ring —OH groups, but the correct assignment is uncertain.

measurements of the dynamics of and interactions between PHCs and water in mixtures should aim at better resolution of the dielectric spectra and nmr data for *both* components. [**68**]

*Hydration and hydroxyl proton nmr* Further indirect, though convincing support for stereospecific hydration was reported by Harvey and Symons who demonstrated that under conditions of slow exchange, the -OH proton nmr signals of sugars could be resolved, even in $H_2O$ solvent. [**70**] This enabled them to avoid the usual $D_2O$ exchange procedure which simplifies the nmr spectrum but also removes all potential information about solute–water interactions which would be contained in the —OH resonances. By careful pH adjustment and operating at low enough temperatures they were able to record and assign some of the —OH signals in anomeric mixtures of aldohexopyranoses; their results are reproduced in diagrammatic form in Figure 11. Two important features emerge: the anomeric β—OH resonance is always found downfield from the corresponding α—OH signal, and for equivalent —OH groups on different sugars, equatorial —OH signals show up downfield form the corresponding axial —OH signal in a different isomer.

The spectra were not well enough resolved to permit estimates of proton exchange rates to be made. However, the spectral assignments formed the basis of more recent glucose–$H_2O$ proton exchange studies in which the temperature range could be extended down to −35 °C in deeply undercooled solutions. [71] The relative line widths of the anomeric —OH proton signals indicate that proton exchange between the equatorial anomeric —OH proton on $\beta$-glucose and water is significantly faster, with a lower activation energy than that between the axial —OH proton on $\alpha$-glucose. Similar observations are on record for xylose in aqueous solution. [72] It is regrettable that such —OH proton studies have not been further pursued, probably because of the obsession of nmr spectroscopists with narrow lines and averaged motions.

*Hydration numbers from spectral intensities* Several attempts to gain information about hydration interactions from near ir or Raman spectra are on record; see for instance ref. 73 which cites $n_h$ values for 2 M solutions of ribose (5.7), glucose (10.3), fructose (12.8), sucrose (25.8), maltose (24.0), lactose (25.1) and raffinose (19.3). The physical significance of such numbers is questionable, because they are obtained from the proportion of the water signal intensity which is *not* detected, ostensibly because 'bound' water does not contribute to the observed intensity. This type of reasoning is also quite frequently applied in spectroscopic studies of water in biological systems. [74]

## 6.4. *Hydration and computer simulation*

The above discussions serve to demonstrate that direct studies of PHC hydration are beset by many problems, arising mainly from the fact that the life times of hydration 'structures' in solution are generally shorter than the time scales of most experimental methods. The incomplete realization of this problem has led to much confusion and many misunderstandings regarding the existence and nature of so-called 'bound' water. Those who interpret experimental observations in terms of bound water seldom define such binding, whether in terms of $g(r)$ (distances), binding energies or life times. Computer simulation, provided the necessary care is taken in the formulation of potential functions, can be of great help in defining hydration geometries and exchange rates, at least to a first approximation.

The only available MD results on sugars are for glucose in water. They demonstrate the effect of water on the conformation and dynamics of the system [35] Some of the details may be due to sampling artefacts, but collectively the results are of great interest. Although the average sugar ring conformation is not much affected by hydration, the solvent acts so as to increase the structural fluctuations. Changes are observed in the —$CH_2OH$ conformation, i.e., in the C(4)—C(5)—C(6)—O(6) angle, as a result of hydration. The radial distribution functions of water in the hydration sphere (nearest neighbour molecules) show up clear differences between —OH and —$CH_2OH$ hydration structures. Thus, apolar hydration is found in the neighbourhood of the $>CH_2$ group, with 12.3 nearest neighbour water molecules at a distance of $< 0.49$ nm. In contrast, the —OH groups have only 3.2 nearest water neighbours, with adjacent —OH groups sharing water neighbours by hydrogen bonding, a structure which may be significant for anomerization. Characteristically, the hydration of O(5) differs from that of the exocyclic oxygen atoms. As expected, water molecules in the oxygen hydration shells are also orientationally restricted. O(2) differs from the other —OH groups in that its water distribution function is flat, with little preference for hydrogen bond orientations. This observation might arise from the particular anomeric structure of glucose.

The simulation provides clear evidence for intramolecular hydrogen bonding which is preserved (most of the time) even in aqueous solution! This finding is in direct contrast to the experimental nmr results (in $D_2O$ solution). If true, such bonds would probably impose very definite —OH group orientations, with correlated —OH rotations. However, the simulation results suggest that —OH rotations are uncorrelated, but slower than *in vacuo*.

Confusing descriptions, like (water) structure making and breaking, found their way into the literature during the 1960s and have gained widespread popularity; PHCs generally are classified as structure makers. [75] The MD simulations on glucose clearly demonstrate that, despite the well-defined and localized hydrogen bonding between sugar —OH groups and water, all first hydration shell water molecules exchange rapidly with bulk water. Water can therefore be regarded as 'bound' in a structural sense, but not in a dynamic sense, since exchange with 'bulk' water takes place on a $< 10$ ps time scale.

MD simulations on mannitol and sorbitol, not being subject to anomeric and tautomeric transitions, provide more exact information regarding PHC interactions with water, albeit 'computer water'. [36] Samoilov defined hydration in terms of a parameter $R = \tau_h/\tau_0$, where the $\tau$ values describe the average life times of a water molecule in the hydration shell (h) and in bulk water (0) respectively. [76] The simulation results show that both polyols have $R < 1$, corresponding to *negative hydration*, by Samoilov's criterion. However, the $R$-values differ for the two diastereoisomers: 0.80 for mannitol and 0.39 for sorbitol, highlighting once again the extreme sensitivity of water to the stereochemical detail of the PHC molecule.

Since dynamic data cannot provide structural information, it is unwise to assign structure making/breaking properties. What is evident, however, is that both hexitols (as well as glucose) generate mobile water environments in their proximities. Hydration numbers can be assigned, being the average number of water molecules at a distance of 1.2 molecular diameters from the hexitol molecule. This yields $n_h = 13.23$ and 11.45 for mannitol and sorbitol, respectively. The ratio, 1.55, is close to the experimental ratio of their calculated hydrodynamic volumes, 1.28. [77]

## 7. Thermodynamics of PHCs in dilute solution

The first systematic thermodynamic studies on simple sugars in solution date from the 1970s when attempts were made to characterize different types of molecular hydration, but even now the available information is fragmentary, nearly all of it confined to aqueous solutions of simple PHCs. No systematic thermodynamic information is as yet available for PHC derivatives, such as aminosugars, sulphates, phosphates and carboxylates, all of which feature prominently in biology. The major aspects of interest are: (1) sugar hydration as affected by stereochemistry, and (2) sugar–sugar interactions, as calculated from thermodynamic excess functions. Wadsö and his colleagues have published high-quality calorimetric data, capable of being extrapolated to infinite dilution. In the laboratories of Wood, Lilley and Barone, work has concentrated on the problems posed by solute interactions in solution. Yet another approach has been adopted by Goldberg and Tewari, who have studied the enzymatically catalysed interconversion thermodynamics of sugars.

The weakness of all thermodynamic studies lies in the chemical heterogeneity of sugar solutions, so that all measured quantities are suitably weighted averages. One method of inhibiting equilibration is to block some or all of the —OH groups by derivatization. Unfortunately such procedures also reduce the information content of the data, because hydration interactions are very susceptible to the —OH group topology of a particular stereoisomer, and derivatization (e.g., acetylation or esterification) partly destroys this sensitivity. Nevertheless, useful information has been obtained from judicious derivatization, e.g., comparisons of $\alpha$- and $\beta$-1-methyl pyranosides. **[78, 79]**

For comprehensive and up-to-date tabulations of thermodynamic properties of PHCs in aqueous solution, the reader is referred to refs. 6 and 19. We have selected data originating from laboratories with reputations for consistent, high-quality experimentation. Only in a few cases can the numbers be confidently compared with corresponding results from other sources. Where this is possible, e.g., homotactic enthalpic pair interaction parameters ($h_{ii}$), derived from heats of dilution, discrepancies become apparent. Whatever the reasons for non-agreement between results from different laboratories, it is advisable not to be too dogmatic in the establishment of universal rules governing the behaviour of PHCs. The carbohydrate literature abounds in such ‘rules’ (some of them critically discussed in refs. 5 and 15), several of which have subsequently had to be discarded or severely qualified in the light of new experimental evidence.

### 7.1. *Limiting thermodynamic quantities*

As regards useful information about hydration details, the limiting partial molar heat capacity $C_p^\circ$, defined in Equation (1), is probably the most sensitive thermodynamic indicator. It is obtained experimentally either from direct heat capacity measurements or from the temperature dependence of the heat of solution of the solid PHC. For purposes of comparisons, several precautions are indicated.

Heat capacity measurements are made on the equilibrium mixture which, in the case of sugars, consists of several isomeric species. For heat of solution measurements, the starting material will normally be a pure crystalline substance. On the assumption that its conversion to the equilibrium mixture is slow, then the measured $C_p$ value refers to a solution of the pure isomer. In order to compare these data with those obtained from equilibrated solutions, it is necessary to correct for the $\Delta C_p$ which accompanies the equilibration. Thus, for the mutarotation of glucose, starting from the $\alpha$-anomer, $\Delta C_p = -9$ J mol$^{-1}$ K$^{-1}$. [**80**] The anomeric mixture contains 37% of the $\alpha$-anomer, and therefore $C_p^\circ$ of the equilibrium mixture should be increased by 5 J mol$^{-1}$ K$^{-1}$ for comparisons with data obtained by heat of solution measurements on the pure $\alpha$-form.

Other sources of discrepancy can arise from differences in the choice of the standard state, especially where a number of isomers and/or polymorphs exist. For instance, in the case of ribose, it must be made clear whether $C_p^*$ refers to $\alpha$- or $\beta$-ribopyranose or furanose. Such information is usually not available.

Equation (1) can also be written for the volumetric properties, i.e., for the evaluation of the limiting partial molar volume $V^\circ$, provided that the necessary $V^*$ data are available. Strictly speaking, these should be density data for the crystalline PHCs. Some authors have calculated $V^*$ from molecular models and/or empirical relationships between $V^\circ$ and a notional (spherical) van der Waals volume, also incorporating a void volume correction. [**81**] Since real differences between $V^\circ$ values for isomeric PHCs tend to be small, uncertainties in $V^*$ may either hide or exaggerate such differences.

The limiting thermodynamic quantities of some PHCs are collected in Table 5. Where several published values exist, we have chosen the value 'preferred' by Lian, Chen, Suurkuusk and Wadsö [**82**] and/or Goldberg and Tewari [**19**] who base their recommendations on a careful analysis of the primary experimental data. In some cases, where the discrepancies appear rather large, we include both the recommended and the discordant values,

Table 5. *Limiting partial molar volumes ($V_2^\circ$), expansibilities ($E_2^\circ$), isothermal compressibilities ($K_2^\circ$) and heat capacities ($C_2^\circ$) of PHCs in aqueous solution at 25 °C. 'Best' values, according to refs.* 6 *and* 19 *are shown, unless otherwise indicated. Asterisks refer to the crystalline state. Units are V: cm³ mol⁻¹, E: cm³ mol⁻¹ K⁻¹, K:* 10⁹ *cm³ mol⁻¹ Pa⁻¹, C: J mol⁻¹ K⁻¹*

| | $V_2^\circ$ | $V^*$ | $E_2^\circ$ | $K_2^\circ$ | $C_2^\circ$ | $C^*$ |
|---|---|---|---|---|---|---|
| β-Arabinose | 93.7 | 92.4 | | −15.2 | 278 (D)<br>270 (L) | 184 |
| Xylose | 95.4 | 98.5 | | −8.8 | 279 | 184 |
| Ribose | 95.2 | 94.4 | 0.12 | −8.4 | 276 | 187 |
| β-Fructopyranose | 110.7 | 112.7 | 0.136 | −16.7 | 338<br>352[c] | 232 |
| α-Galactose | 110.5 | 111.3 | 0.15 | −15.6 | 324 | |
| β-Galactose | | | | | | 217 |
| α-Mannopyranose | 111.5 | 115.2 | | −12.7 | 337 | 216 |
| α-Glucose | | 115.3 | | | 347<br>331[a] | 220<br>224[a] |
| α/β-Glucose | 112.0 | 118.3[b] | 0.09 | −14.8 | 346[a] | |
| β-Glucose | | 116.1 | | | | |
| α-Sorbopyranose | 111.0 | 112.2 | | | | 231 |
| 2-Deoxyribose | 93.8 | | | | 333 | |
| 2-Deoxyglucose | 110.4 | | | | 381 | |
| α-Me glucoside | 133.6 | | | | 435 | |
| α-Me galactoside | 132.6 | | | | 436 | |
| β-Me galactoside | 132.9 | | | | | |
| α-Me mannoside | 132.9 | | | | 445 | |
| Trehalose | 206.9 | | | | | |
| Cellobiose | 213.6 | | | | | |
| Maltose | 208.8 | | | | 614.2 | |
| Sucrose | 211.3 | | | | 649.4 | |
| Lactose | 207.6 | | | | 619 | |
| Melibiose | 204 | | | | | |
| Cellotriose | 309.2 | | | | | |
| Maltotriose | 304.8 | | | | | |
| Melezitose | 305.8 | | | | | |
| Raffinose | 302.2 | | | | 931 | |
| Ethane diol | | | | | 193 | |
| Glycerol | | | | | 240 | |
| Erythritol | | | | | 310 | |
| Ribitol | | | | | 376 | |
| Xylitol | | | | | 346 | |
| Arabinitol | | | | | 374 | |
| Mannitol | | | | | 452 | |
| Glucitol (sorbitol) | | | | | 412 | |
| Galactitol | | | | | 445 | |

[a] F. Kawaizumi, N. Sasaki, Y. Ohtsuka, H. Nomura and Y. Miyahara. *Bull. Chem. Soc. Japan* **57**, 3258(1984).

[b] *Glassy solid*

[c] *Ref.* 82

but only where these have formed part of a wider study, including other PHCs and where the remainder of the data conforms to the recommended values. Table 5 also contains the available crystal data.

Quoted uncertainties range from 0.5 to 3%. It must be remembered that, among other factors, $C_p^\circ$ depends on the accuracy of the heat capacity value for the crystalline sugar. $C_p^\circ$ measures a combination of the effects of hydration (solute–solvent interactions) and any changes in the internal degrees of freedom of the solute molecule when it is removed from the crystal. Thus, $C_p^\circ$ is a measure of the effect of a change in environment on the number of degrees of freedom (mainly intermolecular) of the system.

As expected, $C_p^\circ$ is determined mainly by molecular size and complexity. Nevertheless, second order effects are apparent, e.g. in the $C_p^\circ$ values of the three pentitols listed. In their crystalline forms, ribitol (XIII) and xylitol (XIV) exist with bent carbon chains (sickle-shape), whereas arabinitol (XV) has a planar zig-zag carbon atom arrangement. [**83–5**] It might be expected

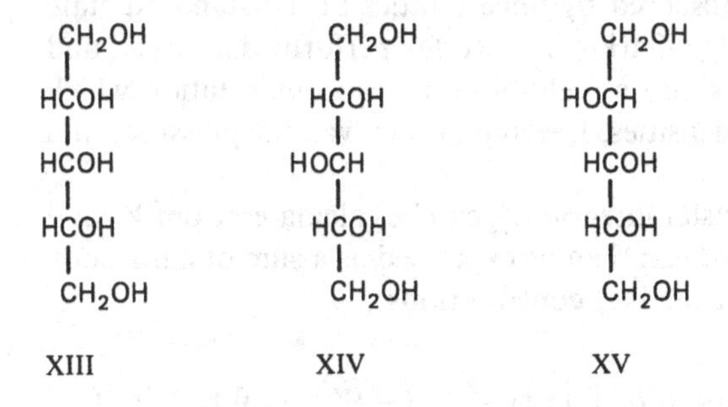


XIII XIV XV

that in solution, ribitol and xylitol should show some similarities in their thermodynamic behaviour, but differ from arabinitol. This is not the case however: xylitol has a $C_p^\circ$ which is significantly lower than those of the other two isomers. A similar difference is observed for the hexitols, where sorbitol (glucitol) has a lower $C_p^\circ$ than its isomers, galactitol and mannitol.

Large positive $C_p^\circ$ values are usually taken as an indicator of thermolabile hydration spheres, and on that basis it appears that the hydration structures surrounding xylitol and sorbitol are of a shorter range and/or less complex than those associated with their isomers, a conclusion which is consistent with the MD results, described in Section 6.4.

Although there is a wealth of limiting partial molar volume ($V^\circ$), expansibility and compressibility data, they have not all been included in Table 5 (but can be found in ref. 19), because measured differences between isomeric sugars are so small that not too much significance can be read into them as regards the influence of sugar stereochemistry on hydration. $V^\circ$ is mainly a function of the molecular weight, with each additional ($C.H_2O$) contributing approx. 16–20 $cm^3\ mol^{-1}$. A general observation is that $V^\circ$ values of PHCs are surprisingly small compared to their non-hydroxylated analogues. Thus, the introduction of a $CH_2$ group into the tetrahydrofuran (I) ring, increases $V^\circ$ by 15 $cm^3 mol^{-1}$ which is almost the same as the increment produced by the addition of —CHOH into the furanose ring, despite the considerable difference in the group molecular weights. [**14**] The comparison of furan and pyran derivatives with furanoses and pyranoses suggests that the density measurement does not 'see' the —OH groups as as part of the solute molecule but as water molecules. [**14**] Observations of this type raise questions whether the $V^*$-values, calculated from van der Waals radii, are realistic reference states, and if not, whether the $V^\circ$ data so obtained can usefully be compared with those referred to crystal densities.

It is regrettable that, despite the considerable experimental effort that has gone into the measurement of limiting volumes, expansibilities and compressibilities of PHCs, little can be said about their interpretation in terms of solute–water interactions, mainly because real differences in $V^\circ$ and its $P$ and $T$ derivatives are obscured by uncertainties in the standard state values. Herrington, Pethybridge, Parkin and Roffey performed a careful and comprehensive experimental study of cellobiose in aqueous solution which comprised measurements of densities, freezing points, vapour pressures and solubilities. [**86**]

The solvent(1)–solute(2) cluster integral $B^\circ_{12}$ can be calculated from $V^\circ$ and other thermodynamic data and can then be expressed as a sum of hard-body (repulsive) and 'chemical' (attractive) contributions:

$$B^*_{12} = \int_0^R [1-\exp(-W_{12}/kT)]\, 4\pi r^2\, dr + \int_R^\infty [1-exp(-W_{12}/kT)]\, 4\pi r^2\, dr \quad (25)$$

where $R$ is the distance of closest approach of the molecules, treated as hard bodies. The two contributions were calculated on the assumption that the water molecule is spherical and cellobiose is a prolate ellipsoid with the same molecular dimensions as those found in its crystalline form. At 25 °C, the repulsive and 'chemical' contributions are, respectively, 405 and $-194$ $cm^3 mol^{-1}$, which can be compared with those of sucrose (476, $-266$) and glucose (357, $-246$). In all three cases the 'chemical' contribution is attractive and increases with decreasing temperature; see Equation (24) and the following discussion.

Despite reservations about the reliance on crystallographic data and modelling the molecules as hard bodies, the above investigation was a useful attempt to relate thermodynamic measurements to molecular interactions through the potential of mean force $W_{12}(r)$. Although the actual numbers must be subject to considerable uncertainty, the results do demonstrate a weak net repulsion between sugars and water. This is surprising, but perhaps the finding is consistent with the MD results, earlier referred to, which suggest a structure-breaking influence of sugars on water.

7.2. *Concentration dependence: solute–solute interactions*

Interest here centres on the effect of stereochemical detail on the virial coefficients in Equation (2), extracted from the measured thermodynamic excess functions. For enthalpic data we rely mainly on the comprehensive studies of Barone and Lilley and their colleagues; most of the relevant results can be found in the compilations of Cesaro [6] and Goldberg and Tewari. [19] In Table 6 we present selected data for the second virial coefficients in Equation (2) for the free energy and enthalpy. While results from calorimetric studies abound, there is little information about $g_{ii}$. The reason is clear: vapour pressure measurements on dilute solutions are difficult to perform with the necessary degree of precision, especially where deviations from ideal behaviour are marginal. By comparison, enthalpy determinations are more straightforward, because of the availability of high-precision calorimeters.

The scattered nature of the data precludes the establishment of patterns in the relationships between the stereochemistry and the viral coefficients. This is particularly true for the $g_{ii}$ data. It was thought earlier that $g_{ii}$ was positive for all PHCs, except *myo*-inositol which is based on the cyclohexane ring and has a high degree of internal symmetry. [87] However, both galactose and mannose appear to have $g_{ii} < 0$. [59]

As regards the available $h_{ii}$ data, mixing appears to be endothermic, except for some hexitols and perseitol (a heptitol) where exothermic mixing has been observed. There is no obvious relationship between molecular size or number of —OH groups and $h_{ii}$, but chemical modification (deoxysugars, methyl pyranosides) leads to increases in $h_{ii}$. As was the case for the limiting thermodynamic properties, measurable differences in $h_{ii}$ exist for diastereoisomers, e.g., ribitol, xylitol and arabinitol, emphasizing once again that sugar–sugar interactions in solution are sensitive to the —OH group topology. Unfortunately, with the present state of our knowledge, no quantitative interpretation of the data can be attempted. Also, as already mentioned earlier, the numbers presented in Table 6 must still be treated with a certain degree of caution, since few of the data have been confirmed by several independent groups of workers.

With a few exceptions, the general rule seems to be that for PHCs, $h_{ii} > g_{ii} > 0$ which contrasts with urea-like compounds for which $h_{ii} < g_{ii} < 0$. It has been suggested that the observed behaviour is to be attributed to a partial reversal of hydration which is speculated to involve only minor changes in the energy and degrees of freedom of the water released from the 'more structured hydration cospheres of the saccharides' to the bulk, with increasing concentration. [75] This interpretation contrasts with the view, discussed in Section 6.4, that PHCs act as water structure breakers.

Barone *et al.* [75] have also attempted to rationalize the observed $h_{ii}$ results in terms of stereochemical detail and, in particular, the specific hydration model. [27] Some correlations can be established between equatorial —OH

Table 6. *Homotactic ($g_{ii}$, $h_{ii}$) and heterotactic ($h_{ij}$) free energy and enthalpic molecular pair interaction parameters, see Equation* (2). *Units: J kg mol*$^{-2}$

| $i$ | $j$ | $g_{ii}$ | $h_{ii}$ | $h_{ij}$ |
|---|---|---|---|---|
| Xylose | | 81.5 | 332 (D)<br>336 (L) | |
| Xylose (L) | Xylose (D) | | | 317 |
| Ribose | | 7.5 | 202 | |
| Arabinose | | 34 | 196 (D)<br>192 (L) | |
| Arabinose (L) | Arabinose (D) | | | 213 |
| Fucose | | | 669 (D)<br>665 (L) | |
| Fucose (L) | Fucose (D) | | | 664 |
| Glucose | | 63 | 342 | |
| Galactose | | 65<br>−139 [59] | 133 | |
| Mannose | | 43<br>−47 [59] | 207 | |
| Fructose | | 83 | 264 | |
| Sorbose | | 148 | 394 | |
| 2-Deoxyribose | | 80 | 464 | |
| 2-Deoxyglucose | | 88 | 592 | |
| 2-Deoxygalactose | | 58 | 442 | |
| 6-Deoxy-L-mannose | | 162 | 685 | |
| 6-Deoxy-L-galactose | | 153 | 665 | |
| 1-Me-α-xyloside | | 114 | 1126 | |
| 1-Me-β-xyloside | | 100 | 1098 | |
| 1-Me-α-xyloside | 1-Me-β-xyloside | | | 1071 |
| 1-Me-α-galactoside | | 33 | 900 | |
| 1-Me-β-galactoside | | 134 | 1081 | |
| 1-Me-α-galactoside | 1-Me-β-galactoside | | | 880 |
| 1-Me-α-glucoside | | 171 | 1097 | |
| 1-Me-β-glucoside | | 82 | 1048 | |
| 1-Me-α-glucoside | 1-Me-β-glucoside | | | 1026 |
| 1-Me-α-mannoside | | 144 | 1206 | |
| Cellobiose | | 138 | 764 | |
| Maltose | | 100 | 483 | |
| Trehalose | | 140 | 595 | |
| Lactose | | 195 | 506 | |
| Sucrose | | 205 | 577 | |
| Melizitose | | | 607 | |
| Raffinose | | 353 | 458 | |
| Ethane diol | | | 362 | |
| Glycerol | | | 251 | |
| Erythritol | | | 358 | |
| Pentaerythritol | | | 395 | |
| Glycerol | Pentaerythritol | | | 431 |
| Ribitol | | | 295 | |
| Xylitol | | | 80 | |
| Arabinitol (D) | | | 187 | |
| Arabinitol (L) | | | 185 | |
| Arabinitol (D) | Arabinitol (L) | | | 195 |
| Mannitol | | | 66 | |

Table 6. (*cont.*)

| $i$ | $j$ | $g_{ii}$ | $h_{ii}$ | $h_{ij}$ |
|---|---|---|---|---|
| Glucitol (sorbitol) | | | −11 | |
| Galactitol | | | −132 | |
| Mannitol | Galactitol | | | 47 |
| Perseitol | | | −299 | |
| *myo*-inositol | | −260 | −800 | |
| *myo*-inositol | Mannitol | | | −220 |
| Mannitol | Cyclohexanol | | | 1267 |
| *myo*-inositol | Cyclohexanol | | | 1487 |

group frequency and the magnitudes of the virial coefficients, but the arguments are tenuous; they would be strengthened by more systematic experimental studies of PHC solution thermodynamics.

Quantitative evaluations of thermodynamic measurements have also been attempted for solute–solute interactions, the relevant virial coefficient $B^*_{22}$ being obtained from activity coefficient data. Thus, Herrington *et al.* calculated the repulsive part in Equation (25) by modelling sugars as a hard ellipsoids. **[86]** The following results are given for the repulsive and attractive contributions to $W_{22}(r)$: cellobiose 622, −335, sucrose 783, −498, and glucose 520, $-403\ cm^3\ mol^{-1}$. Note that for urea the two contributions cancel exactly, so that its aqueous solutions exhibit pseudo-ideal behaviour. **[88]**

The calculations were carried a stage further by Gaffney, Haslam and Lilley who measured the enthalpy of dilution of cellobiose. **[89]** The $h_{22}$ coefficients are included in Table 6. The agreement with the earlier $h_{22}$ coefficients, calculated from freezing point and vapour pressure data is poor, probably because the necessary degree of precision in $g_{22}(T)$ can hardly be achieved with vapour pressure determinations. By combining the cryoscopic and calorimetric results, Gaffney *et al.* obtained $g_{22} = 347\ J\ kg\ mol^{-1}$ at 25 °C. They correctly pointed out that the Helmholtz free energy ($a_{22}$) and the internal energy ($u_{22}$) coefficients were the correct quantities which are related to $W_{22}(r)$ through Equation (25). The attractive contribution to $a_{22}$ was obtained as $-671\ J\ dm^3\ mol^{-2}$, and the hard body (repulsive) contribution as $1542\ J\ dm^{-3}\ mol^{-2}$, with $W_{22}(r) = -71\ J\ mol^{-1}$, i.e., weakly attractive. Just like the earlier authors, Gaffney *et al.* concluded that interactions between sugars in aqueous solution are extremely weak and very sensitive to temperature.

### 7.3. *Group additivity schemes*

Several attempts are on record for the prediction of thermodynamic properties on the basis of group additivity schemes. The particular thermodynamic property in question is regarded as the linear sum of contributions from the various chemical groups which make up the molecule. In the case of PHCs these might be $>CH_2$, $>CHOH$, —O—, —OH, etc. Thus, Nichols *et al.* advanced such a scheme for the prediction of $C_p^\circ$; [**90**] it proved to be successful for hydrocarbons and monofunctional alkyl derivatives. The authors concluded that solvation effects are of a short-range nature, with no 'specific' interactions.

The scheme breaks down for PHCs, with deviations between experimental and predicted values increasing with increasing molecular weight. [**82**] Thus, the group parameter scheme predicts $C_p^\circ = 141$ J mol$^{-1}$ K$^{-1}$ for the aldopentoses, whereas the experimental values average around 270 J mol$^{-1}$ K$^{-1}$. By its very nature, the additivity scheme cannot account for differences between stereoisomers.

Another predictive scheme, for $V^\circ$, was developed by Edward and Farrell [**91**] and tested for PHCs by Shahidi, Farrell and Edwards. [**81**] The agreement with the experimental results is much better than that achieved by the $C_p^\circ$ scheme, but the reason may be that the calculations include an empirical 'shrinkage' parameter which is applied to each solute —OH groups and causes a reduction in $V^\circ$ of 5.9 cm$^3$ (mol —OH groups)$^{-1}$.

Group additivity schemes have also been described for the calculation of the virial coefficients which describe solute–solute interactions; see Equation (2). Savage and Wood first proposed relationships of the type

$$g^{xy} = -(M_1RT/2) + \sum_{ij} n_i^x n_j^y \{G_{ij}\} \tag{26}$$

where $n_i^x$ *and* $n_j^y$ *are the number of functional groups* $i$ or $j$ on species $x$ and $y$, respectively, and $M_1$ is the molecular weight of the solvent (water). [**92**] Similar relationships can be written for $h^{xy}$ etc.

Equations of the type (26) imply that $G_{ij}$ is independent of neighbouring group effects, i.e., steric effects are absent, that the interactions are of a short-range type, and that solvation effects can also be expressed in terms of additive contributions and are of a short-range nature. A further, implicit assumption is that no orientation-specific factors need to be taken into account, so that an orientation-averaged value of $W_{22}(r)$ is adequate. However, we have already discussed that for PHCs, such an assumption is probably unwarranted and that the orientation-dependent part of $W_{22}$ appears to be very important in determining solute–solute interactions.

Group additivity schemes have been elaborated far beyond what appears to be useful or even permissible. Their applicability to PHCs has been critically analysed by Barone *et al.*, [**75**] who demonstrate that, quite apart from the fundamental criticisms, the experimentally determined heats of dilution have led different workers to values for $H_{ij}$ which show poor agreement. For instance, for the OH–OH interaction, estimates for $H_{ij}$ of $-1$, $-26$ and $-59$ J kg mol$^{-2}$ have been reported. On the other hand, for

Table 7. *Enthalpic second virial coefficients $h_{ij}$ for the interactions of PHCs (i) with other solutes (j): glycerol (G), N-methyl formamide (NMF), N-methyl acetamide (NMA), N-acetylglycinamide (NAGA), urea (U) and thiourea (TU); units: J kg mol$^{-2}$. Data from refs.* 79, 154 *and* 155.

| | *j* | | | | | |
|---|---|---|---|---|---|---|
| *i* | G | NMF | NMA | NAGA | U | TU |
| Ribose | −414 | | | 131 | −370 | −559 |
| Arabinose | −370 | | | 152 | −370 | −540 |
| Xylose | −245 | | | | −314 | −402 |
| Glucose | | 397 | 736 | | −317 | |
| Fructose | | | | | −405 | |
| Sucrose | | 506 | 992 | | −598 | |

substantially hydrophobic solutes, the additivity schemes provide a useful, if empirical, method for the prediction of viral coefficients and good agreement exists between the various reported estimates. For instance, for the $CH_2$–$CH_2$ interaction, the proposed $H_{ij}$ values are 43, 40 and 35 J kg mol$^{-2}$. **[75, 93, 94]** In summary, therefore, group additivity schemes do not provide a reliable method for the prediction of PHC interactions in solution, especially where differences in the interactions between different stereoisomers are to be predicted.

## 7.4. *Non-aqueous solutions and mixed solvents*

Some limiting partial enthalpies of PHCs in non-aqueous solvents and mixed aqueous/organic solvents have been measured and interpreted in terms of competing hydrophobic and hydrophilic hydration effects. **[95, 96]**

Despite its shortcomings, the model described by Equation (24) has been applied, although considerably elaborated, to account for the heats of solution of polyols **[95]** and sugars **[96]** in mixed solvents of water and DMF. Unfortunately the treatments contain several unwarranted assumptions. For example, there seems to be no reason why $n_h$ should be set equal to three times the number of —OH groups in the PHC molecule, nor is a good case made for the division of $\Delta H$ into hydrophobic and hydrophilic contributions. It can also not be taken for granted that the sugars exist solely in the C1 conformation. While this may be true for aqueous solutions, in mixed solvents, and especially in pure DMF, furanose forms have been identified. **[97]**

Heats of mixing of PHCs with other water-soluble substances have been reported and $h_{ij}$ virial coefficients extracted. Some representative data for such ternary systems have been collected in Table 7; only data pertaining to common PHCs have been included. Although certain trends in the magnitudes and signs of the enthalpic interaction coefficients seem to exist, the data are too sparse for any useful conclusions to be drawn.

## 8. Mutarotation: equilibria and kinetics in solution

The ability of cyclic PHCs, especially reducing sugars, to undergo a variety of spontaneous stereochemical transformations has for long excited the interest of chemists; it is probably the feature of PHCs which has been studied most intensively. Within the context of this review, interest centres on the role of the solvent in affecting the equilibrium composition in solution and on solvent effects which influence the interconversion kinetics and mechanisms. The simple pentoses and hexoses are taken as typical examples. In all cases, at least two distinguishable species coexist in solution, an indication that such species have similar free energies and that the activation energies for the interconversions are reasonably low.

### 8.1. *The compositions of anomeric equilibrium mixtures*

In the simplest cases, e.g., glucose, the equilibrium mixture only contains the two anomeric forms of one of the possible ring conformers; in this case it is the C1 pyranoid ring. In many other cases several tautomers and their anomeric forms coexist. The equilibrium mixture of ribose contains at least six distinct species: the two anomeric forms of each of the 1C and C1 pyranoses and of the furanose ring. The composition of the equilibrium mixture depends on the solvent. This is shown in Figure 12 which compares the effects of solvent ($D_2O$ and dimethyl sulphoxide (DMSO)) and temperature on the equilibrium mixtures of glucose and ribose. **[98]** For tabulated data of the equilibrium compositions of some representative sugars in solution, see ref. 99. Table 8 provides an abstract of Angyal's compilation which illustrates that the nature of the epimer as well as the solvent determine the equilibrium composition. Actually the situation is somewhat more complex than shown in Table 8, because several pyranose conformers might coexist, as demonstrated for ribose in Figure 12. Furthermore, the sugar concentration also affects the equilibrium composition, as well as its temperature dependence. **[98]**

Shallenberger has discussed the factors associated with different types of mutarotation. **[15]** Thus, all the aldohexoses (with the exception of altrose) which exhibit complex mutarotation, i.e., where more than one tautomer is involved, have an axial —OH on C(4). This is also the case for the two pentapyranoses which exhibit complex mutarotation. These observations have led to the generalization that an axial —OH group on C(4) of a pyranose leads to appreciable proportions of furanose forms, although no reasons have been advanced for this 'rule.' The mutarotation of $\beta$-fructopyranose in water does not produce the corresponding $\alpha$-anomer but

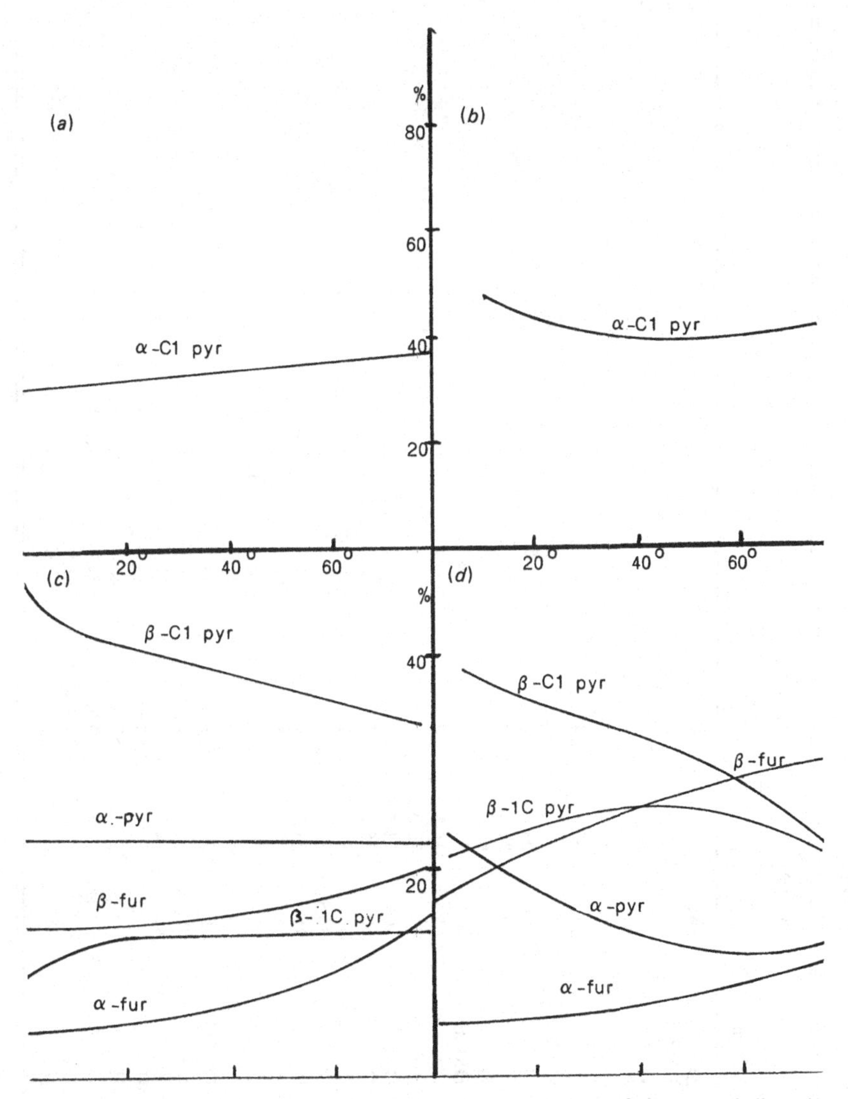


Figure 12. The equilibrium compositions of 10% w/w solutions of glucose and ribose in $D_2O$ and DMSO-$d_6$ as functions of temperature, according to ref. 98: (*a*) glucose/$D_2O$; (*b*) glucose/DMSO-$d_6$; (*c*) ribose/$D_2O$; (*d*) ribose/DMSO-$d_6$.

a mixture of α- and β-fructofuranose. [**100**] This has been rationalized with the aid of a comparison of the favoured $^5C_2$ conformations of the α-pyranose forms of allulose, tagatose and sorbose which do not have an axial —OH at C(5) and β-fructose which does have an axial —OH(5) group in its favoured tautomer, the $^2C_5$ pyranose.

In aqueous solution the pyranose ring of reducing sugars appears to be

Table 8. *Tautomeric and anomeric equilibrium compositions of sugars in solution.* $H = H_2O$, $D = D_2O$, $P =$ *pyridine*-$d_5$, *DMSO* = *dimethyl sulphoxide*-$d_6$. *Data from refs* 98, 99 *and* 122.

| Sugar | Temp. (°C) | Solvent | Pyranose α | Pyranose β | Furanose α | Furanose β | |
|---|---|---|---|---|---|---|---|
| Ribose | 30 | D | 24 | 52 | 9 | 15 | |
| | 30 | DMSO | 16 | 55 | 6 | 23 | |
| Allose | 31 | D | 14 | 77.5 | 3.5 | 5 | |
| | 50 | P | 23 | 61 | 5 | 11 | |
| Altrose | 22 | D | 27 | 43 | 17 | 13 | |
| | 25 | P | 27 | 36 | 24 | 13 | |
| | 25 | DMSO | | | 44 | | |
| Galactose | 31 | D | 30 | 64 | 2.5 | 3.5 | |
| | 25 | P | 33 | 48 | 7 | 12 | |
| Glucose | 30 | D | 34 | 66 | | | |
| | 30 | DMSO | 44 | 56 | | | |
| Mannose | 44 | D | 65.5 | 34.5 | 0.6 | 0.3 | |
| | 116 | P | 78 | 22 | | | |
| | 116 | DMSO | 86 | 14 | | | |
| Arabinose | 31 | D | 60 | 35.5 | 2.5 | 2.0 | |
| | 80 | P | 33 | 33 | 21 | 13 | |
| Xylose | 31 | D | 36.5 | 63 | < 1 | | |
| | 25 | P | 45 | 53 | 1 | 1 | |
| Fructose | 30 | D | 2 | 70 | 5 | 23 | |
| | 80 | D | 2 | 53 | 10 | 32 | Keto: 3 |
| | 33 | P | 5 | 43 | 15 | 35 | |
| | 33 | DMSO | 5 | 26 | 21 | 48 | |
| Sorbose | 27 | D | 98 | | 2 | | |
| | 80 | | 87 | 2 | 8 | 1 | |
| Ribose-5-P | 6 | H | | | 35.6 | 63.9 | Keto: 0.1 |
| Arabinose-5-P | 6 | H | | | 57.3 | 40.4 | Keto: 0.2 |
| Xylose-5-P | 6 | H | | | 70.5 | 24.8 | Keto: 0.3 |
| Fructose-6-P | 6 | H | | | 16.1 | 81.8 | Keto: 2.2 |
| Fructose-1,6-$P_2$ | 6 | H | | | 13.1 | 86.0 | Keto: 0.9 |

more stable than the furanose ring, although this situation is reversed for some sugar derivatives. In non-aqueous solvents appreciable proportions of furanoses exist in the equilibrium mixtures; the reason is not immediately obvious but must reside in the solvation contributions to the conformational free energies of the various species and the ease of interconversion. Accepting then, that water stabilizes the pyranose ring, the next question concerns the relative stabilities of the two anomers. Despite several efforts over the past two decades to calculate conformational free energies, it is still not obvious why the $\alpha/\beta$ ratio for glucose is 36:64, whereas for mannose it is 67:33. Conventionally, such observations have been rationalized (with the benefit of hindsight) by invoking the anomeric effect; for a critique, see Section 4.6.

Alternative explanations of anomeric equilibria take explicit account of solvation as a determining factor. Given that the intrinsic conformational free energy differences between coexisting species, as calculated from vacuum-based potential functions, are small, of the order of $kT$, it is deduced that the solvation contribution plays an important role in determining the equilibrium composition.

### 8.2. *Theoretical 'predictions'*

The fact that $\alpha/\beta$ ratios in aqueous solutions are well established experimentally may be the reason why most theoreticians have concentrated on this particular aspect of sugar stereochemistry, with special emphasis on glucopyranose. Thus, from early days, the achievement of the correct experimental $\alpha/\beta$ ratio for glucose in aqueous solution has been a favourite challenge.

Theoretical studies were first attempted by Angyal, with the aid of empirical force field calculations. **[37]** As mentioned earlier, this type of energy computation has little predictive power. The geometry was assumed to be fixed and equal to cyclohexane, while the atom–atom energies used were obtained from experimental results and calculations on model compounds (mainly cyclitols and acetylated aldoses) in *aqueous* solution. Moreover, the anomeric effect was introduced on an *ad hoc* basis. The results are therefore biased to reflect the properties of aqueous solutions, and it is not surprising that the calculated anomeric ratio matches the experimental value. Since the $\alpha/\beta$ ratio in non-aqueous solvents has a different value (see Figure 12 and Table 8), it follows that the calculated ratio needs to be 'adjusted' to fit. Historically, such adjustments have been performed via the anomeric effect.

The subsequent work by Rao, Vijayalakshmi and Sundarajon was based on a more developed force field, **[101]** including non-bonded, electrostatic and bond angle strain interactions. The partial charges were obtained by MO—LCAO methods. **[44]** The 'effective' dielectric permittivity was set equal to 3.5 to simulate the effect of water as providing a partial screening of the charge interactions. A distortion of the molecule was introduced by tilting the axial —$CH_2OH$ or —OH groups. The —$CH_2OH$ group was tilted by varying symmetrically the angles C(6)—C(5)—O(5) and C(6)—C(5)—C(4), termed $\phi$. These distortions produce changes in the three $\psi$ angles C(6)—C(5)—H(5), C(4)—C(5)—H(5), and O(5)—(C(5)—H(5) which could be adjusted for minimum bond angle strain.

To obtain free energy differences from the calculated potential energies, the contribution of the conformational entropy S was calculated from

$$S = 2.3R\log\Omega \tag{27}$$

where $\Omega$ is the number of possible conformations differing in the orientation of the O—H bonds for the same ring conformation. The conformational entropy computed in this way gives, for $\alpha$-glucose (Cl) 42.3, and for $\alpha$-glucose (1C) 32.2 J $mol^{-1}$ $K^{-1}$; the corresponding values for $\beta$-glucose are 45.7 and 28.8 J $mol^{-1}$ $K^{-1}$. The contribution of the entropy to the free energy ($-TS$) is added to the potential energies. This contribution does not change the qualitative estimates obtained from the potential energy calculations.

The percentage proportion ($p$) of $\alpha$- and $\beta$-anomers in the Cl and 1C conformations present in equilibrium mixtures can be obtained by

$$p = [100\exp(-G_i/RT)]/\sum_i (G_i/RT) \tag{28}$$

where the summation extends to the four states ($\alpha$- and $\beta$-anomers in Cl and 1C conformations) and $G_i$ is the free energy of state $i$. This formula was used to compute the anomeric ratios. **[101]**

The allowance for some flexibility of the ring is an advance. However, the results still suffer from an empirical correction, since some 1.6 kJ $mol^{-1}$ was added for an equatorial anomeric —OH group, i.e., the anomeric effect was used to obtain agreement with the experimental $\alpha/\beta$ ratio.

A more refined treatment was reported by Kildeby, Melberg and Rasmussen, also based on an empirical force field. **[102]** The potential function contained bond, bond angle and and torsional deformation terms and non-bonded interactions. A more complete minimization scheme was used, including all the 14 ring conformers of the pyranose ring system, see Figure 2. The force field was tested against the crystal structure of the Cl ring conformer. Since the computation dealt with an isolated molecules *in vacuo*, crystal lattice forces were not considered. Nevertheless, good agreement with the experimental (solution) $\alpha/\beta$ value was obtained.

Bond lengths and valence angles were found to be independent of the ring conformation. The energies obtained indicate that C- and S-conformers correspond to local minimum conformations, and each B-conformer converts to the closest S-conformer in the topological map. S-conformers have much higher energies than C-conformers, and Cl has the lowest energy. The crystal data are consistent with this result. The main discrepancy between the calculated Cl $\alpha$-glucose structure and the crystallographic data is the length of the anomeric C(1)—O(1) bond. The calculations do not reproduce the abnormally short bond found in the crystal.

We have already discussed the need for keeping the potential functions and the number of parameters for different atoms at their minimum, compatible with reasonable results. The case of the anomeric atoms is interesting. Not only is the anomeric C(1)—O(1) bond sorter than the ring C—O bond, but also, it is possible to assign particular interaction properties to the anomeric atoms. [**103, 104**] Rasmussen has tested different force fields with the aim of observing the influence of the repulsive dispersion terms and the assignment of partial charges. The effect of the presence/absence of torsion potentials on the results was also checked.

The charges used were those obtained from a Mullikan population analysis of *ab initio* results with a reasonably large basis set. [**39**] With one force field, a different charge was assigned to the anomeric C(1) atom from that assigned to the other C atoms. As regards repulsion–dispersion forces, in one case the three-parameter Buckingham 6-exp function, Equation (15), was used, while in others the simpler two-parameter 12–6 Lennard–Jones Equation (7) was preferred. The $\alpha/\beta$ ratios were found to be 0.25:0.75, 0.32:0.68 and 0.35:0.65 for the different force fields used. All these values are in reasonable agreement with experiment (0.36:0.64). The important conclusion was that no differences were observed when more elaborate repulsion–dispersion terms were used, or by treating the anomeric carbon atom differently from the rest. The fact that a reasonable anomeric ratio is obtained for aqueous solutions, albeit based on isolated molecule calculations, is due to the assignment of aqueous solution values to the parameters.

*Mobility* A cursory reading of the literature may suggest that PHC molecules are quite rigid. However, the theoretical results indicate this not to be the case. In the study by Kildeby *et al.* [**102**] no selection of conformers was made *a priori*. Several local minima were found to be associated with any one C- or S-ring conformer. The largest energy difference between local energy minima encountered was 2.5 kJ $mol^{-1}$. Thus, at equilibrium, the local minima will be equally populated. The saddle points between the minima would be of the order of the height of the torsional potential (6–9.5 kJ $mol^{-1}$). At room temperature, the equilibrium would be rapidly established, so that each basic ring conformer almost constitutes a continuum of conformations. An experimental observation would be able to show these different conformers only if the time scale of the experiment is longer than the interconversion rate; otherwise, only an average will be observed.

MD simulations should be able to probe mobility and rigidity of molecules. Brady has simulated $\alpha$- and $\beta$-glucose *in vacuo* and in aqueous solution (SPC computer water). [**35**] *In vacuo*, the pyranoid ring was found to be quite flexible. Numerous spontaneous transitions were observed in 1C $\alpha$-glucose to more favourable geometries, and a spontaneous excursion of $\beta$-

Table 9. *Equilibrium distribution of chair conformers of methyl α- and methyl β-glucopyranosides in vacuo and in several solvents at* 25 °C *(from ref.* 105). *Figures indicate percentages of each conformer*

| Solvent | *A*1 | *A*2 | *E*3 | *E*2 | *E*3 |
|---|---|---|---|---|---|
| '*Vacuo*' | 70.54 | 14.05 | 11.63 | 2.09 | 1.69 |
| *p*-Dioxane | 66.36 | 13.43 | 15.04 | 2.85 | 2.31 |
| Carbon tetrachloride | 67.48 | 14.05 | 13.66 | 2.67 | 2.14 |
| Chloroform | 61.91 | 12.24 | 19.14 | 3.66 | 3.04 |
| Pyridine | 59.16 | 11.74 | 21.41 | 4.20 | 3.50 |
| Acetone | 59.39 | 12.57 | 20.31 | 4.26 | 3.46 |
| Ethanol | 57.08 | 12.13 | 22.21 | 4.74 | 3.84 |
| Methanol | 51.86 | 11.11 | 26.44 | 5.83 | 4.76 |
| Acetonitrile | 54.72 | 11.72 | 24.03 | 5.25 | 4.28 |
| Dimethyl sulphoxide | 57.36 | 12.49 | 21.61 | 4.74 | 3.80 |
| Water | 32.21 | 7.30 | 41.03 | 10.63 | 8.79 |

glucose from C1 to twist-boat and to 1C could also be detected; the transition frequencies are > 4 GHz. The presence of solvent does not restrict the movements; on the contrary, structural fluctuations were found to increase upon solvation.

One can attempt to infer dynamic behaviour from equilibrium data. Five chair conformers of methyl α- and β-glucopyranosides have been studied in different solvents; [**105**] the energies of the isolated molecules were computed by the semi-empirical PCILO method and a solvent contribution was then incorporated, as explained in section 5. The five conformers selected were two for the α- and three for the β-anomer, differing in orientation of the hydroxymethyl group. From the energy differences between conformers, the distributions were computed in the different solvents, as shown in Table 9.

The equilibrium distribution of conformers is seen to be sensitive to the nature of the solvent. It must be recalled that the treatment does not consider any specific solute–solvent interaction, such as might further accentuate the difference. There is no information on the interconversion rate, but the existence of comparable percentages of different species suggests that they are exchanging at high rates, unless the energy barriers are very high. If this situation prevails, however, it would be possible to observe different conformers experimentally; otherwise, only a weighted time average will emerge from the experiments.

One should not rely on theoretical results alone to establish whether a PHC molecule is rigid or flexible. However, the theoretical information can be used to help with the interpretation of experimental results. Present evidence in favour of rigid molecules is not strongly based. We shall return to this subject in the discussion on di- and oligosaccharides.

### 8.3. *Mutarotation kinetics*

As already mentioned, mutarotation is classified according to the observed kinetics. So-called 'simple' mutarotation displays first order kinetics and is characterized by a two-component dynamic equilibrium. This statement is an oversimplification because at least one other, an acyclic species must be involved, even if in very low concentrations. In 'complex' mutarotation, at least three species (pyranoses, furanoses) are involved in producing the final equilibrium mixture. This gives rise to complex kinetics, usually one or several fast reactions, followed by a slow process. Of the pyranoses, glucose, mannose, gulose, xylose and lyxose exhibit simple kinetics, while galactose, talose, altrose, ribose, arabinose, fructose and sorbose undergo complex mutarotation.

Early mutarotation experiments took advantage of available crystalline anomers which were dissolved in solvent and the rate of equilibration measured. Optical rotation was the standard analytical technique and the results have been summarized in a classic review. **[106]** Measurements of optical rotation could, however, lead to ambiguous results, because the specific rotations of the various species themselves depend on the solvent composition and the temperature, perhaps even on pH. A change in the observed optical rotation could therefore not be definitely related directly to a change in the percentage composition of a given anomer/tautomer mixture. Since the advent of nmr, mixtures can be analysed with a high degree of reliability and the interconversion kinetics and mechanisms are now being studied in ever increasing detail.

Mutarotation involves the transfer of a proton from an acid catalyst to the sugar, an opening of the ring, and a molecular rearrangement, followed by ring closure, with the transfer of a proton from the sugar to a base catalyst. Water, being amphoteric, is a most effective catalyst. On the other hand, sugars are also amphoteric and can catalyse mutarotation. Thus, when crystalline sugars are heated close to their melting points, rapid mutarotation takes place. **[107, 108]** Indeed, mutarotation rates in pure sugars resemble those for aqueous solutions, indicating that the catalytic efficiency of sugars approaches that of water. In theory, aprotic solvents should not be able to support mutarotation. However, low, but measurable rates have been reported in DMSO and pyridine, with the eventual establishment of an equilibrium mixture of constant composition. **[109, 110]** Since the solvent cannot donate a proton, presumably the sugar itself acts as cataylst.

Acree suggested that the acyclic intermediate is stabilized by interacting with the solvent, in particular with the tridymite-like intermolecular structure of liquid water. **[111]** This suggestion has been subjected to further examination because both the thermodynamics and the kinetics of mutarotation (and other tautomeric changes) are sensitive to the stereochemical detail of the sugar undergoing the transformation. Anomeric —OH

(a)

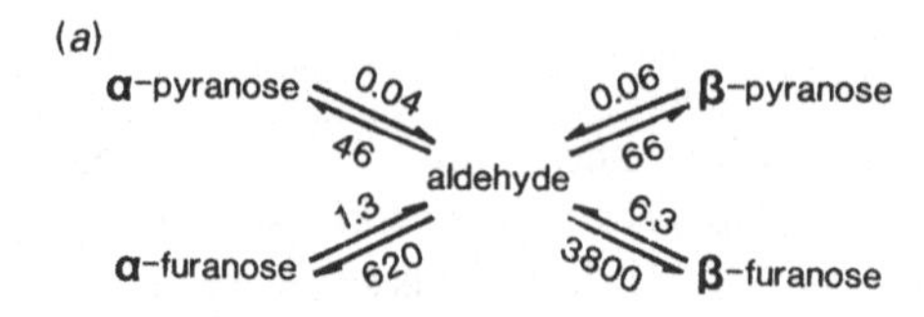


(b)

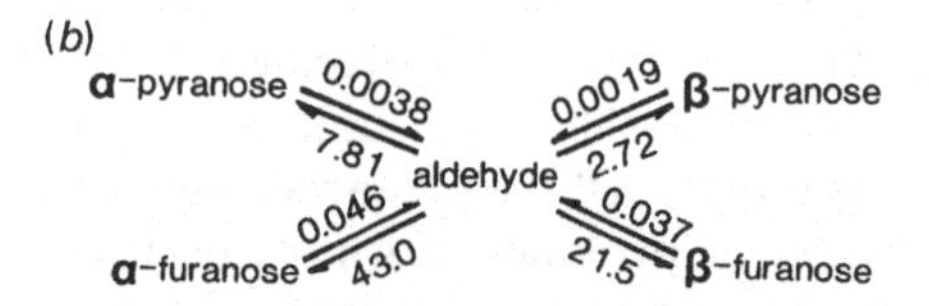


Figure 13. Kinetics of ring opening and closing during the anomerisation and tautomerisation of (*a*) talose and (*b*) idose, according to ref. 112. See also Table 10.

proton nmr relaxation rate measurements have indicated that sugar–water proton transfer involving equatorial anomeric —OH groups is significantly faster than of that observed for axial groups. **[71]** lending support to the specific hydration model.

Emphasis has now shifted to mechanistic investigations. In particular, overall (e.g., α-pyranose → β-furanose) and unidirectional rate constants are being measured and related to sugar structure and solution conditions. **[112]** Eventually such information should shed light on the molecular factors affecting ring formation and ring opening reactions. The available information suggests that these might be quite different. Thus, for ring closure of aldopyranoses, aldofuranoses and ketofuranoses, rate constants differ by factors of 2–7, whereas for the corresponding ring opening reactions they differ by orders of magnitude. This is illustrated in Figure 13 for talose and idose. **[112]**

Serianni and coworkers have developed particularly sensitive analytical techniques, based on sugars enriched with $^{13}C$ at the anomeric carbon, so that even minor components in an equilibrium mixture ($< 0.03$ mol %) can be detected by $^{13}C$ nmr **[113]** They have been able to confirm that an acyclic keto derivative is an obligatory intermediate in mutarotation, and in other tautomerization processes. The intermediate species also interacts with water to yield a hydrate. The reaction scheme is shown in Figure 14. They have also concluded that no correlation exists between thermodynamic and kinetic stabilities. Also, for several sugars studied, $k_{\alpha\to o} \gg k_{\beta\to o}$, but $k_{\alpha\to\beta} < k_{\beta\to\alpha}$, where o denotes the open (acyclic) form of the sugar. Ring opening/closing rates are therefore not correlated with the overall rate constants.

Early studies of the role of the solvent in mutarotation proved to be inconclusive because of several competing solvent effects which could not be resolved by optical rotation methods. They have been described by Shallenberger as: **[15]**

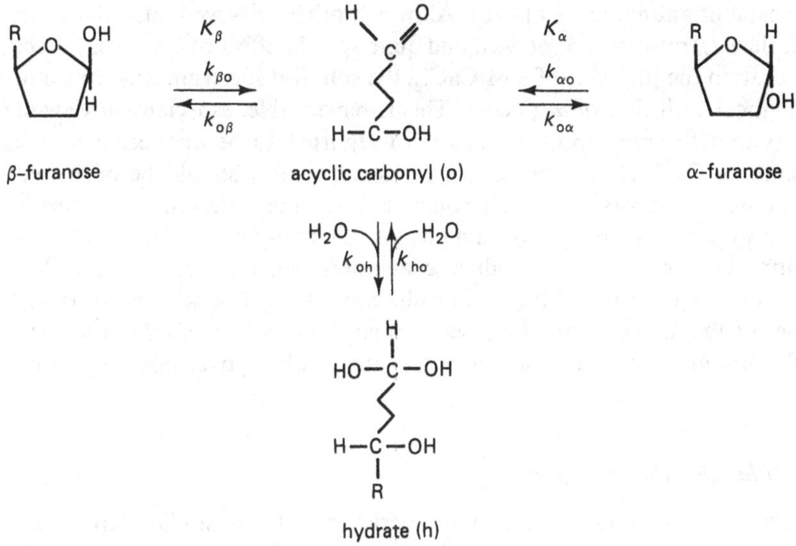

Figure 14. Proposed mechanism of ring opening and closing reactions of furanose sugars. Reproduced, with permission, from ref. 122.

a change in specific rotation of the sugar by virtue of a solvent–sugar interaction, see Table 2;
a change in the tautomer distribution at equilibrium, involving pyranose/pyranose and/or pyranose/furanose shifts;
a change in the conformational equilibrium of each or several given tautomers with subsequent alteration of the rotatory properties.

Mutarotation in mixed solvents is particularly complex, because $k$ does not always exhibit a simple dependence on the solvent composition, i.e., rates do not necessarily follow the 'polarity', as expressed by the dielectric permittivity. Whatever the origins of the solvent effects might be, they can give rise to complicated changes in the mutarotation kinetics. In mixed aqueous/organic solvents, the observed kinetic changes are diagnostic more of the physical properties of the solvent mixture than of the particular sugar employed. **[114]** Now that nmr has become the established technique for studies of PHC transformations, it is to be hoped that a better insight into the details of solvent–PHC interactions will emerge.

The effects of salts on glucose mutarotation have been studied by means of Raman spectroscopy. **[115]** One of the 'rules' of carbohydrate chemistry states that cations interact with only those PHCs that possess the sequence of axial-equatorial-axial —OH groups on consecutive carbon atoms, or *gauche-gauche* in the case of polyols. **[116]** Glucose does not possess that particular —OH arrangement. Contrary to earlier suggestions, **[117]** the authors demonstrated that the 930–830 $cm^{-1}$ region of the spectrum is diagnostic of anomeric $\alpha/\beta$ ratios. Although other salts were also shown to affect the $\alpha/\beta$ ratio, a pronounced and quite specific effect of $Ca^{2+}$ was noted, such that, in the presence of 5 M $CaCl_2$, the solution spectrum was shown to be identical with that of $\alpha$-glucose. This is remarkable, especially in view of reports that the nmr spectrum has been reported to be unaffected by the presence of $Ca^{2+}$. **[118]** Since a general solvent effect should be of a non-specific nature, it was tentatively concluded that the observed $Ca^{2+}$ effect is probably due to preferential ion complexation with $\alpha$-glucose. The conclusion confirms earlier reports that glucose anomers can be separated on $Ca^{2+}$ cation exchange resins. **[119]** The results are also in line with more recent studies of the direct interactions between sugars and ions, see Section 13.1, **[120]** although they run counter to the 'rule' governing sugar–ion interactions.

### 8.4. *Phosphorylated sugars*

In recent years anomeric and tautomeric conversion studies have been extended to phosphorylated sugars. Such compounds should be of particular interest because of their central role in biochemistry. It is curious that, compared to the enormous effort which has been devoted to studies of protein structure and chemistry, little is known about the stereochemical changes of phosphorylated substrates or cofactors. One of the reasons may well be that pure crystalline forms are not readily available.

Compared to their parent sugars, phosphorylated derivatives undergo rapid interconversions. For example, the glucose-6-P mutarotation rate exceeds that of glucose by a factor of 240. The contribution of intramolecular phosphate catalysis to this rate enhancement has been clearly demonstrated. **[12]** The interconversion mechanism resembles that shown in Figure 14. By means of saturation transfer nmr, Pierce, Serianni and Barker have measured ring opening and closing rates and equilibrium compositions of several pentose phosphates. The equilibrium results are included in Table 8. The kinetic data are summarized in Table 10. They refer to measurements made at 6 °C. At higher temperatures the concentration of the carbonyl intermediate shows an increase. **[122]** Note the rather high concentration of the keto intermediate in the fructose phosphate mixtures which is accompanied by complete absence of the acyclic hydrate species. Phosphate catalysis appears to be important in the ring opening reactions, where rates exceed those for parent sugars by factors 6. Intramolecular steric effects have been observed in the ring closure kinetics.

Pierce *et al.* also discuss the biological implications of phosphorylation, especially as applied to glycolysis. The presence of an enzyme with anomerase activity ensures a high metabolic flux, especially where the biologically active sugar tautomers/anomers are present in low concentrations. The phosphorylation of a pentose —OH(1) group serves two functions: (1) the resulting

Table 10. *Sugar ring opening and closing rate constants* ($s^{-1}$); *data from refs* 112 *and* 122

| | | Opening | | Closing | |
|---|---|---|---|---|---|
| Sugar | pH | $k_\alpha$ | $k_\beta$ | $k_\alpha$ | $k_\beta$ |
| Ribose-5-P | 4.2 | 0.86 | 0.44 | | |
| | 7.5 | 100 | 40 | | |
| Arabinose-5-P | 4.2 | 0.64 | 0.49 | | |
| | 7.5 | 50 | 33 | | |
| Xylose-5-P | 4.2 | 0.42 | 0.17 | | |
| | 7.5 | 33 | 9.6 | | |
| Fructose-6-P | 4.2 | 0.20 | 0.22 | | |
| | 7.5 | 18 | 21 | | |
| Fructose-1,6-$P_2$ | 4.2 | 1.2 | 1.9 | | |
| | 7.5 | 28 | 140 | | |
| Erythrose | 4.0 | | | 9 | 12 |
| Threose | 4.0 | | | 5 | 12 |
| 5-Deoxyribose | 4.0 | 0.44 | 0.41 | 80 | 129 |
| 5-Deoxyxylose | 4.0 | 0.53 | 020 | 45 | 15 |
| 5-Deoxyarabinose | 4.0 | 0.13 | 0.23 | 27 | 28 |
| Talopyranose | 4.0 | 0.0038 | 0.0019 | 7.81 | 2.72 |
| Talofuranose | 4.0 | 0.046 | 0.037 | 43.0 | 21.3 |
| Idopyranose | 4.0 | 0.04 | 0.06 | 46 | 66 |
| Idofuranose | 4.0 | 1.3 | 6.3 | 620 | 3800 |

cyclic form is less stable than that of the parent sugar, and (2) the phosphate group catalyses the ring interconversions.

## 8.5. *Isomerization equilibria*

The relative stabilities of isomers in solution can be estimated from enzyme catalysed PHC interconversion studies performed under suitable standard conditions. Goldberg and his colleagues have investigated several aldo–aldo and aldo–keto isomerizations. The results are tabulated in ref. 19. They employed isomerase enzymes, such as xylose isomerase (EC 5·3·1·5), to effect the conversions and determined the equilibrium compositions by HPLC and the enthalpy changes by microcalorimetry. Some representative thermodynamic quantities are summarized in Table 11; the data refer to a standard state based on a hypothetical unit molality ideal solution.

A detailed analysis of the experimental results is difficult because $\Delta H$ values for furanose–pyranose conversions are large, typically of the order of 12 kJ mol$^{-1}$, whereas $\Delta H$ accompanying the sugar anomerization in aqueous solution is small, of the order of 1 kJ mol$^{-1}$. Nevertheless, Tewari and Goldberg comment on the similarities in $T\Delta S°$ between the ribose–ribulose and allose–psicose isomerizations. **[123]** However, they also emphasize the large difference in $T\Delta S°$ between the corresponding pair of processes

Table 11. *Thermodynamics of sugar conversion at 25 °C, according to ref. 19. Units as for other tables*

| | $\Delta G^\circ$ | $\Delta H^\circ$ | $T\Delta S^\circ$ | $\Delta C_p$ |
|---|---|---|---|---|
| Xylose–xylulose | 4389 | 16,090 | −11,710 | 40 |
| Glucose–fructose | 349 | 2780 | 2431 | 76 |
| Ribose–ribulose | 2850 | 11,000 | 8150 | 75±59 |
| Ribose–arabinose | −3440 | −9800 | −6360 | *ca.* 0 |
| Allose–allulose (psicose) | −1410 | 7420 | 8830 | 67 |
| Glucose–mannose | 3070 | 2780 | 290 | −10 |
| Allose–altrose | −2000 | | | |
| Glucose-6-$P^{2-}$–fructose-6-$P^{2-}$ | 3110 | 11,700 | 8590 | 44 |
| Mannose-6-$P^{2-}$–fructose-6-$P^{2-}$ | 25 | 8,460 | 8435 | 38 |

involving glucose–fructose and xylose–xylulose. Although their studies constitute the most complete set of sugar isomerization data, a more rigorous analysis cannot be attempted until a better theoretical basis becomes available for the calculation of tautomeric and anomeric equilibria.

## 9. Solvation and conformation of sugar alcohols

The ambiguities of physico-chemical measurements on unmodified sugars are seen to result largely from the compositional heterogeneity of their solutions. The chemical modifications of sugars which can prevent anomerization and/or tautomerization may also obscure or distort the very properties which are of interest. It might be thought that a useful way of studying the effects of stereochemical detail on solvation, conformation and weak solute–solute interactions is with the aid of acyclic PHCs, such as the sugar alcohols, the crystal structures of which are known. Experimental results could then be related unambiguously to unique stereochemical species.

Since crystal structure data (bond lengths and angles, torsional angles, van der Waals dimensions) usually serve as starting points for experimental and theoretical investigations of molecules in solution, this is an opportune place to discuss the permeability of such extrapolations of solid state data.

### 9.1. *Extrapolation of PHC crystal structure data to molecules in solution*

The problem was addressed by Jeffrey, [**18**] himself a crystallographer, and as far as his analysis relates to sugar alcohols, we shall follow his arguments, in so far as we agree with his exposition. The situation is not as simple as it may appear, because PHCs in the crystalline state are not always stereochemically 'pure'. For example, a recent diffraction and solid state nmr analysis of lactulose has led to the identification of three distinct isomeric forms of the fructose residue: $\alpha$- and $\beta$-furanose and $\beta$-pyranose. [**124**] Similar heterogeneities have also been observed in other PHC crystals.

Jeffrey concludes that, leaving aside other practical complications due to isomerization, hydrolysis or stereospecific solvent interactions, there still remain questions to be answered before the conformation observed in the crystal can be accepted as a valid starting point for calculations or simulations or as a close approximation to one or more possible rotamers which may exist in solution. At this point we part with Jeffrey because of the mounting evidence that stereospecific solvent effects constitute a very central determinant of PHC conformation in solution, especially in aqueous solution.

Other problems highlighted by Jeffrey, but not germane to acyclic molecules, relate to the proper allowance to be made for the conformational flexibility of the furanose ring and to the correct description of intramolecular hydrogen bonds, the latter being difficult to express in terms of limiting van der Waals radii, because of their vectorial character.

A major difference between the crystalline and dilute solution environments is that in the crystal the alditol molecule is extensively hydrogen bonded to several of its neighbours. Each —OH group is linked by two hydrogen bonds (one donor and one acceptor) to —OH groups on other molecules, but the actual hydrogen bond linking schemes differ in different alditol structures. For instance, in xylitol there are finite four-link spirals: O(2)—O(4)—O(2)—O(4) and infinite spirals: —O(5)—O(1)—O(3)—O(5)—O(1)—O(3)—. The torsion angles X—C—C—X of the molecules in the crystal exhibit only minor deviations from the staggered value $n\pi/3$. Using the ethane potential $U_r = \frac{1}{2}U_0\,(1+\cos 3\theta)$, with $U_0 \approx 12.5$ kJ, this amounts to approximately 630 J per C—C bond, substantially less than the energies which are thought to be important in PHC conformational calculations. [**125**] Jeffrey therefore concludes that intermolecular hydrogen bonding contributes little to the shape of the molecule in the crystal.

Crystal structure data have led to the conclusion that the actual conformations of alditols are determined mainly by repulsions between parallel —OH groups on C($n$) and C($n+2$), written as O‖O. Where, for a particular isomer, a linear zig-zag carbon chain would give rise to such a O‖O configuration, rotation about the C($n$)—C($n+1$) bond takes place, producing a sickle-shaped configuration. [**17**] The possibility of parallel C—C and C—O bonds (C‖O) has been dismissed on the grounds that such a configuration would be quite unstable. [**126, 127**]

By 1980, aqueous solutions of most alditols up to the heptitols had been examined by $^{1}$H and $^{13}$C nmr. The accumulated results led to the 'rule' that the molecular configurations in solution and in the crystal are closely similar, if not identical. [**127**] The rule implies: (1) that distinguishable conformations exist in solution and (2) that such conformations are insensitive to solvent effects.

In a detailed reexamination of heptitol conformations, Angyal *et al.* state that the 'rule' that 1,3-parallel C—C and C—O bonds (C‖O) are so unfavourable that they need not be considered at all, is incorrect and that this tacit assumption has led to some wrong assignments of nmr signals. **[128]** They also emphasise that (1) $^{13}$C nmr spectra, taken on their own, can lead to erroneous conclusions, and (2) potential energy calculations based on the summation of group interactions derived from studies of six-membered rings are not necessarily applicable to acyclic molecules. In addition, $^3J_{HH}$ or $^3J_{CH}$ values alone do not necessarily define the alditol shape in solution (or in the crystal), if alternative molecular configurations are also present. Intermediate $^3J_{HH}$ values between 9 Hz (antiperiplanar) and 2 Hz (*gauche*) are usually taken to mean that more than one configuration is present, but they may also arise from a conformation in which the torsional angles differ from the conventional geometry, although this is unlikely for molecules in dilute solution.

### 9.2. *Direct studies of alditol conformations in solution*

The transfer of a molecule, say a pentitol, from the crystal to an aqueous solution takes place with the substitution of ten solute–solute hydrogen bonds by ten (?) solute–water bonds. It is not obvious why this transfer should not affect the stability of the particular molecular configuration as it exists in the crystal. Despite the above-mentioned reservations, solution conformations are often estimated from $^{13}$C nmr spectra alone, because $^1$H spectra of PHCs, in particular alditols, are extremely complex; see Figure 6. At 300 MHz several signals cannot be assigned with any degree of confidence, and even at 400 MHz the $^1$H spectra are very second order and require extensive computer simulation and deuterium labelling. Two recent reports on the $^1$H spectra of sorbitol show up a serious lack of agreement in the $^3J_{HH}$ coupling constants. **[31, 129]**

The 620 MHz spectra of the isomeric pentitols, xylitol, ribitol and arabinitol, in $D_2O$ and pyridine-$d_5$ have now been resolved and all signals unambiguously assigned. **[130]** The $^3J_{HH}$ coupling constants permit comparison to be made between the molecular conformations in the crystal and those in different solvents. The measured *J*-values suggest rapidly interconverting mixtures of *trans* and *gauche* conformers. With the aid of the known crystal structures and the appropriate Karplus equation, the *J*-values for the molecules in the crystalline state have been calculated, as shown in Figure 15 which also includes the measured solution *J*-values plotted against the corresponding torsional angles *in the crystal*. The fractions of each conformer in the *trans* configuration in solution are summarized in Table 12. At high enough temperatures one would expect the molecules to undergo free

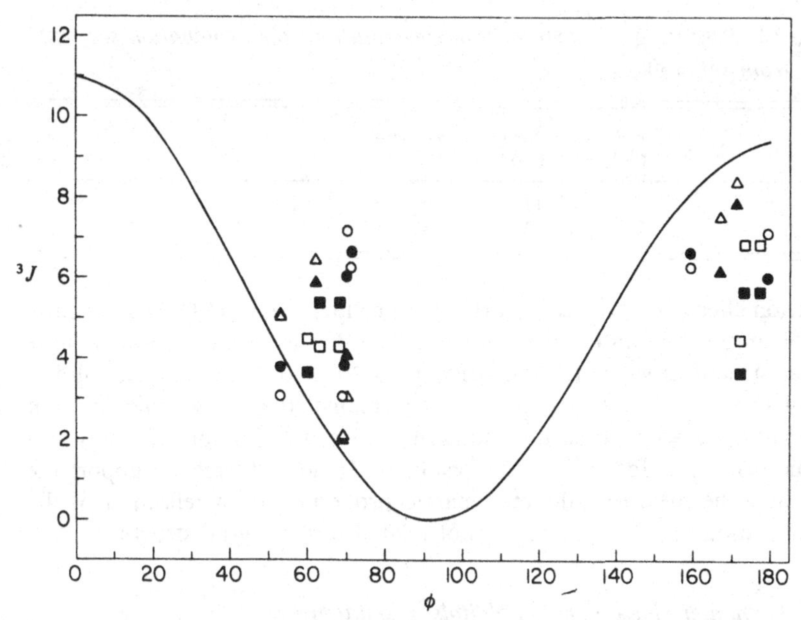


Figure 15. Plot of the Karplus equation $^3J = 0.8 \cos\phi + 10.2 \cos^2\phi$ versus the H—H torsional angles in pentitol crystals. The points denote the experimental $^3J$ values in solutions of: circle, ribitol; square, xylitol; triangle, arabinitol. Open symbols, $D_2O$; filled symbols; pyridine-$d_5$. Reproduced with permission, from ref. 130.

Table 12. *Percentage H—H trans-rotamers in pentitols in solution, at 25 °C, compared to the conformations in the crystal; from ref.* 130

| | Ribitol | | | Xylitol | | | Arabinitol | | |
|---|---|---|---|---|---|---|---|---|---|
| | Cryst. | $D_2O$ (%) | Py-$d_5$ (%) | Cryst. | $D_2O$ (%) | Py-$d_5$ (%) | Cryst. | $D_2O$ (%) | Py-$d_5$ (%) |
| $H_1$—$H_2$ | *g* | 0 | 13 | *g* | 20 | 38 | *g* | 32 | 49 |
| $H_{1'}$—$H_2$ | *g* | 65 | 47 | *t* | 60 | 42 | *t* | 71 | 49 |
| $H_2$—$H_3$ | *t* | 51 | 57 | *g* | 23 | 10 | *g* | −14[a] | −14 |
| $H_3$—$H_4$ | *g* | 51 | 57 | *t* | 23 | 10 | *t* | 85 | 77 |
| $H_4$—$H_5$ | *g* | 0 | 13 | *g* | 20 | 38 | *g* | 0 | 18 |
| $H_4$—$H_{5'}$ | *t* | 65 | 47 | *t* | 60 | 42 | *g* | 54 | 46 |

[a] Negative values result from small $^3J$ values and are indicative of the degree of uncertainty of this type of calculation.

rotation about each bond, and the entries in Table 12 would each be 33%. The average departure from this 'free rotation' value for each isomer is shown in Table 13.

Two effects can be distinguished: In pyridine the alditols have a greater degree of rotational freedom than in $D_2O$, presumably due to decreased

Table 13. *Percentage departure from free rotation; for explanation see text. Data from ref.* 130

| | Ribitol (%) | Xylitol (%) | Arabinitol (%) |
|---|---|---|---|
| $D_2O$ | 28 | 17 | 30 |
| Pyridine-$d_5$ | 19 | 12 | 23 |

solvation effects. Of greater importance is the fact that in $D_2O$, xylitol shows significantly freer rotation than either of the other two isomers. Whereas the crystal structures suggest a similarity in conformation between ribitol and xylitol (both have sickle-shaped carbon chains), thermodynamic solution properties (see below) indicate a similarity between arabinitol and ribitol and unique behaviour for xylitol. The results in Table 13 therefore support the view that the measured thermodynamic properties are a reflection of the conformation in solution and are not related to the crystal structure.

### 9.3. *Computer simulations on alditols in solution*

Although no comparable theoretical or simulation data are available for pentitols, one of us has performed 20 ps MD simulations on the two diastereoisomeric hexitols mannitol and glucitol (sorbitol) which differ only in the position of the —OH group on C(4). [**36**] For the *in vacuo* simulation, no difference could be observed in the end-to-end distances or the radii of gyration (0.54 nm) of the two molecules. However, in aqueous solution ('computer water'), the radii of gyration decreased to 0.417 nm for mannitol and 0.372 nm for sorbitol, irrespective of the starting (crystal) configuration, i.e., planar zig-zig or bent carbon chain. These values were peculiar to the aqueous solvent and differed markedly from those obtained with a Lennard–Jones solvent or an argon-like solvent in which the Lennard–Jones parameters has been chosen to match those of the water oxygen atom. Qualitatively the radii of gyration are consistent with the 'rule' governing O‖O interactions, but quantitatively the results do not correspond to the crystal dimensions.

### 9.4. *Thermodynamic studies of alditol solutions*

Thermodynamic information on alditols is sketchy and confined mainly to $C_p^\circ$ and $h_{ii}$ data. This is unfortunate because polyols are implicated in heat/cold survival mechanisms of plants, insects and microorganisms, [**12**] so that a study of their interactions with one another and with proteins and lipids should prove rewarding.

In contrast to monosaccharides, the $h_{ii}$ coefficients in Table 6 exhibit obvious trends, becoming less endothermic with increasing chain length. Inositol, a cyclitol, shows very different behaviour: the large negative virial coefficients reflect the internal symmetry of the molecule which is also the origin of the low solubility of inositol; for details of the primary references, see refs. 6 and 19. The $h_{ii}$ values of the pentitol isomers reveal similarities between ribitol and arabinitol, with xylitol apparently out of place. The $C_p^\circ$ data show the same pattern which is not related to the molecular configurations in the crystalline state. Ribitol and xylitol which in the planar zig-zag configuration would have O‖O interactions, occur in the sickle form, whereas arabinitol adopts the planar geometry. Although no $g_{ii}$ data are available, it appears that the repulsions between xylitol molecules are of a shorter range than those between the other two pentitols. Similarly, the lower $C_p^\circ$ suggests a more compact hydration shell.

## 10. Di- and trisaccharides

In disaccharides and higher oligomers, conformational complexity is due more to the inherent flexibility of such molecules than to the coexistence of anomeric and tautomeric species. The configurational degrees of freedom are expressed in terms of the torsional angles which measure the rotation of two sugar residues about the glycosidic bond linking them. Although pyranose residues are believed to adopt the same conformations that exist in the monomeric sugars, this is probably not universally true for furanose rings, because of the small energy differences between alternative ring conformers.

For 1–4′-linked sugars the significant angles are $\phi$ (C(1)—O) and $\psi$ (C(4′)—O), as shown in Figure 4. Until the development of very high-resolution nmr methods, these angles were estimated from optical rotation measurements. Despite the interpretative problems, already referred to, Rees and his colleagues used such methods to good effect. They expressed the optical rotatory power of a disaccharide as a sum of contributions from the two sugar residues and from the glycosidic linkage. **[131]** The latter contribution is related to the particular values adopted by $\phi$ and $\psi$. A comparison of the experimentally determined linkage rotations for trehalose, cellobiose and methyl-$\beta$-maltoside in water, DMSO and dioxan with those calculated from crystal structure data and from conformational energy calculations on isolated molecules clearly shows major differences between crystal and solution conformations. The calculated *in vacuo* conformations agree with the experimental values for non-aqueous solvents, but the disaccharide conformations in aqueous solution differ markedly from those calculated from crystal structures and those experimentally determined for organic solvents, as shown in Table 14. Where crystal and solution data diverge, this may be due to inter- and/or intramolecular hydrogen bonds in the crystal which are replaced in solution by sugar–water bonds (but see below for claims to the contrary).

Table 14. *Glycosidic linkage optical rotation [Γ] (deg) at 25 °C and its temperature dependence; after Rees and Thom.* [**131**]

| Sugar | State | [Γ] | d[Γ]/d$T$ |
|---|---|---|---|
| Me-β-maltoside | Crystal | −110 | |
| | Calculated *in vacuo* | −12 and +83 | |
| | Water solution | +46 | −0.12 |
| | Dioxan solution | −18 | +0.22 |
| | DMSO solution | −24 | +0.27 |
| α,α-Trehalose | Water solution | +18 | −0.18 |
| | DMSO solution | −15 | +0.15 |

## 10.1. *Extrapolation of crystal structure data to molecules in solution*

We now return to Jeffrey's analysis, as it applies to di- and trisaccharides. [**18**] From the available crystal structure data it is possible to conclude that pyranoses usually occur in the C1 form, rarely in the 1C form, and that other possible conformations (Figure 2) need not be considered at all. The ring torsion angles seldom lie outside the range of 50–65°, as compared with the strain-free range of 56–62°. [**132**] Sucrose provides an interesting case: the existence of two intramolecular hydrogen bonds between the glucose and fructose rings causes an extreme flattening of the glucopyranose ring. Other disaccharides only contain one intramolecular hydrogen bond.

Furanose rings, because of their greater flexibility, occur in crystals in a variety of twist and envelope forms, shown in Figure 1. Such minor conformational differences are greatly amplified in the —OH group topology, as shown in Figure 16 for a series of trisaccharides, all of which contain the sucrose residue. [**18**] It is reasonable to assume that the —OH geometry is sensitive to the solvent environment, especially where the solvent itself possesses donor and acceptor properties. Conversely, since the energy differences between the different furanose conformers are small, the solvation contribution to the free energy may be significant in determining the most 'stable' ring conformation(s).

Intramolecular hydrogen bonds are generally thought to be unstable in hydroxylic solvents; they are rare even in crystalline PHCs. Exceptions are sucrose, maltose monohydrate, cellobiose and α-lactose monohydrate, but not α,α-trehalose or any trisaccharides (for primary references, see ref. 124).

Jeffrey concluded that the most serious reservations concerning the extrapolation from crystal structures to PHC solution conformations are for molecules which contain furanose rings and for solvents which do not have both donor and acceptor properties. In order to test these conclusions, he constructed conformational maps for sucrose and for the sucrose residue in the four trisaccharides shown in Figure 16. The main differences in the starting coordinates were the shape of the furanose ring, the torsion angles

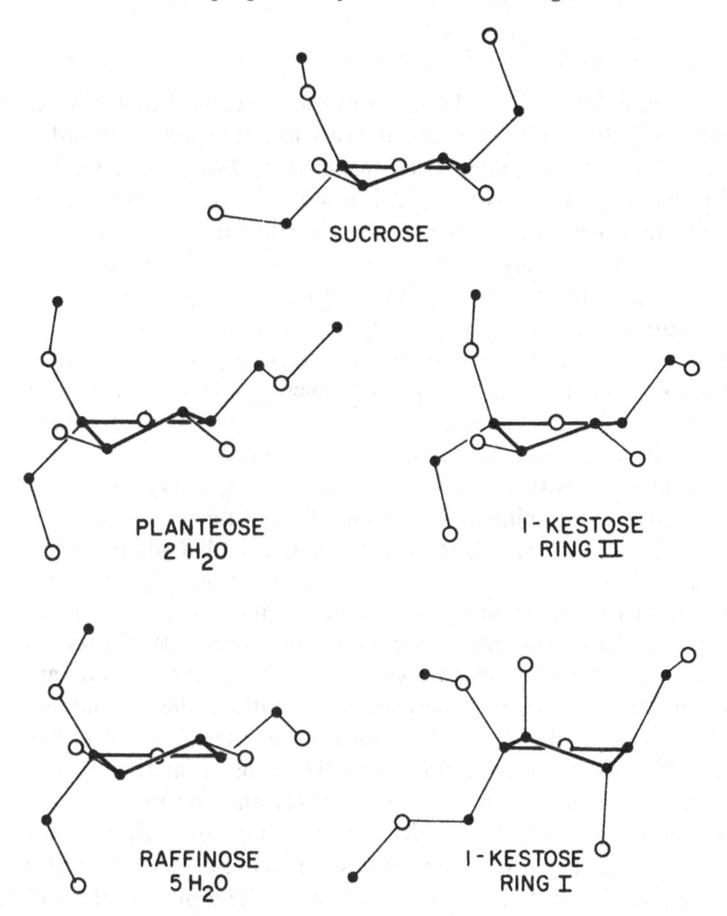


Figure 16. Conformations of the furanose rings in sucrose and the sucrose-containing trisaccharides, raffinose, planteose and 1-kestose. Reproduced, with permission, from ref. 18.

of the glycosidic linkages and, in the case of sucrose, the intramolecular hydrogen bonds. A Buckingham potential function, Equation (15), was used. Sucrose was found to be the abnormal case: it appears that the molecular conformation in the crystal places constraints on the rotations which are not observed in the sucrose residues of the trisaccharides. The constraints clearly arise from the positions of the hydrogen bonds involved which are only present in the disaccharide.

When these two bonds are removed, the $\phi,\psi$ map of sucrose closely resembles that of kestose. On the assumption that intramolecular hydrogen bonds have no permanent existence in aqueous solutions of disaccharides, the conformational maps suggest a complete rotameric freedom about $\psi$, with a restricted ($\pm 60°$) rotation about $\phi$.

## 10.2. *Theoretical calculations on disaccharides*

The theoretical treatments of disaccharide conformations have to deal with relatively large molecules. Some simplifications are usually required, in order to keep the problem within manageable proportions. For $\beta$-maltose, for instance, Melberg and Rasmussen, [**133**] using empirical force fields, performed conformational calculatinos, allowing a relaxation of the internal coordinates and searching only for *skeletal* minima (i.e., those excluding exocyclic C—O bonds). In accordance with other reports on glucopyranoses by the same research group, any local energy minimum is associated with a group of minimum energy conformations, corresponding to the exocyclic torsion angles. In terms of the minimization problem, as depicted in Figure 7, a search limited to skeletal minima is equivalent to a partial 'smoothing' of the energy surface, although still allowing for the existence of substates.

The above-mentioned work has led to some interesting results. After the (skeletal) minimization, five minima were found. One of them (denoted as 5 in ref. 133), is considered as unrealistic. It is argued that such a minimum has an unrealistically high potential energy due to strong bond deformation. After discarding this conformation, four 'realistic' minima remain, from which the one with the lowest energy is in good agreement with the results of X-ray diffraction. The equilibrium distribution of the different minima, based on an estimation of their free energies, gives a ratio of the populations as $M1:M2:M3:M4 = 59:19:15:7$, where $Mi$ denotes the minimum $i$. It is relevant to consider the interconversion rates. The rate of interconversion between the conformer of lowest free energy ($M1$) and the next ($M2$) is $k_{12} \approx 4$ GHz, with $k_{21} \approx 9$ GHz in the reverse direction. These rates correspond to fast exchange on the nmr time scale, being of the same order of magnitude as, or faster than molecular tumbling. The nmr results will therefore indicate a weighted average conformation of all minima, although on a static interpretation of the data one may be tempted to consider only *one* solution conformation. In the present case, $M1$ is the most highly populated minimum, and the average will lie close to it.

It is important to note that in this calculation the locations of the minima are almost independent of the force field used, which is not the case for the heights of the barriers, This dependence is so important that for a given value of $K_\theta$(C—O—C), the 'unrealistic' minimum $M5$ does not appear at all. In any comparison of the calculations with experimental results, the barrier heights are important, because they are related to the rates of interconversion. X-ray results apply to crystals; in solution, many states may be accessible, but only experimental measurements on a time scale faster than the interconversion rates of the conformers can show distinct populations. The calculations do not explicitly include the solvent. Any solvent effect is incorporated in the force field, which is selected to reproduce the experimentally determined conformations.

Table 15. *Mol fraction of β-maltose conformers in different solvents (from ref.* 135).

| Solvent | *M*1 | *M*2 | *M*3 | *M*2 |
|---|---|---|---|---|
| '*Vacuo*' | 59.5 | 19.2 | 15.1 | 6.2 |
| *p*-Dioxane | 53.3 | 20.2 | 17.7 | 8.7 |
| Carbon tetrachloride | 50.7 | 22.8 | 18.3 | 8.2 |
| Benzene | 49.6 | 23.0 | 18.7 | 8.6 |
| *tert*-Butylamine | 41.2 | 25.6 | 22.1 | 11.2 |
| Chloroform | 50.4 | 19.3 | 19.0 | 11.3 |
| Pyridine | 47.1 | 19.4 | 20.3 | 13.2 |
| Acetone | 43.8 | 20.9 | 21.5 | 13.8 |
| Ethanol | 42.0 | 20.7 | 22.1 | 15.2 |
| Methanol | 38.8 | 19.8 | 23.2 | 18.2 |
| Acetonitrile | 39.8 | 20.8 | 22.9 | 16.5 |
| Dimethyl sulphoxide | 40.1 | 21.8 | 22.8 | 15.3 |
| Water | 25.7 | 17.5 | 26.3 | 30.5 |

Minima 1 and 2 (those of the lowest energies) were further analysed by *ab initio* methods. [134] These computations also refer to an isolated molecule. The energy difference between *M*1 and *M*2 is 15.879 kJ $mol^{-1}$, as against 2.371 kJ $mol^{-1}$ obtained with the force field calculation (FF300 parameter set). An important feature that emerges from the calculation is the distinction between different atoms considered as equivalent from the point of view of non-bonded interactions in the force field calculations. The most important difference concerns the anomeric carbon atom. The information derived by *ab initio* calculations is useful for modifications of the force field. However, the introduction of new atom types increases the number of parameters in the calculations, and thus, also the degree of uncertainty.

A further step in the theoretical studies of β-maltose involves the inclusion of the solvent, in terms of a continuum. [135] The evaluation of the *free* energy of conformers has been attempted by dividing the overall free energy into intramolecular $G_u$ and intermolecular $G_{solv}$ interactions, as described in Section 5.

In the evaluation of $G_{solv}$, the term corresponding to the specific interaction is disregarded. This approximation means that a solvent is characterized only by its dielectric permittivity, molecular size and density, and that dispersion and electrostatic interactions are of a symmetrical nature. Any other particular property, such as the possibility (or impossibility) of hydrogen bonding, is neglected.

With the four 'best' conformers described by Melberg and Rasmussen as starting points [133] and equating their potential energies to the free energies of conformers in free space, the contributions to the free energy of 12 different solvents have been added and the mol fraction of each conformer obtained, as shown in Table 15. Several points deserve emphasis. Even knowing that the non-specificity of the solute–solvent interaction may be significant, we note that the relative populations of conformers change with the solvent polarity in such a way that in water the order of populations is almost reversed, compared to the isolated molecule *in vacuo*. Moreover, the energy differences are not large. The rates of interconversion between the four conformers should be very high. To compare the results with experimental findings is hardly worthwhile. The starting point of the calculation was limited to four conformers, obtained from *in vacuo* force field calculations. Many other conformers may exist in solution. The comparison with experimental data should be based on an average conformation. If many conformers are excluded from the theoretical data, the resulting average will be unreliable. However, by using as a source only the four conformers included in the calculation, we find that the average torsional angles of the glycosidic linkage in β-maltose are shifted by about 30–50° from the corresponding solid state values. This result is consistent with the glycosidic angles obtained from the optical rotation measurements described earlier. [**131**]

Once again one cannot but conclude that the conformations of disaccharides are sensitive to the solvent in such a way that the crystal data cannot be taken as representative of solution conformations. The difference between aqueous and non-aqueous solvents, even neglecting 'specific' interactions, is also quite remarkable, but firmly established by experiment.

β-Cellobiose has been studied along similar lines. [**136–8**] Although the populations of the conformers identified again depend on the solvent, the average of the five conformers is close to the conformation in the solid state.

It is clear that no general conclusion about disaccharides can be extracted from a single theoretical study. The information obtained to date emphasizes the need to take solvent effects into account and also the possibility of mobility about the glycosidic linkage. No MD simulations have yet been performed on disaccharides. It is to be expected that a full simulation will give a better description of the conformational state(s). For a sufficiently long simulation time, the probability of sampling the relevant regions of phase space is high. Moreover, the results obtained will give a weighted average of different substrates which is also observed experimentally.

Sucrose has been studied by a combination of nmr spectroscopy and theoretical calculations based on the HSEA method, discussed in Section 4.6. [**139**] We have already criticized the arbitrary introduction of the anomeric (or exoanomeric) effect as part of the potential. The authors interpret their results as showing extreme conformational rigidity. Their *in vacuo* force field calculation suggests a strong conformational preference for the glycosidic linkage that is close to that of sucrose in the crystalline state and retains one of the intramolecular hydrogen bonds which exist in the crystal. [**140**] The calculated potential energy surface shows a deep and narrow well, coinciding with the conformation obtained from X-ray and neutron diffraction. We suggest that little confidence can be attached to the theoretical results. As pointed out earlier by the same research group, the calculation '...involves no effort to judge the influence of either dipole–dipole interactions or of the solvent water.' [**33**] It is by now clear that the solvent exerts a major influence on intramolecular mobility and flexibility; for water, the influence may also extend to a modification of the hydrogen bond lattice.

More serious is the finding that the nmr $^3J$ coupling constants and nuclear Overhauser enhancements (nOe) are also said to be indicative of the persistence of the intramolecular hydrogen bond in solution. Our preliminary (unpublished) nmr studies at 620 MHz of sucrose in water and pyridine are not fully consistent with these conclusions, but the matter requires resolution.

### 10.3. *Thermodynamic studies of disaccharide solutions*

The available thermodynamic information for di- and trisaccharides is included in Table 5 and 6. The $g_{ii}$ coefficients are positive. Of interest is the marked dissimilarity in $h_{ii}$ of the three diglucose isomers. Such subtle differences in sugar–sugar interactions, clearly depending on molecular shape, demonstrate the futility of hard sphere models, orientational averaging procedures and group additivity schemes as devices to obtain a better understanding of such interactions.

## 11. Oligosaccharides: rigidity versus flexibility

The brief review of disaccharide studies demonstrates that there is little agreement on the issue of rigidity of flexibility of such compounds. However, we may be dealing with a subject which is not yet clearly established; in some cases the interpretations are biased by preconceptions and prejudices which will require revision. The same situation applies to the oligosaccharides in general. Leaving aside the mobility or rigidity of the individual sugar residues, already discussed, let us examine the possible mobility about the glycosidic bonds.

Detailed conformational information on complex PHCs can only be obtained from a combination of experiment and calculation. A discussion of high molecular weight, linear polysaccharides is outside the scope of this review; it is a popular subject which continues to be well covered by frequent reviews. Such molecules, derived from plant or microbial sources, can be considered as 'regular' in the sense that the numbers of monomers and/or linkages types are limited. This gives rise to structural regularity within the polymer chain. We here consider oligosaccharides of complex geometry due to a high degree of chain branching, as found in glycoproteins.

The biological relevance of secondary oligosaccharide structures, as they occur in glycoconjugates, is difficult to assess, and no correlation appears to exist between the primary structural type and the secondary structure,

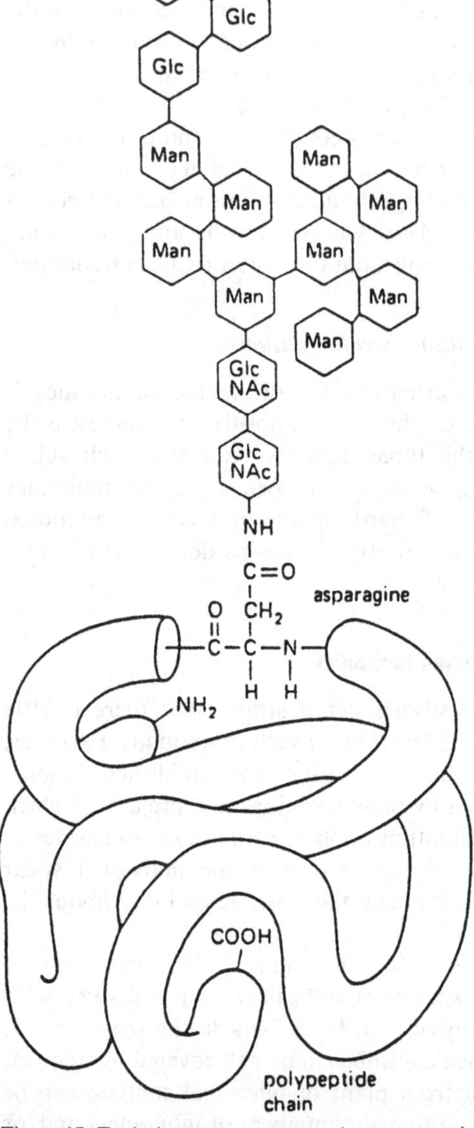


Figure 17. Typical sugar arrangement in an asparagine-linked membrane glycoprotein.

although it is likely that minor primary sequence changes can produce significant changes in shape (*viz.* protein mutants) and/or biological activity.

The N-linked oligosaccharides are implicated in biological processes, ranging from cell–cell recognition and receptor function to targeting, malignant transformations and microorganism–plant interactions. The features of glycoconjugates which are currently under study relate to their stability, folding, antigenicity, charge, solubility and function/activity relationships. In size and complexity the saccharide moieties range from simple disaccharides to the 14-sugar oligosaccharides in mammalian cell membranes, shown in diagrammatic form in Figure 17.

### 11.1. *Fish antifreeze glycoprotein*

An interesting example of the combination of nmr and energy calculations is the investigation of the conformation of Antarctic fish antifreeze glycoprotein (AFGP). [**141**] After assigning all peptide and carbohydrate proton resonances and nOe enhancements and testing the resulting possible torsional angles against calculated conformations, the authors conclude that the polymer which is based on an ala—ala—thr tripeptide repeat, with threonine O-linked galNAc—gal disaccharide units, takes up a polyproline-II type helix, 'coated' with a carbohydrate layer, the —OH groups of which face the aqueous phase. This configuration, so they speculate, is responsible for the ability of AFGPs to inhibit ice crystal growth. While this *in vitro* ability has been clearly demonstrated, its physiological or ecological relevance remains open to question. To render the animal freeze resistant, one would expect the AFGP to be able to prevent ice nucleation at the subzero environmental temperatures, rather than retard the growth of already existing crystals. Even if the solution conformation, as determined by Bush and Feeney, is correct and corresponds to the *in vivo* molecular shape, it is not clear how the collective —OH group topology of the disaccharides interacts with the aqueous medium so as to *prevent* freezing.

### 11.2. *Blood group oligosaccharides*

More complex saccharide structures are also receiving increasing attention by theoreticians and experimentalists alike. Particular interest centres on the *N*-asparagine-linked high mannose clusters of the blood group glycoproteins. In this context it must again be emphasized that most so-called proteins are in fact *glyco*proteins, even though the functions of the oligosaccharide residues are not yet fully understood in many cases. [**1**] A typical example is provided by the blood group A tetrasaccharide shown below which has been studied by this combination of methods. [**142**]

$$\left.\begin{array}{r}\text{Fuc}(\alpha 1 \rightarrow 2)\\ \text{GalNAc}(\alpha 1 \rightarrow 3)\end{array}\right\rangle \text{Gal}(\beta 1 \rightarrow 3)\text{—GalNAc—ol}$$

Model calculations were performed on the two disaccharides Fuc($\alpha$1 $\rightarrow$ 2)Gal—OMe and GalNac($\beta$1 $\rightarrow$ 3)Gal$\beta$—OMe. Figure 18 shows the non-bonded potential energy map for the former model dissacharide. The three

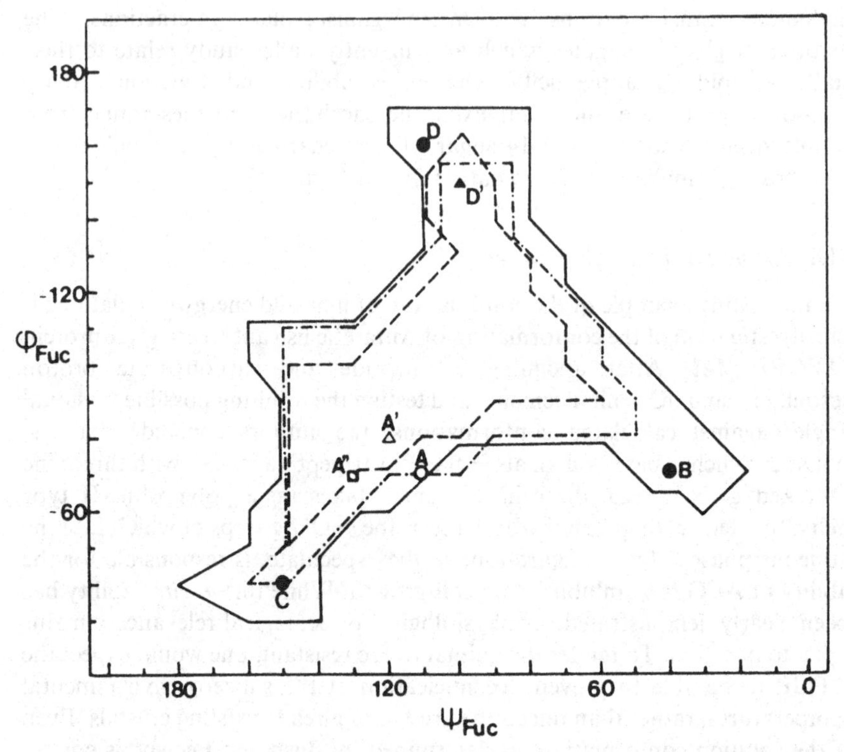


Figure 18. Non-bonded energy for Fuc($\alpha$1 $\rightarrow$ 2)Gal-OMe as a function of the glycosidic torsional angles $\phi$ and $\psi$. Contours at 17 kJ mol$^{-1}$ are drawn for three different model potentials. Global minima are shown as A, with susidiary minima at B, C and D. Reproduced, with permission, from ref. 142.

contours (17 kJ mol$^{-1}$ above the minimum energy) were arrived at by three different sets of potential functions. Points marked A, A′ and A″ represent the *global* energy minima according to the three calculations, with other *local* energy minima designated as B, C and D. Although the three methods used in the calculations differ markedly in the types of interactions included, the global minima coincide reasonably well. On the other hand, the fact that subsidiary minima are identified raises questions whether a unique 'native' conformation exists *in vivo*. When the energy data are combined with those for the other model disaccharide, the conformation of the trisaccharide Fuc($\alpha$1 $\rightarrow$ 2)[GalNAc($\beta$1 $\rightarrow$ 3)]Gal$\alpha$—OMe can be deduced. Here the different model potentials do yield different low-energy conformations; the authors conclude that '...it is difficult to judge what is the conformation, if indeed there is a unique fixed conformation.'

With the aid of nOe and $T_1$ measurements, the area on the $\phi$,$\psi$ map can be delineated where all the experimental results agree with each other and with those computed from the calculated conformations. The results suggest that the oligosaccharide residue is quite rigid, but the various empirical methods used in the calculations of molecular shapes are inadequate and subject to uncertainties amounting to $> 3kT$. The possibility that observed experimental results could arise from an average of several conformations was ruled out as '... that seems quite unlikely', without further discussion. However, the interconversion rate between conformers can be faster than the nmr time scale. Although details of the height of the barriers are not available, we can estimate their values from the contour plots, as yielding rates ranging from 70 MHz to 1 GHz.

Many questions are seen to remain: which energy contributions are important in determining molecular shape; is there indeed a unique conformation or are the results diagnostic of an averaging over several rotamer conformations; does saccharide function depend on unique 'native' configurations and are they subject to cooperative transitions, analogous to those of proteins?

The complexity of the problems posed by small oligosaccharides is well illustrated by Homans *et al.* **[143]** in their structural analysis of high mannose derivatives such as occur in cell-surface glycoproteins, e.g.,

```
                Manα1↘
                      3
Manα1↘                 Manα1 → 4GlcNAcβ1 → 4GlcNAc
      3               6
       Manα1↗
      6
Manα1↗
```

By a combination of several refined nmr techniques and quantum mechanical calculations the rotamer distributions at the Manα1—6Manβ— or Manα1—6Manα— linkages for six related compounds were analysed. The observed values of $J_{56}$ and $J_{56'}$ for hydroxymethyl groups are sensitive to the rotamer distribution about the C(5)—C(6) bond. Calculations based on molecular mechanics and semi-empirical quantum mechanical (MNDO) methods of energy profiles about this linkage are shown in Figure 19 for β-glucose and β-mannose. Different constraints have been imposed in each case.

The analysis eliminates the energy profiles (*a*), (*b*) and (*c*) in Figure 19. However, it is worth noting that the theoretical prediction can give quite different (and misleading) results, unless great care is exercised. From the analysis of the possible values of both $J$-values for the structures under study, it was concluded that the energy profiles shown in Figure 19(*d*) and (*e*), for β-glucose and β-mannose, respectively, are compatible with the experimental information. Therefore, the rotamers with $\omega \approx 170°$ and $-30°$ are considered as predominant in aqueous solution.

The presence of three rotatable bonds between 1 → 6-linked sugars provides for considerable flexibility. Figure 20 shows different conformational transitions within oligosaccharides, according to the results pre-

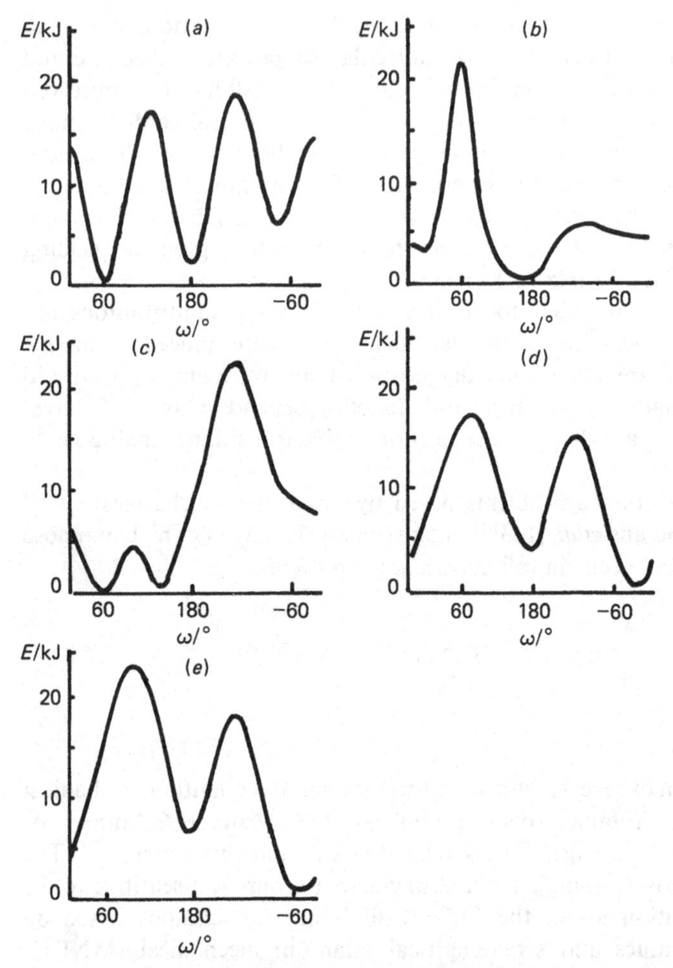


Figure 19. Energy profiles about C(5)—C(6) linkages for $\beta$-glucose, calculated by the following methods: (*a*) molecular mechanics with the torsional angles H′(4)—O(4)—C(4)—C(5) and H′(6)—O(6)—C(6)—C(5) set at 180°; (*b*) by semi-empirical quantum mechanics (MNDO) methods for a fixed, optimised structure with the above torsional angles set at 55° and 60°, respectively; (*c*) as in (*b*), with H′(4)—O(4)—C(4)—C(5) = 180°; (*d*) as in (*b*), with both torsional angles = 180°; (*e*) as in (*d*) for the optimised geometry of $\beta$-mannose. Only profiles (*d*) and (*e*) are compatible with the experimental evidence. [143]

viously discussed. The double headed arrows indicate different possible conformations which, from the experiments, are observed as averaged conformations. It is seen that some possible multiple conformations are not shown. This is due to the lack of experimental evidence of the actual existence of averaged conformations between such states.

From the probable biological significance of multiple conformations in N-

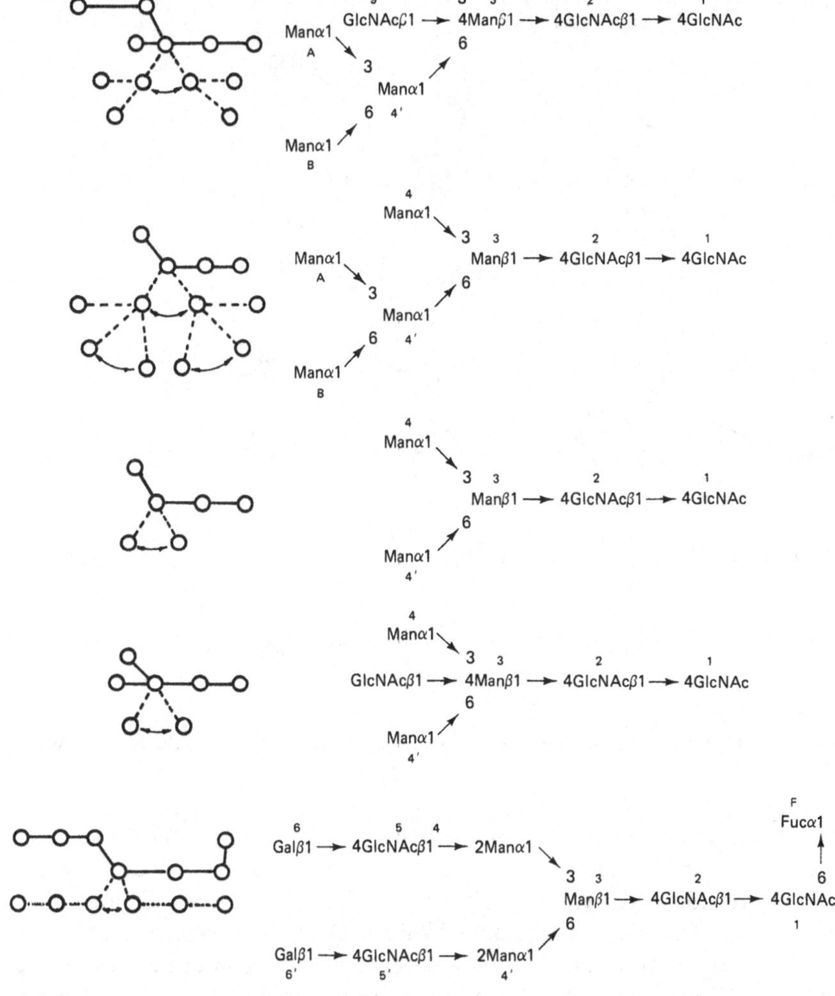


Figure 20. Conformational transitions in oligosaccharides (shown on the right hand side). Averaging between $\alpha 1 \rightarrow 6$ linkages of two conformations is shown by broken lines and double-headed arrows. Redrawn, with modifications, from ref. 143.

linked oligosaccharides, it can be argued that the appropriate conformation is selected when the oligosaccharides are bound to proteins, either covalently or as ligands. It has been suggested **[40]** that the stabilization of individual conformational subsets is due to different glycosylation sites of glycoproteins; this may help to explain the differential site-specific activities of glycosyl transferases at the various glycosylation sites.

In summary, it must be concluded that the arguments against flexibility in

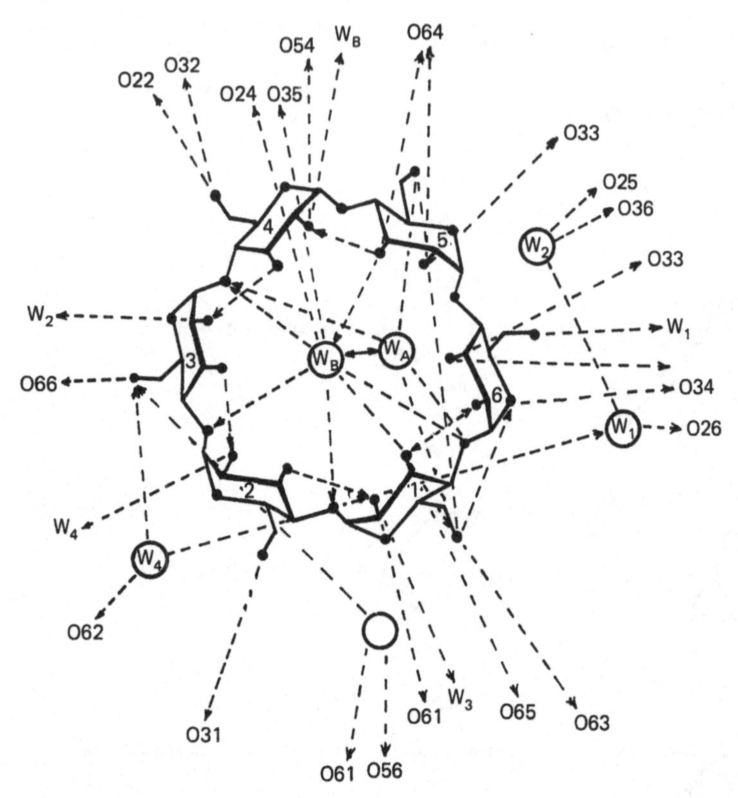


Figure 21. Hydrogen bonding pattern of $\alpha$-cyclodextrin$\cdot 6H_2O$ (form I) at 193 K. Redrawn from ref. 41.

di- and oligosaccharides are not convincing. There is strong evidence in favour of bond mobility in a variety of compounds. In all cases where interconversion rates between different conformers have been estimated, they occur faster than can be measured by nmr. Thus, the predicted mobility is high and the nmr results are indicative of average conformations. Some results [**143**] are in clear favour of a relatively high mobility, while in other cases, where the authors favour the rigidity hypothesis, the possibility of a time-averaged structure is ignored, without justification. The MD simulations of cyclodextrin, see below, predict a rather mobile structure, although the ring structure may reduce the degree of flexibility which can be achieved.

## 11.3. *Cyclodextrins*

The sizes of oligosaccharides make their theoretical study difficult, but new computing facilities are a great help in this respect. Kohler has studied $\alpha$- and $\beta$-cyclodextrin by MD simulations. [**41**] Cyclodextrins are torus-shaped

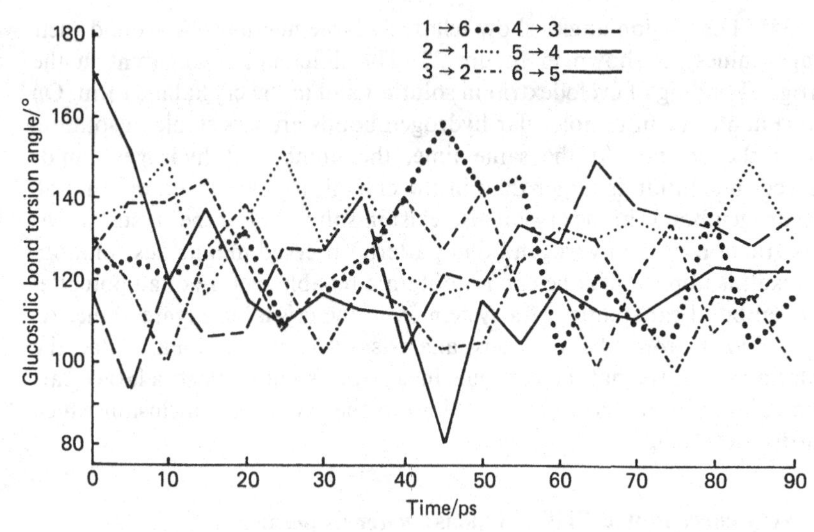


Figure 22. Time evolution over 90 ps of the six glucosidic bond torsion angles C(3)—C(4)—O(4)—C(1) of α-cyclodextrin in solution. Redrawn from ref. 41.

oligosaccharides, consisting of six, seven or eight glucose units (α, β and γ-cyclodextrin, respectively) that are covalently linked to each other by 1—4 glucosidic bonds. Simulations were performed for the molecule in the crystal and in solution. In each case the starting configuration was that obtained from neutron diffraction. For the simulation of the hydrated crystal, the experimental crystalline structure was maintained. The hydrogen bond network observed by MD agrees with that obtained from neutron diffraction, although some other short-lived bonds are apparent. Figure 21 shows the MD structure of α-cyclodextrin with its hydrogen bond network. Two included water molecules can be seen with hydrogen bonds towards the ring, and others pointing to other cyclodextrin molecules. The experimental and simulated average crystal structures of α-cyclodextrin have root mean square differences for all atoms of only 0.023 nm and for the non-hydrogen atoms of 0.015 nm. These small differences are taken as proof of the goodness of the force field used. The difference between the average atomic positions for the crystal and the molecule in aqueous solution obtained in the simulation are about four times larger than the above values, indicating that the differences are indeed significant. The mobility of the α-cyclodextrin atoms in solution is about twice as large as in the crystal. The glucosidic bond torsion angles show an interesting conformational difference between the crystal and solution. Experimentally it is observed that the values lie in the range 115–135°, with the exception of the torsion angle connecting gluose 5 and 6 which has a value of 171° (the simulation yields 158°). In solution, the twist of glucose 5, which is necessary to produce that value, disappears, and the six glucosidic bond torsion angles exhibit comparable behaviour, lying between

115–135°. The torsion angles show relatively large fluctuations around their average values, as shown in Figure 22. The difference is apparent in the hydrogen bonding of cyclodextrin in solution and in the crystalline form. On the one hand, the intramolecular hydrogen bonds are less stable in solution than in the crystal. At the same time, the number of hydrogen bonds observed in solution is larger than in the crystal. It is worth noting that all the hydrogen bonds in the crystal also exist in solution, but the crystal shows only a fraction of the hydrogen bond pattern observed in aqueous solution. This result is important: one cannot obtain a reliable picture of all potential hydrogen bond capabilities of a system from the crystal structure alone. As we concluded from the previous analysis of conformational data, the simulations indicate that the molecule in aqueous solution visits a larger part of the conformational space than it does in the crystal, a conclusion which is hardly surprising.

## 12. Very concentrated PHC solutions: water as plasticizer

Most information on PHC interactions and conformations has been obtained from studies of their dilute solutions and under equilibrium conditions. On the other hand, the practical applications lie in the area of very concentrated solutions, bordering even on the solid state. In such situations equilibrium conditions cannot be expected to apply, bearing in mind particularly the high viscosities of concentrated PHC solutions.

In the accompanying chapter on nucleotide hydration, stress is laid on the correspondence between the crystalline state and the aqueous solution. Indeed, much useful information about nucleotide hydration has come from the study of crystals. Similar arguments apply to protein hydration because the protein molecule in the crystal exists in an aqueous environment. The situation is entirely different for PHCs. Crystals are mainly anhydrous, although some few stoichiometric hydrates do exist. The molecules in the crystals are extensively hydrogen bonded and complex intermolecular hydrogen bonding patterns are the rule, see ref. 124.

Few PHCs crystallize readily from aqueous solution. This presents problems in several technologies, for instance in the refining of sugar. As can be seen from Table 6, the second virial coefficients $g_{tt}$ which are indicative of solute–solute interactions tend to be small and positive, suggesting that there is very little driving force for crytallization. Also, in view of the complex crystal structures, crystal nucleation presents severe problems. Furthermore, sugars in solution do not occur in the 'pure' state but as mixtures of isomers. Even when seeded, crystallization may not go to completion so that the resulting mixture contains crystalline and amorphous states.

When a PHC crystal is melted and the resulting melt is cooled, amorphous states are common, and cannot be treated by equilibrium thermodynamics. Amorphous PHCs have a high affinity for water with which they form solid solutions. In a study of such systems it is difficult to define a useful reference state for comparison purposes, because the physical behaviour is determined by kinetics and cannot be referred to equilibrium. In most cases kinetics are extremely slow so that, what appears to be an equilibrium, is really a stationary state.

A common property of such materials is the glass–rubber transition which is in the nature of a relaxation process and is characterized by the onset of intra- and/or intermolecular mobility. In terms of thermodynamics, the glass transition is defined as a second order transition, in the sense that the second derivatives of free energy undergo discontinuities. Experimentally, therefore, glass transitions can be studied by changes in the specific heat or the coefficient of expansion.

Operationally a glass transition is often defined in terms of the temperature dependence of mechanical relaxation properties, e.g., diffusion, viscosity, elastic modulus. Whereas the properties of the glass and the mobile fluid obey Arrhenius kinetics, the intermediate state, referred to as 'rubber', exhibits a different behaviour, such that the relaxation rates follow an equation of the type

$$\log k_{\mathrm{T}} = -C_1(T-T_{\mathrm{g}})/[C_2+(T-T_{\mathrm{g}})] \quad (29)$$

where $k_T$ is a rate constant which can be expressed in terms of the mechanical or transport properties. For viscosity, $k_T = \eta/\eta_{\mathrm{g}}$, where the subscript refers to the value at $T_{\mathrm{g}}$, the glass transition temperature. $C_1$ and $C_2$ are 'universal' constants, related to functions of the free volume of the system above and below $T_{\mathrm{g}}$. The numerical values of the constants have been obtained from measurements on synthetic amorphous polymers: $C_1 = 17.44$ and $C_2 = 51.6$. **[144]**

Equation (29) shows that in the intermediate temperature range, relaxation rates are much more sensitive to temperature than is predicted by the Arrhenius equation. Indeed, in the 'rubber' range of a material, the relaxation rate may well change by 12 orders of magnitude.

The mechanical properties, and hence the glass transition of a material, are sensitive to the presence of low molecular weight species which are described as diluents or plasticizers. Thus. the plasticizer depresses $T_{\mathrm{g}}$ which can be described in terms of the mobility (relaxation rate), as shown in Figure 23 for sucrose/fructose and glucose/fructose blends in which the fructose acts as diluent. **[145]**

Surprisingly, low molecular PHCs behave like high polymers, possibly because of supramolecular chains formed by hydrogen bonding. Table 16 summarizes the glass transitions and other related physical properties of PHCs. There is as yet a scarcity of data on the effect of water and other diluents on $T_{\mathrm{g}}$; Figure 24 shows the depression of $T_{\mathrm{g}}$ of PHCs, resulting from the addition of moisture. **[146, 147]** Various semi-empirical treatments have been advanced to account for properties of amorphous polymers. The

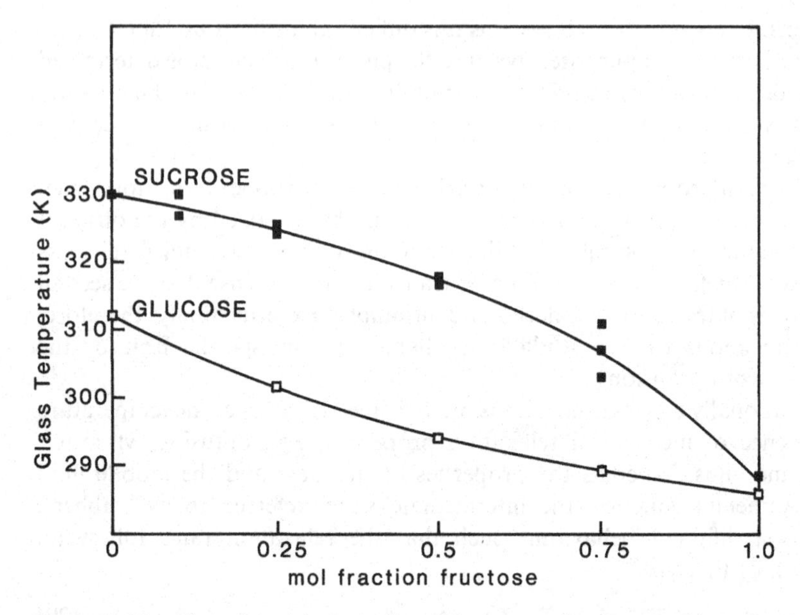


Figure 23. The diluent (plasticisation) effect of fructose on the glass transitions of sucrose and glucose. [**145**]

Table 16. *Glass/rubber transitions of anhydrous and freeze-concentrated PHCs; data from ref.* 11

| | $T_g$ (K) | $T_g'$ (K) | Wt % water |
|---|---|---|---|
| Glycerol | 180 | 208 | 46 |
| Xylitol | 234 | 226.5 | 42.9 |
| Ribose | 263 | 226 | 33 |
| Xylose | 282 | 225 | 31 |
| Glucitol (sorbitol) | 270 | 229.5 | 18.7 |
| Glucose | 304, 312[b] | 230 | 29.1 |
| Mannose | 303 | 232 | 25.9 |
| Galactose | 383[a] | 232.5 | 45 |
| Fructose | 373[a], 286[b] | 231 | 49 |
| Maltose | 316 | 243.5 | 20 |
| Cellobiose | 350 | | |
| Trehalose | 350 | 243.5 | 16.7 |
| Sucrose | 325, 330[b] | 241 | 35.9 |
| Glucose/fructose (equimolar) | 293 | 230.5 | 48 |
| Maltotriose | 349 | 249.5 | 31 |

[a] The higher of two observed glass transitions which, according to ref. 11, determines the onset of translational diffusion. For a discussion of the significance of two observed $T_g$ values, see ref. 145.
[b] Ref. 145.

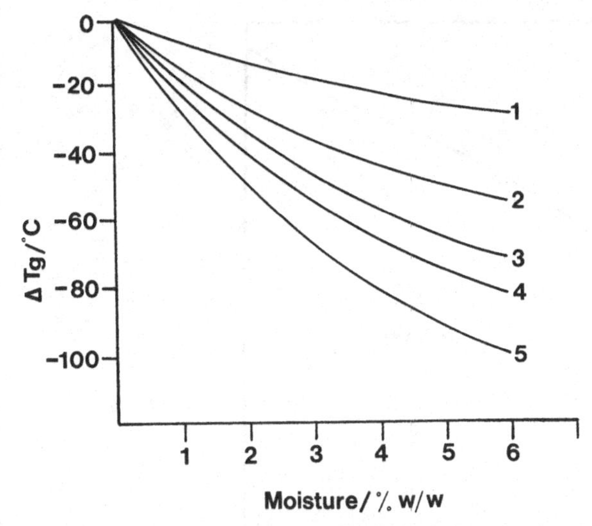


Figure 24. Plasticisation by water of PHCs, as monitored by the depression of their glass temperatures $\Delta T_g$. 1: sorbitol, 2: sucrose, 3: maltose, 4: maltotriose, 5: maltohexose.

equations, developed for high polymers, for the $T_g$-depressant effect of diluent in terms of the glass temperatures of the pure components and the specific heat of glasses and rubbers, have been applied to PHCs and have been found to be reasonably successful in accounting for glucose and fructose diluents in blends with sucrose. [**145**] Although it is not yet generally recognized, such information is of great importance for the prediction of shelf stability of low moisture food and pharmaceutical products, especially freeze-dried preparations. [**148**]

The water contents at which glass transitions in aqueous PHC systems occur at temperatures of practical importance ($> -40$ °C) can only be achieved either by moisture sorption by the amorphous PHC or, more commonly, by the freeze-concentration of PHC solutions. Because of the reluctance of most PHCs to crystallize from aqueous solution, eutectic phase separation is not normally detected when dilute solutions are frozen. Instead, the solution becomes supersaturated and the liquidus curve approaches the glass–rubber transition curve. Ice crystallization slows down as the viscosity of the mixture increases rapidly with decreasing temperature and increasing concentration. The observed behaviour follows Equation (29). The intersection of the liquidus and glass transition curves occurs at a temperature $T_g'$, where the freeze-concentrated, amorphous PHC contains $w_g'$ g water/g PHC. By definition, ice crystallization practically ceases at $T_g'$, where the viscosity of the mixture has reached the value of $\eta_g \approx 10^{14}$ Pa s, which corresponds to a crystallization time of several years per $\mu$m. A typical state diagram is shown in Figure 25 for the sucrose/water system; a selection of $T_g'$ and $w_g'$ data for PHCs is included in Table 16.

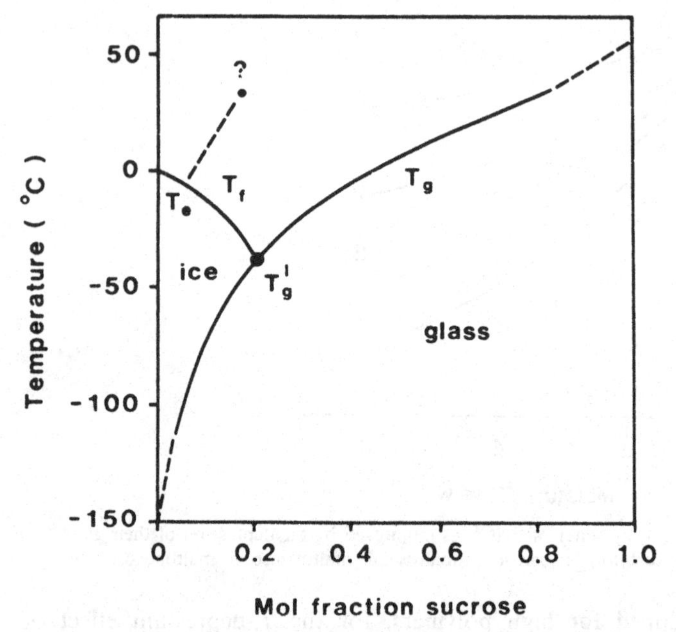


Figure 25. State diagram for sucrose water mixtures, showing liquidus ($T_f$) and glass ($T_g$) curves. $T_e$ denotes the eutectic temperature, but sucrose *cannot* be made to crystallise from the freeze-concentrated mixture. Instead the solution becomes supersaturated and freezing becomes progressively slower until it practically ceases altogether at $T_g'$. The system then consists of ice embedded in a sucrose–water glass, containing approx. 0.5 g water/g sucrose. Reproduced, with permission, from ref. 12.

Until very recently, the measurement of $T_g'$ and $w_g'$ relied on differential thermal analysis of frozen solutions. The difference between the known water content calculated from the area of a DSC melting endotherm as taken as a measure of the 'unfreezable water' which was equated to $w_g'$. The temperature $T_g'$ was taken to be the onset of the melting endotherm. Recent improvements in DSC methodology as applied to PHCs have made it possible to measure $T_g'$ independently and it is found to occur several degrees below the onset of the melting endotherm. Instrumental improvements have also resulted in more reliable estimates of the 'unfrozen' water. We distinguish between 'unfreezable' and 'unfrozen' water in the sense that the former description implies that a certain quantity of water is incapable of freezing, whereas the latter suggests that the water could freeze, if given sufficient time, once again emphasizing that the observed properties of amorphous states are governed by slow kinetics rather than by equilibrium thermodynamics.

The data in Table 16 show that, whereas $T_g'$ is a function of the PHC molecular weight, the relationship, if any, existing between $w_g'$ and the stereochemistry of the PHC is obscure. Note, for instance, the abnormally low $w_g'$ of freeze-concentrated trehalose: $T_g'$ does now show a corresponding anomaly, compared to other disaccharides.

In complex aqueous mixtures, containing PHCs, proteins, salts, pH buffers, etc., the crystallization behaviour of all the components is affected by their concentrations in the freeze concentrate. Thus, although salt/water mixtures commonly display eutectic phase separation, in the presence of PHCs, salt precipitation may be inhibited partially or completely. The salt will then be included in the supersaturated solution. $T_g'$ and $w_g'$ values are then determined by the concentration ratios of the various components in the freeze concentrate. A knowledge of the crystallization/vitrification characteristics of such freeze-concentrated aqueous systems is a prime requirement for the specification of effective freeze-drying processes, **[148]** a fact which is not at all widely recognized. Instead, freeze-drying is treated as a 'black art' where each process and formulation are arrived at *de novo*.

The technological and biological implications of amorphous aqueous systems, including PHCs, are enormous. It is gradually beginning to be accepted that attributes such as the microbiological safety, shelf life and texture of many foods are determined by glass–rubber transitions, **[149]** where formerly concepts such as water activity ($a_w$) and 'bound water' had been employed to rationalize the properties of low and intermediate moisture products. Even more remarkable, the ability of some living organisms, particularly insects, to survive for long periods at subzero temperatures, is now being linked to the possible vitrification of their body fluids which, in a state of cold acclimation, often contain high concentrations of low molecular weight PHCs. **[150]**

An issue of great interest but also of great complexity concerns the elucidation of the structures of amorphous PHCs and their molecular packing and intermolecular hydrogen bonding geometries which together determine their bulk mechanical properties and moisture sensitivities. Attempts to compare Raman spectra of stereoisomeric alditols in crystalline, amorphous and concentrated solution states were abandoned because of the complexity of the spectra and the lack of normal mode analytical data which might have assisted in the interpretation of observed spectral differences. **[151]** On the other hand, Mathlouti and coworkers have made an encouraging start to address these problems. **[152]** Their chosen techniques, FTir and MAS-nmr, provide data which require careful analysis. Their comparison of sucrose spectra in the crystal, in the glass and in dilute and concentrated solution and in the freeze-dried state is likely to be a useful way forward.

It is to be hoped that more attention will in future be paid to aqueous PHC glasses and rubbers and that theoretical descriptions of such systems can be developed to relate $T_g'$/composition profiles to molecular structure, intermolecular hydrogen bonding and packing. Recent studies of the specific heats of sugars and their blends in these various physical states may lead the way. **[145]**

## 13. PHCs in ternary and more complex systems

In this final section we briefly touch upon the roles and interactions of PHCs in even more complex situations. The discussions are not meant to be exhaustive, nor is the literature coverage. The aim is to indicate to the reader unresolved issues which merit closer study and to demonstrate that the still widely propagated view that PHC solutions are ideal (and therefore of little interest) is far from the truth. **[153]**

### 13.1. *Interactions of PHCs with metal ions*

Since the solvent medium is now recognized as playing a role in determining the positions of anomeric and tautomeric equilibria of PHCs, it is hardly surprising that solvent perturbations also exert an influence. The commonest method of perturbing aqueous solvents is by means of salts, and salt effects on chemical reactions involving PHCs have been a subject of study for many years. Apart from general solvent effects, some ions also form stoichiometric complexes with PHCs; it is not always easy to distinguish between these two effects. Angyal has reviewed the chemistry of sugar–metal complexes and proposed that for such complexation to occur, the sugar must possess an axial-equatorial-axial (ax-eq-ax) arrangement of —OH groups on three consecutive carbon atoms. **[116]** Many complexes have been identified, some of them, e.g., the $CaCl_2$ complex of mannose, **[156]** have been isolated in the solid state.

In principle, three vicinal oxygen atoms in every alditol could take up the required *gauche-gauche* (*threo-threo*) conformation for metal complex formation. However, it must be remembered that this might involve O‖O interactions which are said to be unfavourable. By means of lanthanide shift nmr and electrophoresis, Angyal, Greeves and Mills have studied the propensity for complex formation by pentitols and hexitols and have concluded that the energy required to form this arrangement, by the rotation about C—C bonds, determines the extent of complexation. **[157]** On the other hand, studies on ethane diol and its derivatives have demonstrated that there are few generalized limits that can be placed on complexation. **[158]**

In an effort to distinguish general solvent effects from possible complex formation, Franks *et al.* studied the Raman spectra of $\alpha$- and $\beta$-glucose solutions in the presence of mono- and divalent metal ions. **[115]** These conditions allow a more definite assignment of spectral features to a particular anomer in the equilibrated mixture than is possible from a comparison with the spectra of the polycrystalline solids. **[113]** All ions were found to produce changes in the anomeric Raman bands, but the $Ca^{2+}$ ion had a particularly marked effect on the anomeric equilibrium, shifting it in favour of the $\alpha$-anomer. Surprisingly, the C—H stretching regions (2903 and 2951 $cm^{-1}$) were also found to be sensitive to the presence of salts. From a

Table 17. *Thermodynamic functions associated with the transfer of* 1-*phenacyl*-3-*phenyl*-1,2,4-*triazole from water to* 4.0 *molal aqueous PHC solutions at* 25 °*C*; *units*: *kJ mol*$^{-1}$. *According to ref.* 163

| | $\Delta G_t$ | $\Delta H_t$ | $T\Delta S_t$ |
|---|---|---|---|
| Glucose | −0.68 | 1.6 | 2.3 |
| Ribose | −2.5 | −11.1 | −8.6 |

thorough analysis of the spectral changes, the authors favoured an interpretation of weak complexing, rather than a medium-induced shift of the anomeric equilibrium which would be expected to be much less ion specific.

Similar Raman studies were performed on the interactions of $Ca^{2+}$ with arabinitol, xylitol and ribitol. [**151**] The pentitol:water:$Ca^{2+}$ mol ratios were adjusted to test for specific complexes. Here, too, intensity changes, proportional to the salt concentration, were particularly marked in the C—H stretching bands around 2900 cm$^{-1}$. However, no stoichiometric pentitol:calcium complexation ratios could be detected.

The thermodynamics of sugar/ion mixtures have been investigated by ion specific electrodes and calorimetry. [**120, 159**] Pair interaction parameters $g_{ij}$ etc. have been determined and corrected for anion effects and for non-specific cation–sugar interactions. Here again it has been found that various sugars interact very differently with the same cation, $Ca^{2+}$, but not necessarily following the rule that ax-eq-ax —OH arrangements are required. According to the structure of the sugar and also the nature of its hydration, $g_{ij}$ can be positive (salting-out) or negative (salting-in). This underlines the fact that the interactions in dilute solutions are very different from those in concentrated solutions or in the solid amorphous state.

It might be expected that studies on sugar acids and their salts would reveal interesting, conformation-specific effects, but to the authors' knowledge, no such physico-chemical studies have yet been performed. This is surprising, especially in view of the important biochemical role played by sugar sulphates, phosphates and carboxylates.

## 13.2. *Kinetic solvent effects*

There exists a vast literature on the effects of mixed aqueous solvents on the kinetics and mechanisms of organic rections; see, for instance, ref. 160. However, until recently, such reports did not include aqueous solutions of PHCs. Engberts and his colleagues are in the process of performing detailed studies on PHC solvent effects, particularly on hydrolysis reactions involving non-polar substrates, e.g., variously substituted 1-acyl-1,2,4-triazoles, [**161**] for which the activation parameters $\Delta H^*$ and $\Delta S^*$ have been reported. Apart from a general rate retarding effect of sugars, it has been found that $T\Delta S^*$ is insensitive to substitution of glucose by xylose or maltose, but decreases significantly in the presence of ribose and arabinose.

Differences in the solvent environments were also tested by measuring the thermodynamic functions associated with the transfer of a non-hydrolyzable substrate analogue from water to aqueous glucose and ribose solutions, as shown in Table 17. The differences are unexpected, startling and, so far, without a convincing explanation.

### 13.3. *PHC-induced stabilization of proteins*

Extensive studies by Timasheff and his coworkers, extending over several years, of the influence of cosolvents on the stabilities of globular proteins, have clearly demonstrated the ability of PHCs to stabilize proteins against thermal denaturation. **[162–4]** The phenomenon had been observed earlier, **[165]** but without explanation. Timasheff first pointed out that, in order to study preferential interactions in a rigorous manner, it was necessary to make two series of measurements under conditions where (1) the molal concentrations of solvent components were identical in the solvent and in the protein solution, and (2) the chemical potentials of diffusible components were kept identical in the solvent and in the protein solution. Operationally, the latter condition can be (almost) achieved by equilibrating the protein solution with the solvent by dialysis. The methods employed for the processing of the experimental data are based on a formal thermodynamic treatment of ternary (and multicomponent) mixtures, developed by Casassa and Eisenberg. **[166]**

It is now well established that most PHCs *raise* the chemical potentials of proteins, i.e., PHCs repel proteins, even to the point where proteins can be precipitated from solution. However, the native form of the protein is affected to a greater degree than the denatured (inactive) form, so that the net effect produced by the PHC is one of conformational stabilization. The stabilization is not a linear function of PHC concentration and at high concentrations it levels off or even reverses. To a first approximation, the stabilization per mol PHC is a function of the number of —OH groups, but within a group of stereoisomers, specific effects can be observed.

The stabilization of proteins by PHCs is utilized technologically in freeze-drying operations, and ecologically by microorganisms, insects or plants when they are subjected to osmotic or chill stresses. **[12]** There is also accumulating evidence that some PHCs, particularly trehalose, are able to protect native proteins, **[167]** and even some low forms of life, **[168]** against the otherwise damaging effects of complete desiccation.

### 13.4. *PHC-induced stabilization of lipid phases*

Similar stabilizing effects to those described above have also been observed in lipid systems. Thus, PHCs have the ability to depress the lyotropic and thermotropic phase transition temperatures commonly observed in lipid/water systems. [**169**] Here again, the physical stabilization extends to the dry lipid/PHC system, where the PHC molecules appear to substitute for water in their interactions with the polar headgroups of the lipids, as observed by ir spectroscopy. [**170**] A more stringent test is the preservation of biological function during desiccation, and its subsequent recovery upon rehydration. Both laboratory studies on isolated membrane systems and field studies have shown convincingly that PHCs have the ability to stabilize dry membranes in anhydrobiotic organisms and that full functional activity is recovered on hydration. [**172**] Such remarkable properties of PHCs, which are clearly related to their propensity, described in Section 12, to form aqueous glasses, are now the subject of intense investigation and patenting activity. [**171, 172, 173**] In this connection it may be of relevance that, unlike proteins, nucleotides and supramolecular lipid structures, PHC polymers are able to resist freezing and drying without damage.

## 14. Conclusions and future prospects

The history of investigations into the solution physical chemistry of PHCs demonstrates that early progress was stopped in its tracks because of the widely held, but mistaken view that such solutions were ideal and their behaviour could be adequately accounted for by simple mass action hydration equilibria. Concepts such as hydration numbers and bound water were used indiscriminately and are only now being replaced by more realistic approaches. The extreme solvent sensitivity of low molecular weight PHCs is being recognized and, with the availability of advanced analytical techniques, it is to be hoped that their conformational behaviour will now receive closer examination. The paucity of reliable data, as witnessed by the gaps in the table entries shows that there is much leeway to be made up. In particular, phosphorylated, carboxylated and sulphated sugars and amino-sugar derivatives, e.g., *N*-acetylamino sugars, deserve much closer study.

As regards the theoretical treatment of PHC conformation, important advances have been made, dating from the initial empirical attempts. However, many questions remain to be answered. Most available results pertain to a static view of the problem, with petrified single molecules *in vacuo* or in an artificial solvent continuum. Nevertheless, these representations were significant starting points and the results formed the basis for future refinements. An important reflection is that we must not become attached to methodologies which, although useful in the past, are not likely to provide new important insights in the future. The representation of PHCs as hard (or soft) spheres or ellipsoids of revolution is an example of such an outdated concept.

Clearly, any new efforts should include a full treatment of the solvent. MC and MD techniques are candidates for further exploration, but not exclusively so. New methodologies may come to help with conformational predictions. The very active and fruitful area of protein structures will, without doubt, make a contribution. It is within that area that the major advances in the simulation of large molecules have been made. Paradoxically, there exist many more simulation studies, elaborate and expensive, of large proteins and polysaccharides than of low molecular weight peptides and PHCs, for which one could expect valuable information for a modest investment. It is also a case of trying to run before we are able to walk.

It is the use of simulation techniques which we consider will lead to the most useful results. It is worth emphasizing that simulations are *not* experiments or theories. The results will always have to be contrasted with the most appropriate experimental information. The ignoring of this caveat, which also applies to other computational techniques, is sure to produce large quantities of data but little useful information.

## Acknowledgements

We wish to thank the following for helpful criticism or for making available their results prior to publication: Ross Hatley, Louise Slade, Harry Levine, Jenny Green, Austen Angell and David Davies.

## Dedication

The authors wish to dedicate this chapter to Stephen J. Angyal and George A. Jeffrey in recognition of their pioneering studies into the intricacies of 'simple' sugar solution and crystal chemistry which laid the foundation for most of the work discussed in this review.

## References

1. T. W. Radedmacher, R. B. Parekh and R. A. Dwek. *Ann. Rev. Biochem.* **57**, 785(1988).
2. A. Suggett. In *Water – A Comprehensive Treatise* (ed. F. Franks), Vol. 4. Plenum Press: New York, 1975, p. 519.
3. D. A. Rees. *Carbohydrates* (ed. G. O. Aspinall) MPT Internat. Rev. of Science: Organic Chemistry, Series One, Vol. 7. *Butterworths*: *London*, 1973.
4. F. Franks. In *Polysaccharides in Foods* (eds J. M. V. Blanshard and J. R. Mitchell). Butterworths: London, 1979, p. 33.
5. F. Franks. *Pure Appl. Chem.* **59**, 1189(1987).
6. A. Cesaro. In *Thermodynamic Data for Biochemistry and Biotechnology* (ed. H.-J. Hinz). Springer: Berlin, 1986, p. 177.

7. D. A. Rees, E. R. Morris, D. Thom and J. K. Madden. In *The Polysaccharides* (ed. G. O. Aspinall), Vol. 1. Academic Press: New York, 1982, p. 195.
8. N. Schuelke and F. X. Schmid. *J. Biol. Chem.* **263**, 8827, 8832(1988).
9. U. Heber, J. M. Schmitt, G. H. Krause, R. J. Klosson and K. A. Santarius. In *Effects of Low Temperatures on Biological Membranes* (eds G. J. Morris and A. Clarke). Academic Press: London, 1981, p. 263.
10. J. R. Mitchell. In ref. 4, p. 51.
11. L. Slade and H. Levine. *Pure Appl. Chem.* **60**, 1841(1988).
12. F. Franks. *Biophysics and Biochemistry at Low Temperatures.* Cambridge University Press: Cambridge, 1985.
13. G. M. Fahy, D. I. Levy and S. E. Ali. *Cryobiology* **24**, 196(1987).
14. F. Franks, J. R. Ravenhill and D. S. Reid. *J. Solution Chem.* **1**, 3(1972).
15. R. S. Shallenberger. *Advanced Sugar Chemistry*, Ellis Horwood: Chichester, 1982.
16. G. Barone, P. Cacace, G. Castronuovo and V. Elia. *Carbohydrate Res.* **91**, 101(1981).
17. G. A. Jeffrey and S. H. Kim. *Carbohydrate Res.* **14**, 207(1970).
18. G. A. Jeffrey. *Adv. Chem. Ser. No. 117,* 177(1973).
19. R. N. Goldberg and Y. B. Twari. *J. Phys. Chem. Reference Data* **18**, 809(1989).
20. H. Hoiland. in ref. 6, p. 129.
21. W. G. McMillan and J. E. Mayer. *J. Chem. Phys.* **13**, 276(1945).
22. H. L. Friedman. *J. Solution Chem.* **1**, 387, 413, 419(1972).
23. J. L. Finney, J. E. Quinn and J. O. Baum. *Water Sci. Rev.*, **1**, 93(1985).
24. K. Freudenberg. *Stereochemistry*, Franz Deuticke: Leipzig, 1932.
25. D. A. Rees. *J. Chem. Soc. B*, 877(1970).
26. D. A. Rees and P. J. C. Smith. *J. Chem. Soc. Perkin II*, 830(1975).
27. M. J. Tait, A. Suggett, F. Franks, S. Ablett and P. A. Quickenden. *J. Solution Chem.* **1**, 131(1972).
28. A. Suggett and A. H. Clark. *J. Solution Chem.* **5**, 1(1976).
29. A. Suggett. *J. Phys. E.* **8**, 327(1975).
30. J. L. Markley. In *Protein Engineering* (eds. D. L. Oxender and C. F. Fox). Alan R. Liss: New York, 1987, p. 15.
31. R. E. Hoffman and D. B. Davies. Personal communication.
32. C. M. Preston and L. D. Hall. *Carbohydrate Res.* **37**, 267(1974).
33. R. U. Lemieux, K. Bock, L. T. J. Delbaere, S. Koto and V. S. Rao. *Canad. J. Chem.* **58**, 631(1980).
34. R. J. Abraham and E. Bretschneider. In *Internal Rotation in Molecules* (ed. J. Orville-Thomas). Wiley: London, 1974, p. 481.
35. J. W. Brady. *J. Amer. Chem. Soc.* **108**, 8153(1986). *Carbohydrate Res.* **165**, 306(1987).
36. J. R. Grigera. *J. Chem. Soc. Faraday Trans. I.* **84**, 2603(1988).
37. S. J. Angyal. *Austral. J. Chem.* **21**, 2737(1968).
38. S. Diner, J. P. Malrieu and P. Claverie. *Theoret. Chim. Acta* **13**, 1(1969).
39. K. Rasmussen. *Acta Chem. Scand.* **A36**, 323(1982).
40. A. Mardsen, B. Robson and J. S. Thompson. *J. Chem. Soc. Faraday Trans. I.* **84**, 2519(1988).
41. J. Kohler. *On the Structure and Dynamics of Cyclodextrins in the Crystal and in Solution. A Theoretical Study.* Dissertation, Berlin, 1987.

**42.** H. J. C. Berendsen, J. P. M. Postma, W. F. van Gunsteren and J. Herman. In *Intermolecular Forces* (ed. B. Pullman). D. Reidel Publ. Co: Dordrecht, 1981, p. 331.

**43.** H. J. C. Berendsen, J. R. Grigera and T. Straatsma. *J. Phys. Chem.* **91**, 6269(1987).

**44.** G. Del Re. *J. Chem. Soc.*, 4031(1958).

**45.** L. Dosen-Micovic, D. Jermic and L. Allinger. *J. Amer. Chem. Soc.* **105**, 1716(1983).

**46.** L. Dosen-Micovic, D. Jermic and L. Allinger. *J. Amer. Chem. Soc.* **105**, 1723(1983).

**47.** J. T. Edward. *Chem. Ind.* 1102(1955).

**48.** R. U. Lemieux. In *Molecular Rearrangement* (ed. P. de Mayo), Vol. 2. Interscience: New York, 1964, p. 709.

**49.** S. J. Angyal. *Angew. Chem.* (internat. edn.) **8**, 157(1969).

**50.** K. Bock. In *Gangliosides and Neuronal Plasticity.* (eds. G. Tettamanti, R. W. Ledeen, K. Sandhoff, Y. Nagai and G. Toffano). Fidia Res. Series, Vol. 6, 47. Liviana Press: Padova, 1986.

**51.** A. I. Kitaygorodsky. *Tetrahedron* **14**, 230(1961).

**52.** A. I. Kitaygorodsky. *Chem. Soc. Rev.* **7**, 113(1961).

**53.** R. U. Lemieux, K. Bock, L. T. J. Delbaere, S. Koto and V. S. Rao. *Canad. J. Chem.* **58**, 631(1980).

**54.** I. Tvaroska and T. Kozar. *J. Amer. Chem. Soc.* **102**, 6029(1982).

**55.** L. Onsager. *J. Amer. Chem. Soc.* **58**, 1486(1936).

**56.** R. H. Stokes and R. A. Robinson.. *J. Phys. Chem.* **76**, 2126(1966).

**57.** J. J. Kozak, W. S. Knight and W. Kauzmann. *J. Chem. Phys.* **48**, 675(1968).

**58.** K. Miyajima, M. Sawada and M. Nakagaki. *Bull. Chem. Soc. Japan* **56**, 1620(1983).

**59.** H. Uedaira and H. Uedaira. *J. Chem. Thermodynamics* **17**, 901(1985).

**60.** W. Kauzmann. *Adv. Protein Chem.* **14**, 1(1959).

**61.** Papers and discussions in *J. Chem. Soc. Faraday Symp. No. 17*, 1982.

**62.** R. Jaenicke. *Prog. Biophys. molec. Biol.* **49**, 117(1987).

**63.** W. L. Jorgensen and J. Gao. *J. Amer. Chem. Soc.* **110**, 4212(1988).

**64.** K. Miyajima, K. Machida, T. Taga, H. Komatsu and M. Nakagaki. *J. Chem. Soc. Faraday Trans. I* **84**, 2537(1988).

**65.** M. A. Kabayama and D. Patterson. *Canad. J. Chem.* **36**, 568(1958).

**66.** J. C. Dore, in ref. 23, p. 1.

**67.** A. Suggett, S. Ablett and P. J. Lillford. *J. Solution Chem.* **5**, 17(1976).

**68.** A. Suggett. *J. Solution Chem.* **5**, 33(1976).

**69.** W. Derbyshire. In *Water – A Comprehensive Treatise* (ed. F. Franks), Vol. 7. Plenum Press: New York, 1982, p. 339.

**70.** J. M. Harvey and M. C. R. Symons. *J. Solution Chem.* **7**, 571(1978).

**71.** S. Bociek and F. Franks. *J. Chem. Soc. Faraday Trans. I* **75**, 571(1979).

**72.** E. Lai and F. Franks. *Cryo-Letters* **1**, 20(1979).

**73.** J. L. Hollenberg and D. O. Hall. *J. Phys. Chem.* **87**, 695(1983).

**74.** P. T. Beall. In *Water Biophysics* (eds. F. Franks and S. F. Mathias). John Wiley & Sons: Chichester, 1982, p. 323.

**75.** G. Barone, G. Castronuovo, D. Doucas, V. Elia and C. A. Mattia. *J. Phys. Chem.* **87**, 1931(1983).

**76.** O. Ya. Samoilov. *Structure of Aqueous Electrolyte Solutions and the Hydration of Ions*. Consultants Bureau: New York, 1965.
**77.** A. Mayaffre, R. Bury, M. Chemla and M. H. Mannebach. *J. Chim. Phys.* **83**, 637(1986).
**78.** G. Barone, P. Cacace, G. Castronuovo, V. Elia and U. Lepore. *Carbohydrate Res.* **115**, 15(1983).
**79.** G. Barone, G. Castronuovo, V. Elia and V. Savino. *J. Solution Chem.* **13**, 209(1984).
**80.** J. M. Sturtevant. *J. Phys. Chem.* **45**, 127(1941).
**81.** F. Shahidi, P. G. Farrell and J. T. Edwards. *J. Solution Chem.* **5**, 807(1976).
**82.** Y.-N. Lian, A.-T. Chen, J. Suurkuusk and I. Wadsö. *Acta Chem. Scand.* **A36**, 735(1982).
**83.** S. H. Kim, G. A. Jeffrey and R. D. Rosenstein. *Acta. Cryst. Sect. B* **25**, 2223(1969).
**84.** H. S. Kim and G. A. Jeffrey. *Acta Cryst. Sect. B* **25**, 2607(1969).
**85.** F. D. Hunter and R. D. Rosenstein. *Acta Cryst. Sect. B.* **24**, 1652(1968).
**86.** T. H. Herrington, A. D. Pethybridge, B. A. Parkin and M. G. Roffey. *J. Chem. Soc. Faraday Trans. I* **79**, 845(1983).
**87.** G. Barone, P. Cacace, G. Castronuovo and V. Elia. *Carbohydrate Res.* **119**, 1(1983).
**88.** H. S. Frank and F. Franks. *J. Chem. Phys.* **48**, 4746(1968).
**89.** S. H. Gaffney, E. Haslam and T. H. Lilley. *Thermochim. Acta* **86**, 175(1985).
**90.** N. Nichols, R. Skold, C. Spink, J. Suurkuusk and I. Wadsö. *J. Chem. Thermodyn.* **8**, 1081(1976).
**91.** J. T. Edward and P. G. Farrell. *Canad. J. Chem.* **53**, 2965(1975).
**92.** J. J. Savage and R. H. Wood. *J. Solution Chem.* **5**, 733(1976).
**93.** B. Y. Okamoto, R. H. Wood, J. E. Desnoyers, G. Perron and L. Delorme. *J. Solution Chem.* **10**, 139(1981).
**94.** G. Barone, B. Bove, G. Castronuovo and E. Elia. *J. Solution Chem.* **10**, 803(1981).
**95.** R. W. Balk and G. Somsen. *J. Phys. Chem.* **89**, 5093(1985).
**96.** R. W. Balk and G. Somsen. *J. Chem. Soc., Faraday Trans. I.* **82**, 933(1986).
**97.** A. Reine, J. A. Hveding, O. Kjolberg and O. Westbye. *Acta Chem. Scand.* **828**, 690(1974).
**98.** F. Franks, P. J. Lillford and G. Robinson. *J. Chem. Soc. Faraday Trans. I.* **85**, 2417(1989).
**99.** S. J. Angyal. *Adv. Carbohydrate Chem. Biochem.* **42**, 15(1984).
**100.** S. J. Angyal and G. S. Bethell. *Austr. J. Chem.* **29**, 1249(1976).
**101.** V. S. R. Rao, K. S. Vijayalakshmi and K. S. Sundarajan. *Carbohydrate Res.* **17**, 241(1971).
**102.** K. Kildeby, S. Melberg and K. Rasmussen. *Acta Chem. Scand.* **31**, 1(1977).
**103.** G. A. Jeffrey, J. A. Pople, J. S. Binkley and S. Vishveshwara. *J. Amer. Chem. Soc.* **100**, 373(1978).
**104.** M. D. Newton. *Acta Cryst.* **839**, 104(1983).
**105.** I. Tvaroska and T. Kozar. *Chem. Papers* **41**, 501(1987).
**106.** H. S. Isbell and W. Pigman. *Adv. Carbohydrate Chem. Biochem.* **24**, 13(1969).
**107.** A. Broido, Y. Houminer and S. Patai. *J. Chem. Soc. B* **411**(1966).

**108.** F. Shafizdeh, G. D. McGinnis, R. A. Susott and H. W. Tatton. *J. Org. Chem.* **36**, 2813(1971).
**109.** N. M. Ballash and E. B. Robertson. *Can. J. Chem.* **51**, 556(1973).
**110.** A. S. Hill and R. S. Shallenberger. *Carbohydrate Res.* **11**, 541(1969).
**111.** T. E. Acree. Ph.D. Thesis. Cornell University: Ithaca, NY, 1968.
**112.** A. S. Serianni and J. R. Snyder. 8th Internat. Symposium on Solute–Solvent Interactions, Regensburg. Abstracts (Section 6), 1987, p. 134.
**113.** R. Barker and A. S. Serianni. *Acc. Chem. Res.* **19**, 307(1986).
**114.** G. Livingstone, F. Franks and L. J. Aspinall. *J. Solution Chem.* **6**, 203(1977).
**115.** F. Franks, J. R. Hall, D. E. Irish and K. Norris. *Carbohydrate Res.* **157**, 53(1986).
**116.** S. J. Angyal. *Chem. Soc. Rev.* **9**, 415(1980).
**117.** M. Mathlouthi and D. V. Luu. *Carbohydrate Res.* **81**, 203(1980).
**118.** S. J. Angyal. *Austr. J. Chem.* **25**, 1957(1972).
**119.** R. W. Goulding. *J. Chromatrogr.* **103**, 229(1975).
**120.** J.-P. Morel, C. Lhermet and N. Morel-Desrosiers. *J. Chem. Soc. Faraday Trans. I.* **84**, 2567(1988).
**121.** J. M. Bailey, P. G. Fishman and P. G. Pentchev. *Biochem.* **9**, 1189(1970).
**122.** J. Pierce, A. S. Serianni and R. Barker. *J. Amer. Chem. Soc.* **107**, 2448(1985).
**123.** Y. B. Tewari and R. N. Goldberg. *Appl. Biochem. Biotechnol.* **11**, 17(1985).
**124.** G. A. Jeffrey, R. A. Wood, P. E. Pfeffer and K. B. Hicks. *J. Amer. Chem. Soc.* **105**, 2128(1983).
**125.** J. F. Stoddart. *Stereochemistry of Carbohydrates.* Wiley-Interscience: New York, 1971, chap. 3.
**126.** J. A. Mills. *Austr. J. Chem.* **27**, 1433(1974).
**127.** S. J. Angyal and R. Le Fur. *Carbohydrate Res.* **126**, 15(1984).
**128.** S. J. Angyal, J. K. Saunders, C. T. Grainger, R. Le Fur and P. G. Williams. *Carbohydrate Res.* **150**, 7(1986).
**129.** G. E. Hawkes and D. Lewis. *J. Chem. Soc., Perkin Trans. II*, 2073(1984).
**130.** F. Franks, R. L. Kay and J. Dadok. *J. Chem. Soc., Faraday Trans. I*, **84**, 2595(1988).
**131.** D. A. Rees and D. Thom. *J. Chem. Soc., Perkin Trans. II* 1919(1977).
**132.** S. H. Kim and G. A. Jeffrey. *Acta Crystallogr.* **22**, 537(1967).
**133.** S. Melberg and K. Rasmussen. *Carbohydrate Res.* **69**, 27(1979).
**134.** S. Melberg and K. Rasmussen. *Carbohydrate Res.* **76**, 23(1979).
**135.** I. Tvaroska. *Biopolymers* **21**, 1887(1982).
**136.** D. A. Rees and R. J. Skerret. *Carbohydrate Res.* **7**, 334(1968).
**137.** S. Melberg and K. Rasmussen. *Carbohydrate Res.* **71**, 25(1979).
**138.** I. Tvaroska. *Biopolymers* **23**, 1951(1984).
**139.** K. Bock and R. U. Lemieux. *Carbohydrate Res.* **100**, 63(1982).
**140.** G. M. Brown and H. A. Levy. *Science* **141**, 921(1963).
**141.** C. A. Bush and R. E. Feeney. *Int. J. Peptide Protein Res.* **28**, 386(1986).
**142.** C. A. Bush, Z. Y. Yan and B. N. Narasinga Rao. *J. Amer. Chem. Soc.* **108**, 6168(1986).
**143.** S. W. Homans, R. A. Dwek, J. Boyd, M. Mahmoudian, W. G. Richards and T. W. Rademacher. *Biochem.* **25**, 6342(1986).
**144.** H. Levine and L. Slade. *Water Sci. Rev.* **3**, 79(1988).

**145.** L. Finegold, F. Franks and R. H. M. Hatley. *J. Chem. Soc. Faraday Trans. I.* **85**, 2945(1989).
**146.** S. Quinquenet, C. Grabielle-Madelmont, M. Ollivon and M. Serpelloni. *J. Chem. Soc. Faraday Trans. I.* **84**, 2609(1988).
**147.** R. H. M. Hatley, T. Webb and F. Franks. Unpublished data.
**148.** F. Franks. *Cryo-Letters* **11**, 93(1990).
**149.** L. Slade and H. Levine. In *Food Structure – its Creation and Evaluation* (eds. J. R. Mitchell and J. M. V. Blanshard). Butterworths: London, 1988, p. 115.
**150.** J. M. Waslyk, A. R. Tice and J. G. Baust. *Cryobiology* **25**, 451(1988).
**151.** F. Franks, D. E. Irish and L. Rajadhyax. Unpublished results.
**152.** M. Mathlouti, A. L. Cholli and J. L. Koenig. *Carbohydrate Res.* **147**, 1(1986).
**153.** A. Hvidt. *Ann. Rev. Biophys. Bioeng.* **12**, 1(1983).
**154.** G. Barone, G. Castronuovo, V. Elia and K. Stassinopoulou. *J. Chem. Soc. Faraday Trans. I.* **80**, 3095(1984).
**155.** G. Barone, G. Castronuovo, P. Del Vecchio, V. Elia and M. T. Tosto. *J. Solution Chem.* **17**, 925(1988).
**156.** D. C. Craig, N. C. Stephenson and J. D. Stevens. *Carbohydrate Res.* **22**, 494(1972).
**157.** S. J. Angyal, D. Greeves and J. A. Mills. *Austr. J. Chem.* **27**, 1447(1974).
**158.** R. M. Williams and R. H. Atalla. *J. Chem. Soc. Perkin II* 1155(1975).
**159.** N. Morel-Desrosiers and J.-P. Morel. *J. Chem. Soc. Faraday Trans. I*, **85**, 3461(1989).
**160.** J. B. F. N. Engberts. In *Water – a Comprehensive Treatise* (ed. F. Franks), Vol. 6. Plenum Press: New York, 1979, p. 139.
**161.** J. J. H. Nusselder and J. B. F. N. Engberts. *J. Org. Chem.* **52**, 3159(1987).
**162.** K. Gekko and S. N. Timasheff. *Biochemistry* **20**, 4667(1981).
**163.** K. Gekko and T. Morikawa. *J. Biochem.* **90**, 51(1981).
**164.** T. Arakawa and S. N. Timasheff. *Biochemistry* **21**, 6536(1982).
**165.** S. Y. Gerlsma. *J. Biol. Chem.* **243**, 957(1968).
**166.** E. F. Casassa and H. Eisenberg. *Adv. Protein Chem.* **19**, 287(1964).
**167.** J. F. Carpenter and J. H. Crowe. *Cryobiology* **25**, 459(1988).
**168.** J. H. Crowe and L. M. Crowe. *Cryobiology* **19**, 317(1982).
**169.** L. M. Crowe, J. H. Crowe and D. Chapman. *Arch. Biochem. Biophys.* **236**, 289(1985).
**170.** L. M. Crowe, R. Mouradian, J. H. Crowe, S. A. Jackson and C. Womersley. *Biochim. Biophys. Acta* **769**, 141(1984).
**171.** J. L. Green and C. A. Angell. *J. Phys. Chem.* **93**, 2880(1989).
**172.** B. J. Roser. Europ. Patent Applic. No. 88305979.2, 1988.
**173.** F. Franks and R. H. M. Hatley, Europ. Patent Applic. No. 90.301561.8, 1990.